OXFORD ENGINEERING SCIENCE SERIES

SERIES EDITORS

THE OXFORD ENGINEERING SCIENCE SERIES

---

1 D. R. RHODES: *Synthesis of planar antenna sources*
2 L. C. WOODS: *Thermodynamics of fluid systems*
3 R. N. FRANKLIN: *Plasma phenomena in gas discharges*
4 G. A. BIRD: *Molecular gas dynamics*
5 H.-G. UNGER: *Planar optical waveguides and fibres*
6 J. HEYMAN: *Equilibrium of shell structures*
7 K. H. HUNT: *Kinematic geometry of mechanisms*
8 D. S. JONES: *Methods in electromagnetic wave propagation* (*Two volumes*)
9 W. D. MARSH: *Economics of electric utility power generation*
10 P. HAGEDORN: *Non-linear oscillations* (*Second edition*)
11 R. HILL: *Mathematical theory of plasticity*
12 D. J. DAWE: *Matrix and finite element displacement analysis of structures*
13 N. W. MURRAY: *Introduction to the theory of thin-walled structures*
14 R. I. TANNER: *Engineering rheology*
15 M. F. KANNINEN and C. H. POPELAR: *Advanced fracture mechanics*
16 R. H. T. BATES and M. J. McDONNELL: *Image restoration and reconstruction*
17 K. HUSEYIN: *Multiple-parameter stability theory and its applications*
18 R. N. BRACEWELL: *The Hartley transform*
19 J. WESSON: *Tokamaks*
20 P. B. WHALLEY: *Boiling, condensation, and gas-liquid flow*
21 C. SAMSON, M. LeBORGNE, and B. ESPIAU: *Robot control*
22 H. J. RAMM: *Fluid dynamics for the study of transonic flow*
23 R. R. A. SYMS: *Practical volume holography*
24 W. D. McCOMB: *The physics of fluid turbulence*
25 Z. P. BAZANT and L. CEDOLIN: *Stability of structures: principles of elastic, inelastic, and damage theories*
26 J. D. THORNTON: *Science and practice of liquid–liquid extraction* (*Two volumes*)
27 J. VAN BLADEL: *Singular electromagnetic fields and sources*

Science and Practice of Liquid–Liquid Extraction

Volume 2

Process Chemistry and Extraction Operations in the Hydrometallurgical, Nuclear, Pharmaceutical, and Food Industries

JOHN D. THORNTON

Emeritus Professor of Chemical Engineering,
University of Newcastle upon Tyne

CLARENDON PRESS • OXFORD

This book has been printed digitally and produced in a standard specification in order to ensure its continuing availability

OXFORD
UNIVERSITY PRESS

Great Clarendon Street, Oxford OX2 6DP

Oxford University Press is a department of the University of Oxford.
It furthers the University's objective of excellence in research, scholarship,
and education by publishing worldwide in

Oxford New York

Auckland Cape Town Dar es Salaam Hong Kong Karachi
Kuala Lumpur Madrid Melbourne Mexico City Nairobi
New Delhi Shanghai Taipei Toronto
With offices in
Argentina Austria Brazil Chile Czech Republic France Greece
Guatemala Hungary Italy Japan South Korea Poland Portugal
Singapore Switzerland Thailand Turkey Ukraine Vietnam

Published in the United States
by Oxford University Press Inc., New York

Reprinted 2011

ISBN 978-0-19-856237-5

Printed and bound in Great Britain by CPI Antony Rowe,
Chippenham and Eastbourne

TO

THE LATE F.H. GARNER OBE

One time Professor of Chemical Engineering
and Dean of the Faculty of Science
University of Birmingham

PREFACE
VOLUMES 1 AND 2

Since the adoption of liquid extraction as a separative procedure towards the end of the last century, many industries have employed this technique as a means of separating the constituents of liquid mixtures. In the early days, the petroleum industry used extraction as a means of refining kerosenes and, from this beginning, developed a wide range of other extraction operations for hydrocarbon processing.

The use of liquid extraction for separating metal ions in solution is a later development which received great impetus from the need of the nuclear industry to extract uranium ores, to reprocess spent reactor fuels, and to purify other materials of nuclear importance. The reprocessing of nuclear fuels presented unprecedented problems not only in the field of extraction chemistry but also in the areas of contactor design and engineering. These problems were solved and, as a result, new advances were made in the technology of the process as a whole. Today, extraction provides viable procedures for separating and purifying many non-nuclear metal systems, and has, for example, been proposed for the production of high-purity materials for electronic applications.

Just as the advent of the nuclear age catalysed developments in both the theory and practice of extraction, so the emergence of the biologically based industries is generating new requirements for novel extractants as well as short residence time contactors capable of processing a wide range of fermentation liquors with a minimum of solute degradation.

The present volumes are an attempt to bring together some of the more important developments of recent years and to present them in a way that might be helpful to those engaged in research in the extraction field, as well as to engineers and chemists concerned with the industrial aspects of extraction processes.

The presentation assumes a working knowledge of the subject and is therefore not designed as an undergraduate text, although it is hoped that many post-graduate students will find much that is useful to them. The chapters have been contributed by writers with a first-hand knowledge of their respective fields and each is supplemented by references to the published literature.

Any work of this kind involves an element of selection if the text is not to become unduly long and expensive. I have therefore restricted the material to those topics which illustrate either the fundamental ideas behind the subject or new requirements and developments at the process level. Wherever possible, the treatment of the various topics has been as fundamental as current knowledge allows with a minimum of empiricism.

For the convenience of readers, the work has been divided into two volumes. The first is primarily concerned with the fundamental aspects of extraction leading up to extractor selection and design while the second volume is devoted to process chemistry and extraction operations in a number of selected industries where new developments are rapidly emerging.

Volume 1 contains much that is likely to be of interest to research workers and design engineers and opens with a discussion of the measurement, correlation, and prediction of equilibrium data from a thermodynamic standpoint. Although vapour–liquid equilibria have long been interpreted in terms of activity coefficients, the same cannot be said of liquid–liquid systems and it is hoped that this presentation will help to stimulate renewed interest in a more fundamental approach to the interpretation of liquid–liquid equilibrium measurements rather than a continued reliance upon essentially empirical procedures.

The subject of mass transfer has been dealt with at some length and the fundamental ideas behind the diffusion of a solute across a liquid–liquid interface are first described in detail for 'simple' systems in which complicating factors such as interfacial turbulence are absent. It is not always appreciated that the transfer of certain solutes across a liquid–liquid interface can set up intense turbulence in the vicinity of the phase boundary with the result that mass transfer coefficients calculated purely on the basis of molecular diffusion can considerably underestimate the observed transfer rates.

It might be thought that such effects are a bonus in so far as actual transfer rates will be higher than those predicted by diffusion theory. Unfortunately this is not necessarily true since a concomitant effect is enhanced droplet coalescence rates which in turn result in a smaller interfacial area for solute transfer. Thus whilst the mass transfer coefficient may be enhanced, the interfacial area may be reduced to an extent that is not yet quantifiable so that the overall effect remains uncertain at the present time. The nature of these so-called Marangoni effects has therefore been dealt with in some detail and possible mechanisms by which they can influence the mass transfer rates are discussed. Although the prediction of interdroplet coalescence rates can be notoriously unreliable when traces of surface active materials are present, an account of the underlying theories of droplet coalescence in the absence of surfactants has been included with particular reference to coalescence mechanisms and their application to settler design.

The material dealing with contactor design has been divided into three separate chapters. Thus Chapter 6 deals with the analysis of contactor performance based upon plug flow considerations, whereas the more complex situation involving axial mixing is dealt with in Chapter 7. Here we are concerned with methods of modelling axial mixing together with appropriate methods of solving the equations involved. Finally, in Chapter 8, the selection, design, and scale-up of contactors is discussed with reference to a range of typical units.

Volume 2 is concerned with the process chemistry and extraction operations carried out in a number of key industries. These are industries which either demonstrate novel extraction technology or else have generated a need for unconventional extractants and/or high efficiency contactors with unusually short residence times.

The applications of extraction in the hydrometallurgical field are described in detail for a wide range of metal systems including the extraction of uranium from ores. The more circumscribed use of liquid extraction in a radioactive environment where special requirements prevail has been dealt with separately in the context of nuclear fuel reprocessing.

The extraction of complex labile molecules poses a number of problems to both the pharmaceutical and the foodstuffs industries where there is frequently a need for new types of extractants which are compatible with the solutes to be extracted coupled with high efficiency low residence time contactors in order to minimize solute degradation. Thus, for example, the possible use of two-phase aqueous–aqueous systems and supercritical extraction may well offer attractive process routes in the foreseeable future. The potential of such techniques has been examined, as have the future requirements for extractants and equipment in the biologically based industries.

The volume concludes with an account of some current developments in the extraction field including supercritical extraction, liquid membranes, biochemical processes, and electrically enhanced extraction.

When this work was first envisaged, I had the task of persuading the many contributors to give of their time and undertake the many hours of labour involved in the preparation of each chapter. This they all agreed to do and I should like to record my appreciation and thanks to each and every author for so generously agreeing to be part of this project.

For my part, I must accept the responsibility for the contents of the two volumes and if my readers feel that I have omitted anything of moment, I can only plead that I have had to bear production costs in mind continually and therefore set a limit to the ultimate number of pages if the volumes were not to become prohibitively expensive.

Finally, I should like to thank Oxford University Press for all their help in the production of this work and for so kindly producing the Index to each volume.

Worcestershire J. D. T.
August 1991

CONTENTS

VOLUME 2

CONTRIBUTORS

VOLUME 2

W. BATEY — Dounreay Nuclear Power Development Establishment, Thurso, Caithness, UK

MICHAEL COX — Division of Chemical Sciences, Hatfield Polytechnic, Hatfield, Hertfordshire, UK

W. HAMM — Consulting Engineer, Harpenden, Hertfordshire, UK

MICHAEL S. VERRALL — SmithKline Beecham Pharmaceuticals, Chemotherapeutic Research Centre, Brockham Park, Betchworth, Surrey, UK

L. R. WEATHERLEY — Department of Chemical and Process Engineering, Heriot-Watt University, Riccarton, Edinburgh, UK

CONTENTS

VOLUME 1

1

LIQUID–LIQUID EXTRACTION IN HYDROMETALLURGY

MICHAEL COX

1.1 Hydrometallurgy

Hydrometallurgy involves a series of interlinked unit processes for the recovery and purification of elements using aqueous solutions. One of the essential features of present flowsheets is the inclusion of a solution treatment step to purify and concentrate the desired element before recovery. Several alternative techniques are available but liquid–liquid extraction has the greatest versatility for the separation of elements and also the range of solution concentrations which can be treated. This process is therefore now seen as the preferred technique for the treatment of all except the most dilute leach liquors, such as in gold and some uranium operations where resin ion exchange and carbon adsorption processes offer distinct advantages (Streat and Naden 1987). Although a detailed discussion of hydrometallurgy is outside the remit of this chapter, it is appropriate to consider the demands made by such a flowsheet (Fig. 1.1) as these will influence the choice and operation of the liquid–liquid extraction system.

Following mining and preliminary ore treatment, the feed is subjected to leaching with an aqueous lixiviant, chosen on the basis of the mineralogy of the ore, availability of chemicals, and overall flowsheet demands. Three major types of lixiviant are commonly used: acids, normally sulphuric; bases, for example ammonium carbonate or sulphate; and oxidizing solutions, such as iron(III) chloride. These produce solutions containing the metal as aquated cations, or complex species such as ammines, for example $[M(NH_3)_x]^{n+}$, or chloroanions, for example $[MCl_x]^{(x-n)-}$, where n represents the oxidation state of the metal. This variation in solution species largely determines the nature of the extractant to be used. Also extractant selection is influenced by the nature and concentration of the elements present as impurities in the leachate. Thus in most flowsheets it is necessary to remove iron prior to solution purification, by precipitation as a complex sulphate, jarosite, or a hydrated oxide species, goetite, or to select an extractant which will preferentially extract the desired element leaving iron in the raffinate. Also it should be recognized that it is generally more satisfactory to purify a solution by removing the impurities rather than extracting the major component. Because of the complexity of leach solutions this is seldom possible, but it can be used for the purification of subsequent liquid

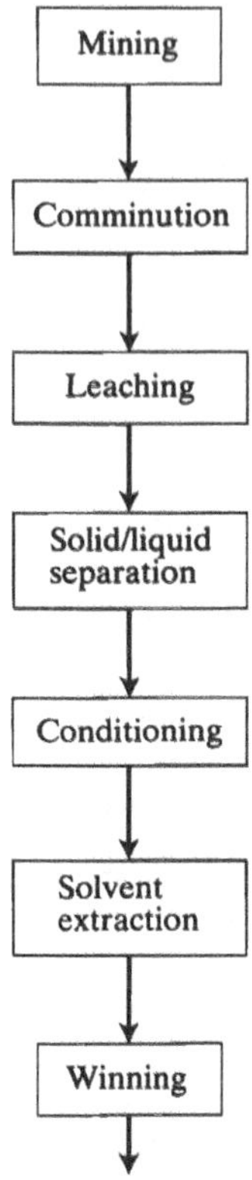

Fig. 1.1. Hydrometallurgical route using solvent extraction for extraction of metals from ores.

streams (as, for example, in electrolytic tank house liquor purification). The choice of extraction process is also influenced by processes downstream of solution purification, especially where the operation is intended to fit into an existing flowsheet. In particular, electrowinning processes impose rigorous demands both on the level and nature of impurities in the liquor, and on the presence of organic compounds entrained in the extraction process. However, it should be realized that liquid–liquid extraction also imposes restrictions on the upstream unit operations. In particular, the presence of surfactants, whether used as flocculants to aid solid–liquid separation or as mist supressants in tankhouses, can have a deleterious effect on extractant performance. Thus the liquid–liquid extraction operation must not be considered in isolation, but should be based on the fullest knowledge of both the upstream and downstream processes with which it will be interfaced. This is particularly true of the leaching circuit, because as a result of the precise mineralogy of the feed ore body each leachate, although nominally containing the same mineral, will have a different composition.

This is why it is essential in the early stages of development of a liquid–liquid extraction circuit that the actual leach liquor is used, rather than a simulated leach liquor made to match the analytical data or, worse still, pure laboratory solutions containing just the major elements.

1.2 Liquid–liquid equilibria

[See also Newsham, Vol. 1]

The basic solvent extraction circuit in hydrometallurgy consists of a series of interconnecting loops (Fig. 1.2) that allows the desired element to be transferred first into an immiscible organic phase, which may then be contacted with a series of aqueous solutions designed to strip, or back extract, any impurities coextracted with the desired element. This process is termed scrubbing, and the resulting aqueous scrub raffinates may then be further treated to remove any valuable elements. Finally the purified organic phase is contacted with an appropriate strip solution to return the chosen element to an aqueous phase from which it can be recovered, thereby regenerating the organic extractant solution for recycling. The ability to recycle a number of these process streams is a desirable feature of the flowsheet. Inevitably

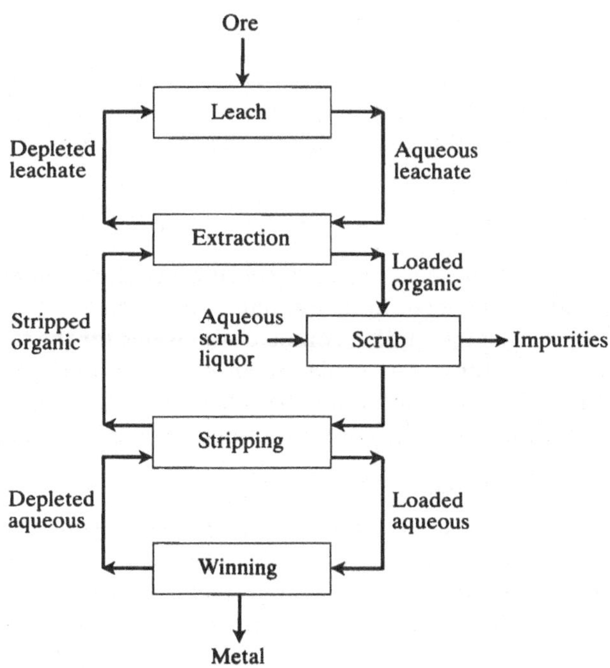

Fig. 1.2. Typical solvent extraction flow sheet for the recovery of metals from a leach solution.

with extensive recycling, some impurities do build up in the process streams, so it is usual to incorporate purification procedures into bleed streams off the main process stream.

Although it is not intended to cover exhaustively the theory of liquid–liquid extraction equilibria and kinetics in this chapter, to appreciate the criteria involved in the design and selection of extraction reagents for hydrometallurgical processes, it is necessary to outline the underlying principles.

As stated earlier, elements can exist in aqueous solution in a number of different forms, which may be charged, for example as aqueous cations or chloroanions, or uncharged, as in some aquated complex species such as $UO_2(NO_3)_{2,aq}$. All these species have little or no tendency to transfer to a non-polar organic phase as used in liquid–liquid extraction. Thus, to achieve the desired extraction, the aqueous species have to be modified in some way to make them compatible with the organic phase. This modification can be carried out by neutralization of any ionic charge and/or replacement of water of hydration by other coordinating species. These requirements determine the nature and mode of operation of the organic extractants used in hydrometallurgy, which can be conveniently divided into the following three classes.

1. Acidic extractants which possess ionizable hydrogen atoms that can be replaced by metal ions to give an overall neutral species (eqn (1.1))*;

$$M^{n+}_{aq} + \overline{nRH} \rightleftharpoons \overline{MR_n} + nH^+_{aq}. \tag{1.1}$$

2. Basic, or anion exchangers, which are normally the protonated form of high molecular weight amines or quaternary compounds. These form ion-pairs with anionic species, thus reducing the effective ionic charge of the latter (eqn (1.2));

$$(p-n)\overline{R_3HN^+} + MX^{(p-n)-}_{p,aq} \rightleftharpoons \overline{(R_3NH^+)_{p-n}(MX_p^{(p-n)-})} + H_2O. \tag{1.2}$$

This equation also represents an ion association reaction, hence the alternative names of ion-association or ion-pair extractants.

3. Solvating extractants, which compete with water for the first solvation shell of the metal atom. The replacement of water molecules by these reagents, which often contain donor oxygen atoms, facilitates the transfer of a metal-containing complex into a non-polar organic phase (eqn (1.3))

$$MX_{p,aq} + \overline{qS} \rightleftharpoons \overline{MX_pS_q} + H_2O. \tag{1.3}$$

All the above reactions are equilibria so the essential requirement that the extraction process is easily reversible is observed. Thus, with acidic extractants, change of acidity will cause the equilibrium to be shifted, whereas stripping for the other two extractants may be carried out with water. Before

*In this and all subsequent extraction equations the barred species occur in the organic phase.

considering these extraction types in more detail it is appropriate to list some general requirements that a reagent should satisfy for hydrometallurgical operation. The following alliterative list (Price and Tumilty 1975) summarizes the main features.

1. Strength—a measure of the ability of an extractant to transfer the metal both in extraction and stripping. This includes such properties as metal loading capacity, which relates the mass of metal extracted to the mass of reagent used and determines the number of stages to meet the design capacity, and also the amount of reagent required.

2. Selectivity—the ability to provide a pure product stream free from deleterious impurities.

3. Solubility—high solubility in the chosen organic diluent of both the extractant and the metal complex. This determines the required volume of the organic phase and hence the size of the plant. Also, low aqueous solubility of the extractant is desirable to minimize loss from the circuit to the environment.

4. Speed—kinetics of extraction and stripping which determine mixer size and stage efficiency.

5. Separation—good coalescing properties of the aqueous and organic phases after contacting either in extraction or stripping, with little tendency to form stable emulsions.

6. Stability—the extractant should have good stability with regard to, for example, hydrolysis and oxidation, thereby allowing extensive recycling without degradation. In addition, any degradation products should preferably be inactive towards extraction, especially of impurities.

7. Synthesis—the extractant should have a consistent quality with little batch-to-batch variation, thereby allowing consistent plant performance following any make-up operations.

8. System—the ability to interface with the remainder of the hydrometallurgical flowsheet and provide a product in the form and purity required by the industry.

An addition should be made to this list, that is, Sterling or $, the cost of the reagent which is important for the initial outlay and for make-up. Also, because it is unrealistic to assume that no organic phase will be released into the environment (Flett *et al.* 1983), the extractant and diluent should be non-toxic and biodegradable.

1.2.1 Acid extractants

These extractants react with metal cations to form neutral complexes that are preferentially dissolved by the organic phase, and the extent of extraction depends on the acidity of the aqueous phase (eqn (1.4)). In addition, the extraction will also depend on the nature of the metal, the pK_a of the extractant, its concentration, and any specific interactions between the metal and extractant arising from consideration of the appropriate coordination chemistry.

The simple extraction (eqn (1.4)) can be regarded as a series of step-wise equilibria with associated equilibrium coefficients, involving: partition of the extractant (eqn (1.5)); ionization (eqn (1.6)); formation of a series of extractant–metal complexes (eqn (1.7)); and partition of the neutral complex between the aqueous and organic phases (eqn (1.8)).

$$MR^{n+}_{aq} + n\overline{RH} \rightleftharpoons \overline{MR_n} + nH^+ \tag{1.4}$$

$$RH_{aq} \rightleftharpoons \overline{RH} \qquad p_{RH} = [\overline{RH}]/[RH] \tag{1.5}$$

$$RH_{aq} \rightleftharpoons R^- + H^+ \quad K_a = [R^-][H^+]/[RH] \tag{1.6}$$

$$M^{n+}_{aq} + R^- \rightleftharpoons RM^{(n-1)+}$$

$$\begin{array}{ccc} \cdot & \cdot & \cdot \\ \cdot & \cdot & \cdot \\ \cdot & \cdot & \cdot \end{array}$$

$$MR^+_{(n+1)} + R^- \rightleftharpoons MR_n \qquad \beta = [MR_n]/[M][R]^n \tag{1.7}$$

$$MR_n \rightleftharpoons \overline{MR_n} \qquad p_{MR_n} = [\overline{MR_n}]/[MR_n]. \tag{1.8}$$

With the definition of distribution coefficient, D, as the ratio of the total analytical concentration of the metal in the two phases (eqn 1.9), then the application of the law of mass action to eqn (1.4) yields eqn (1.10).

$$D = [\overline{M}]/[M] \tag{1.9}$$

$$\log D = \log K + n\text{pH} + n\log[\overline{RH}] \tag{1.10}$$

where $K = p^{-n}_{RH} \cdot K^n_a \cdot \beta_n \cdot p_{MR_n}$.

Thus the metal distribution should increase with the aqueous solubility and acidity of the extractant, increasing pH and reagent concentration. The term β_n, overall formation coefficient for the metal–extractant complex, reflects specific interactions between the metal and extractant thereby including factors such as chelation, steric effects, and hard and soft acid–base considerations.

This particular model implies that the relevant chemical reactions occur in the aqueous phase and thus a reasonable water solubility of the extractant. This conflicts with the requirements for an extractant listed above, and so must be considered as representing one limiting case for a broad spectrum of possible models. The other limiting case is the situation where the extractant is completely water insoluble and the reactions occur solely at the aqueous–organic interface. In this situation, the partition equilibria, represented by eqns (1.5) and (1.8), should be replaced by the appropriate adsorption isotherms relating interfacial to bulk phase concentrations. However, in most circumstances the former scheme, in spite of the obvious limitations, has advantages in describing the extraction chemistry. In addition the equilibrium constants should involve activities of the species present rather than concentrations. However, it is extremely unlikely that such activity data are available. In the laboratory this problem can be circumvented if the concentrations are low, since activity and concentration may then be regarded as being numerically identical. It should be remembered in practical situations, such as in leach liquors where salt concentrations can be high, substantial differences exist between activity and concentration (Gokcen 1982; Muir and Parker 1982). Although it may be possible in a particular study to circumvent the problem by maintaining a constant ionic strength, it is important to realize that errors can arise in practical hydrometallurgical applications, when using data from laboratory studies at low metal and salt concentrations. This caveat is true for all types of extractant.

1.2.1.1 *Types of acid extractant*

These extractants can be further subdivided into chelating and non-chelating species. The latter include carboxylic, phosphoric, and sulphonic acids (Table 1.1) and provide a very versatile series of reagents. The extent of extraction follows a combination of metal ion basicity and extractant acidity, so that for any particular extractant a selectivity series of elements can be constructed which tends, in the absence of any specific metal–extractant interaction, to follow the lyotropic series.

However, because these reagents are so versatile and extract so many metals their use in hydrometallurgical flowsheets is minimal. Thus although a considerable amount of data have been published on carboxylic acid extractions (Yamada and Tanaka 1985; Rice 1978; Flett and Jaycock 1973) no current industrial application is known. Similarly, with organosulphonic acids, here the extractants are more acidic than either the carboxylic or organophosphoric acids, which tends to shift the pH range for extraction to lower values that may make integration with the rest of the flowsheet difficult. The condensing of the overall pH range for metal extraction also makes selectivity for a particular metal a problem (Markovits and Choppin 1973). Finally, both the carboxylic and organosulphonic acids do tend to

TABLE 1.1 *Commercial acidic extraction reagents*

Carboxylic acids	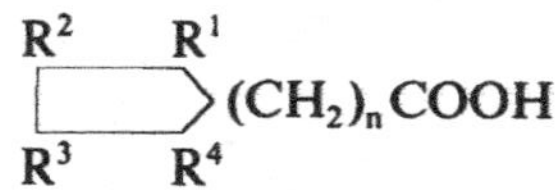	Naphthenic acids (9,13)
	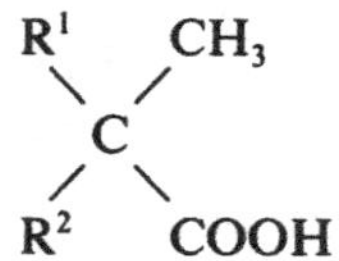	Versatic acids (9)
	$R^1 = R^2 = C_6$ (Versatic 10) $R^1 = R^2 = C_4 - C_5$ (Versatic 911)*	
Phosphorus acids	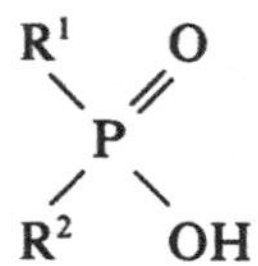	
Phosphoric acids	$R^1 = OH$; $R^2 = C_4H_9CH(C_2H_5)CH_2O-$	mono-2-ethylhexylphosphoric acid (4,7)
	$R^1 = R^2 = C_4H_9CH(C_2H_5)CH_2O-$	di-2-ethylhexylphosphoric acid (DEHPA) (3,4,7,11)
	$R^1 = R^2 = p-CH_3(CH_2)_6CH_2-C_6H_5O$	di-p-octylphenylphosphoric acid (OPPA) (7)
Phosphonic acids	$R^1 = C_4H_9CH(C_2H_5)CH_2-$ $R^2 = C_4H_9CH(C_2H_5)CH_2O-$	2-ethylhexylphosphonic acid mono-2-ethylhexyl ester (PC88A) (3); (SME418) (9); (P-507) (12)

Phosphinic acids	$R^1 = R^2 = CH_3C(CH_2)_2CH_2CH(CH_3)CH_2-$	di-2, 4, 4-trimethylpentylphosphinic acid (Cyanex 272) (2)
	$R^1 = R^2 = C_4H_9CH(C_2H_5)CH_2-$	di-2-ethylhexylphosphinic acid (P-229) (12)
Thiophosphorus acids	$(RO)_2P(S)OH$ $R = C_4H_9CH(C_2H_5)CH_2-$	monothiophosphoric acid di-2-ethylhexyl ester (HoE F 3787) (4)
	$R_2P(S)OH$ $R = CH_3C(CH_3)_2CH_2CH(CH_3)CH_2-$	di-2, 4, 4, -trimethylpentyl monothiophosphinic acid (Cyanex 302) (2)
	$(RO)_2P(S)SH$ R unknown (2-ethyl hexyl ?)	dithiophosphoric acid diester (HoE F 3541) (4)
	$R_2P(S)SH$ $R = CH_3C(CH_3)_2CH_2CH(CH_3)CH_2-$	di-2, 4, 4, -trimethylpentyldithiophosphinic acid (Cyanex 301) (2)
Sulphonic acid	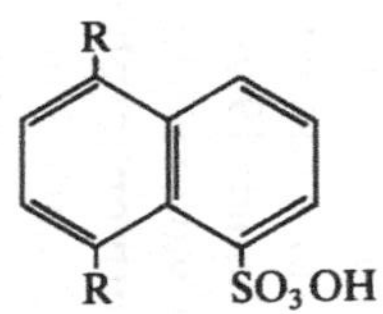 $R = C_9H_{19}$	5, 8-dinonylnaphthalenesulphonic acid (DNNSA)

have an appreciable aqueous solubility, which can, under operating conditions, become too high for economic and environmental considerations. The organophosphorus acids however are used commercially in a number of processes. Their use is generally confined to the monobasic ester derivatives as, although some dibasic reagents are available, the formation of polymeric metal complexes with the latter causes solubility problems. The most widely used reagent is di-2-ethylhexylphosphoric acid, and its chemical stability, low aqueous solubility, and good loading and stripping characteristics make it one of the most versatile extractants. It has been shown to exist largely as dimers in organic diluents leading to the formation of polymeric metal species at low metal loadings (eqn (1.11))

$$M^{n+} + \overline{n(RH)_2} \rightleftharpoons \overline{M(R-HR)_n} + nH^+. \quad (1.11)$$

However, as metal loading increases these polymers dissociate,giving species such as MR_n, although these too may be polymeric with bridging oxygen atoms (Kolarik and Grimm 1976). Recently interest has been revived in the commercial use of organophosphonic and organophosphinic extractants (Table 1.1), especially for the separation of cobalt and nickel (see later). These reagents also exhibit interesting variations in the selectivity series for zinc and the alkaline earth elements (Table 1.2), which would allow zinc to be extracted at a lower pH than normally possible with DEHPA and with a greater selectivity over calcium. Also for Cyanex 272 cobalt can be extracted at a lower pH than calcium which has advantages in eliminating any possibility of precipitating gypsum during acid stripping of a cobalt-loaded organic phase (Rickelton *et al.* 1984.) The irregular behaviour of calcium when compared with the other elements may be attributed to specific interactions between the metal and extractant.

Recently commercial reagents based on thiophosphinic acids have become available. This type of extractant has been studied for the extraction of second and third row elements for several years (Sekine and Hasagawa 1977; Cote and Bauer 1986) but have only just excited commercial interest. American Cyanamid have now produced two such reagents, Cyanex 301, a dialkyl dithiophosphinic acid, and Cyanex 302, the dialkyl monothio analogue. Cyanex 301 has been developed as an extractant for zinc in the presence of calcium (Brown *et al.* 1987) and this type of compound can also

TABLE 1.2 *Selectivity series for organophosphorus acid extractants (Cox and Flett 1987)*

Phosphoric acid (DEHPA)	Zn > Ca > Cu > Mg > Co > Ni
Phosphonic acid (PC88A)	Zn > Cu > Ca > Co > Mg > Ni
Phosphinic acid (Cyanex 272)	Zn > Cu > Co > Mg > Ca > Ni

be used to extract cadmium from, for example, wet process phosphoric acid (Bierman *et al.* 1985).

Cyanex 302 has a similar performance to the dithio compound but has a higher hydrolytic stability (American Cyanamid). Hoechst also supply two similar extractants in HOE F 3541, a dialkyldithiophosphoric acid ester, and HOE F 3787, di-2-ethylhexyl monothiophosphoric acid ester.

1.2.1.2 *Chelating acid extractants*

During the past 20 years the commercial availability of these extractants has provided a particular stimulus to the use of liquid–liquid extraction in hydrometallurgy. This was largely due to the development of reagents which were selective for copper over ferric iron in acidic leach solutions. The background to this development has been widely documented (Ritcey and Ashbrook 1984) and many papers have been published especially on the copper–hydroxyoxime system (Hanson and Whewell 1981). In addition to hydroxyoximes other reagents have been proposed for copper extraction from both acidic and ammoniacal solutions. The former category includes extractants based on 8-quinolinol, for example: Kelex 100 and 108, LIX 26; 8-sulphonamidoquinoline, LIX34, whereas the latter are mainly derivatives of 1,3-diketones, LIX54, DK16. All these chelating reagents (Table 1.3) are capable of extracting other metals, but under conditions which normally allow adequate separation factors for copper to be achieved. The separation of copper and iron with the 8-quinolinol derivatives is interesting in that it is based on kinetic rather than equilibrium factors (Fleming *et al.* 1980). All these extractants rely on a combination of an acidic function and a specific chelation effect with the metal to achieve selectivity and extraction. In addition to these groupings other substituents are incorporated to convey compatability with the organic phase and satisfy particular steric requirements (Dalton and Price 1982). Thus this type of reagent offers the greatest potential for the design and production of selective extractants for any desired element; whether or not these will become available commercially in the future will depend mainly on economic factors.

1.2.2 Basic, anionic extractants

In hydrometallurgical processes these basic extractants are limited to the alkylamines and quaternary ammonium compounds. Although it is possible to use other cationic 'onium species in analytical applications (Sekine and Hasagawa 1977) their cost and potential toxicity make them unsuitable for hydrometallurgical use.

Extraction depends on the ability of metal ions to form anionic species in the aqueous phase which are then extracted as ion-pairs by amine salts in an ion-exchange process (eqn (2.12)).

TABLE 1.3 *Commercial chelating acid extractants*

α-hydroxyoximes	$R^1C(NOH)CH(OH)R^2$ $R^1 = R^2 = C_4H_7CH(C_2H_5)-$	5,8-diethyl-7-hydroxydodecan-6-oxime (LIX63) (5) *
β-hydroxyaryloximes	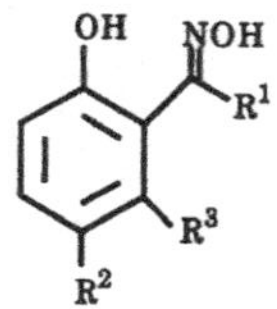	
R^1 = phenyl	$R^2 = C_{12}H_{25}$; $R^3 = H$	LIX64 (5) *
	$R^2 = C_9H_{19}$; $R^3 = H$	LIX65N (5)
	LIX65N + LIX63 (1%)	LIX64N (5)
	$R^2 = C_9H_{19}$; $R^3 = Cl$	LIX70 (5) *
	LIX70 + LIX65N	LIX71 (5)
	LIX70 + LIX65N + LIX63	LIX73 (5)
R^1 = benzyl ($C_6H_5CH_2-$)	$R^2 = C_9H_{19}$; $R^3 = H$	P17 (1) *
$R^1 = CH_3$	$R^2 = C_9H_{19}$; $R^3 = H$	SME529 (9) LIX84 (5)
$R^1 = H$	$R^2 = C_9H_{19}$; $R^3 = H$	P1 ; P50 (1)
	P50 + nonylphenol modifier (1 : 1 ratio P5100) (1 : 3 ratio P5300)	P5000 series (1)

	P50 + tridecanol modifier	PT5050 (1)
	P50 + unknown ester modifier	M Series (1)
	$R^2 = C_{12}H_{25}$; $R^3 = H$	LIX860 (5)
	LIX84 + LIX860	LIX984 (5)
	LIX860 + tridecyl alcohol modifier	LIX622 (5)
Quinolines	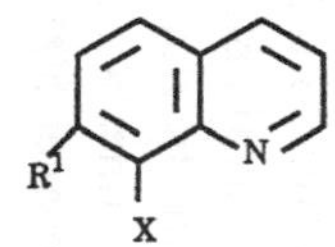	
X = OH	R^1 = 5,5,7,7, tetramethyl-1-octenyl = 4-ethyl-1-methyloctyl (post 1976) Demopoulos and Distin (1983)	Kelex 100 (10) (pre 1976)
	R^1 = unknown side chain structure	Kelex 108 (10)
	R^1 = unknown alkyl	LIX26 (5)*
$X = NHSO_2C_6H_5R^2$		
	R^1 = H ; R^2 = unknown alkyl ($C_{12}H_{25}$?)	LIX34 (5)*
β-diketones	$R^1COCH_2COR^2$	
	$R^1 = R-C_6H_5$; $R^2 = CH_3(CH_2)_5-$	LIX54 (5)
	R^1 and R^2 unknown	DK16 (4)*
	R^1 and R^2 unknown but containing fluorine substituents	LIX51 (5)*

$$(p-n)\overline{R_4N^+Y^-} + MX_p^{(p-n)-} \rightleftharpoons \overline{(R_4N)_{p-n}(MX_p)} + (p-n)Y^-. \quad (1.12)$$

An alternative mode of extraction involves an addition reaction, (2.13):

$$\overline{R_4N^+X^-} + MX_n \rightleftharpoons \overline{R_4N^+MX^-_{(n+1)}}. \quad (1.13)$$

There is little doubt that both types of reaction occur even when the anion is the same in both reactants. The controlling factor is the free concentration of the anion in the aqueous phase which determines the nature and concentration of the possible anion–metal complexes. Thus low anion concentration would favour the addition reaction and high concentrations the exchange reaction.

In the case of extraction by primary, secondary, or tertiary amines these require protonation before they can act as extractants (eqn (1.14)):

$$\overline{R_3N} + HY \rightleftharpoons \overline{R_3HN^+Y^-}. \quad (1.14)$$

This indicates that these reagents also behave as extractants for acids in their free base form.

In general the ease of extraction for nitrato- and halo-complexes follows the order quaternary > tertiary > secondary > primary amine salts, although for economic reasons the quaternary species are seldom used in commercial operations. The nature of the alkyl group is also important, thus it is found that aliphatic amines are generally better extractants and that aromatic groups, especially when attached directly to the nitrogen, weaken the extraction ability. Also too much chain branching tends to reduce extraction, probably as a result of steric hindrance impeding the solvation of the ion-pair thus reducing its stability relative to the free amine in the organic phase. Inspection of the commercially available extractants (Table 1.4) shows that they have molecular weights in the range of 250–600, with the majority lower than 450. This is because amines with molecular weights less than 250 are too water soluble, and those over 600 have reduced solubility in the common organic diluents. In both aqueous and organic solution at low concentrations the amine salts are present as monomers, but as the concentration increases micelle formation occurs. The concentration at which this takes place depends on the nature of the alkyl substituents, the organic diluent, and temperature, but is likely to be less than that used in hydrometallurgical operations. This means that problems with solubility of the extractant and metal–extractant complex do occur and third-phase formation is quite common. This can be eliminated by the addition of diluent modifiers such as long chain alcohols, which are able to solvate the amine salts and ion-pair complexes.

Metals are capable of forming a number of different types of anionic complex which can be extracted by these reagents, for example, the simple and condensed oxo-anions of molybdenum, tungsten, rhenium, technecium, and manganese. However, most systems involve complex anions of the mineral acids and their salts used in leaching. The extent of complexation by these common anions is nitrate < chloride < sulphate. Thus nitrate only forms weak complexes in aqueous solution but with some cations, e.g. uranyl (VI), they are sufficiently stable to allow extraction. In this particular case it has been suggested that extraction occurs via the neutral complexes such as $(R_3N.HNO_3)UO_2(NO_3)_2$ (Lloyd and Mason 1964). Sulphato complexes, although more common than nitrato, are still rather uncommon and suggested processes are confined to uranium(VI), thorium, hafnium, and rare earths (Cattrall and Slater 1973). In this system the sulphate–bisulphate equilibrium becomes important and tends to impose a limit on the range of sulphate concentration and pH for extraction. Thus at high acid concentration bisulphate is the dominant species and extraction is reduced, presumably because bisulphate forms a weaker complex and also the HSO_4^- ion competes for the amine extractant. Conversely, as the acid concentration decreases the sulphate concentration increases but the pH now becomes too high for the amine to protonate readily and again extraction is reduced. Finally, in the sulphate system the extraction depends strongly on the type of amine with the order primary > secondary > tertiary being followed. This is the reverse of that found for nitrato- and halo-complexes but is the same as that for sulphuric acid extraction.

The majority of hydrometallurgical applications of ion-pair extraction involves the formation of chloro-anions. Extraction from alkali chloride media is usually preferred to mineral acid not only because metal chlorides are used as lixiviants, but also the HCl_2^- ion formed in high mineral acid concentrations competes for the extractant.

The selectivity series for metals depends on the ease of formation of stable aqueous chloroanions, and for base metals follows the order: zinc > iron (III) > copper > cobalt > iron (II) > manganese (II). Because nickel does not form chlorocomplexes under these conditions easy separation of cobalt from nickel can be achieved with this system. Again it should be emphasized that as very high aqueous concentrations of salts are used, activity cannot be equated to concentration, and also, because of non-ideal behaviour of the extractant in the organic phase, great care must be taken in extrapolating from laboratory data obtained in dilute solutions. Consequently work on these systems should rely on experimental data using the actual operating conditions of the proposed process rather than predictions from published data.

The stripping of the loaded organic phase can be achieved by a wide variety of salt solutions such as chloride, sulphate, or carbonate, the choice depending on the subsequent recovery process used for the metal. An alternative, which can be used for primary, secondary, and tertiary amines, is to use basic solutions which reverse the amine protonation step. This generally gives a more complete strip in fewer stages.

TABLE 1.4 *Commercial basic extractants*

Primary amines	RNH_2	
	$R = (CH_3)_3C(CH_2C(CH_3)_2)_4-$	Primene JMT (8)
Secondary amines	R^1R^2NH	
	$R^1 = C_9H_{19}CH=CHCH_2-$ $R^2 = CH_3C(CH_3)_2-(CH_2C(CH_3)_2)_2-$	Amberlite LA-1 (8)
	$R^1 = CH_3(CH_2)_{11}-$ $R^2 = CH_3C(CH_3)_2-(CH_2C(CH_3)_2)_2-$	Amberlite LA-2
	$R^1 = R^2 = CH_3(CH_2)_{12}-$	Adogen 283 (10)
	$R^1 = R^2 = CH_3CH(CH_3)CH_2(CH_2)_6-$	HOE F 2562 (4)
Tertiary amines	$R^1R^2R^3N$	
	$R^1 = R^2 = R^3 = CH_3(CH_2)_7-$	tri-octylamine, TOA (13)
	R^1, R^2, R^3 or C_8-C_{10} mixture with C_8 predominant about 2:1	Alamine 300 (5) Alamine 336 (5)
	$R^1 = R^2 = R^3 = (CH_3)_2CH(CH_2)_5-$	tri-iso-octylamine TiOA (13) Adogen 381 (10) Alamine 308 (5) Hostarex A 324 (4)

	mixture of octyl and decylamines (1:1)	Hostarex A 327 (4)
		Adogen 364 (10)
	$R^1 = CH_3(CH_2)_7-$; $R^2 = CH_3(CH_2)_9-$; $R^3 = CH_3(CH_2)_{11}-$	Adogen 368 (10)
	$R^1 = R^2 = R^3 = (CH_3)_2CH(CH_2)_7-$	tri-isodecylamine
		Alamine 310 (5)
	$R^1 = R^2 = R^3 = CH_3(CH_2)_{11}-$	Adogen 363 (10)
		Alamine 304 (5)
	$R^1 = R^2 = R^3 = CH_3(CH_2)_{12}-$	Adogen 383 (10)
	$R^1 = R^2 = R^3 = C_{28}H_{57}-$	Amberlite XE204 (8)
Quaternary amine salts	$R^1R^2R^3N^+(CH_3)Cl^-$	
	$R^1R^2R^3$ as Alamine 336	Aliquat 336 (5)
	$R^1 = R^2 = R^3 = (C_8 - C_{10})$	Adogen 464 (10)
	Mixture of octyl and decylamines	HOE S2706 (4)

1.2.3 Solvating extractants

In general a solvating extractant tends to be rather unselective, extracting a wide range of neutral complexes and also under certain conditions free acid. Like the anionic extractants, the level of metal extraction by solvating extractants depends on the extent of complex formation in the aqueous phase. Extraction is by replacement of water in the primary solvation shell by the solvating extractant thus rendering the complex more compatible with the organic phase, (eqns (1.15), (1.16)).

$$MX_p(H_2O) + \overline{qS} \rightleftharpoons \overline{MX_pS_q} + qH_2O \qquad (1.15)$$

or

$$HMX_{(n+1),aq} + \overline{xS} \rightleftharpoons \overline{HS_x^+ \, MX_{(n+1)}^-} + H_2O. \qquad (1.16)$$

An example of the latter reaction is the extraction of ferric iron from strong hydrochloric acid, where it is present as $HFeCl_4$, by carboxylic acid esters. The extent of extraction depends on a number of factors including the nature and concentration of the complexing anionic ligand, which in turn influences the type of metal complex formed; the degree of hydration of these metal complexes in the aqueous phase; and the relative strength of the metal–water and the metal–extractant bonds, which leads to competition between water and the extractant for the solvation sites. The nature of the extracted complex is determined by a balance between the electropositivity of the metal and the polarity of the solvating ligand compared with water. Thus with a strongly electropositive metal the strongly polar organophosphorus compounds compete favourably with water for the solvation sites and replace all the coordinated water in the extracted complex. On the other hand the extracted complex formed between this same metal and the less polar ethers and ketones still retains some water molecules. Similarly, the extracted complex of a less positive metal and an organophosphorus extractant would still include some water. A further factor which influences the degree of extraction is the ionic strength and composition of the aqueous phase. This is because the species in this phase will compete for water of solvation, and if cations are present which are more easily hydrated than the metal complex being extracted then the degree of hydration of the latter will be reduced, making formation of bonds to the extractant easier. This process is termed 'salting out' and is exemplified by an increase in the distribution coefficient of the extractable species in the presence of an electrolyte. In these situations the extractable complex may be regarded as a non-electrolyte and hence be described in terms of appropriate thermodynamic interactions and models (Marcus and Kertes 1969). The most extensive studies relate to the extraction of metal nitrates (in particular uranium), which are only poorly extracted in the absence of salting agents. However, in some cases the extractable salt can act as its own salting agent. Empirically it has been shown that for good extraction the aqueous phase should contain 2–3 mol/dm^3 of a non-extractable nitrate or 3–4 mol/dm^3 of an extractable species (Jenkins and McKay 1954). The cation of the salting agent does have a major effect on extractability with the distribution coefficient increasing as the cationic radius, in any one periodic group, decreases and as the charge on the cation increases. Thus, for example, the extraction of europium by tri-n-butylphosphate from 5.6 mol/dm^3 neutral nitrate media decreases in the order: Al > Mg > Zn > Li > Cu > Na, Ca > NH_4 (Brown 1959). As this phenomenon can be related to the availability of 'free water' in the aqueous phase it follows that metal nitrates which crystallize with water are better salting out agents than those that have no water of crystallization (Scott 1949).

The phenomenon of salting-in is less common. In this case the extractable species has a lower activity in the electrolyte than in pure water. A typical example is the depression of the extraction of nitric acid by tri-n-butylphosphate in the presence of uranyl nitrate or certain actinide or high valent nitrate species. This is a feature of situations where the salting agent, i.e. uranyl nitrate, has a higher affinity for the extractant than the extractable species, i.e. nitric acid.

As implied above there are two major types of commercially available oxygen-donating extractants together with several more specialist reagents (Table 1.5). The use of carbon–oxygen extractants like ethers and ketones is well established for halide and nitrate systems. However, aqueous solubility and flammability limits their hydrometallurgical use to a few specialist and low tonnage operations like gold recovery with dibutyl carbitol (Morris and Khan 1968) and niobium/tantalum separation with methyl isobutylketone (Soisson *et al.* 1961). In both these examples the reagent is used in an undiluted form, a rather rare occurrence.

A range of organophosphorus extractants are commercially available, relying on the donor properties of a phosphorus–oxygen bond for extraction. It is found that the extractive power of these reagents is reflected by the polarity of this P=O bond, which increases with increasing number of carbon-phosphorus bonds, i.e. $(RO)_3PO < (RO)_2RPO < (RO)R_2PO < R_3PO$. This order also reflects the aqueous solubility, with phosphine oxides being the least soluble. The situation becomes more complex in acid solution when generally the aqueous solubility of all compounds increases with increasing acidity, sometimes to prohibitive amounts.

Like the amine extractants these compounds will also extract mineral acids, in this case by solvation of the proton. As metal extraction is often carried out from acid media the possibility of coextraction of acid can make the overall situation complex especially at high acidities. The stripping reaction can involve a simple re-aquation of the metal cation or, if the

TABLE 1.5 *Commercial solvating extractants*

Carbon-oxygen extractants		
Amides	$R^1CONR^2_2$	
	$R^1 = R^2 =$ unknown	A101 (12)
	$R^1 = CH_3$; $R^2 = CH_3(CH_2)_5CH(CH_3)-$	N503 (12)
Ethers	$CH_3(CH_2)_3OCH_2CH_2OCH_2CH_2O(CH_2)_3CH_3$	dibutylcarbitol (13) Butex
Ketones	$(CH_3)_2CHCH_2COCH_3$	isobutylmethylketone (13) MIBK
Phosphous-oxygen extractants	$R^1R^2R^3PO$	
Phosphorus esters	$R^1 = R^2 = R^3 = CH_3(CH_2)_2CH_2O-$	tri-n-butylphosphate (13) TBP
	$R^1 = R^2 = CH_3(CH_2)_2CH_2O-$; $R^3 = CH_3(CH_2)_2CH_2CH_2-$	dibutylbutylphosphonate (7,11) DBPP
	$R^1 = R^2 = CH_3O-$; $R^3 = CH_3-$	dimethylmethylphosphonate (7)
Phosphine oxides	$R^1 = R^2 = R^3 = CH_3(CH_2)_6CH_2-$	tri-n-octylphosphine oxide TOPO (13) Cyanex 921 (2)
	Mixed alkyl groups n-octyl and n-hexyl	Cyanex 923 (2)
	$R^2 = CH_3C(CH_3)_2CH_2CH(CH_3)CH_2-$ $R^3 = CH_3(CH_2)_6CH_2-$	Cyanex 925 (2)

Phosphine sulphides	$R^1R^2R^3PS$	
	$R^1 = R^2 = R^3 = (CH_3)_2CHCH_2-$	tri-isobutylphosphine sulphide Cyanex 471 (2)
Alkyl sulphides	R^1R^2S	
	$R^1 = R^2 = C_6H_{13}-$	dihexyl sulphide (13)
	$R^1 = R^2 = C_7H_{15}-$	diheptyl sulphide (13)
Nitrogen extractants	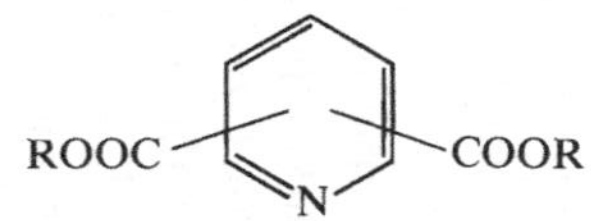	
	(Precise structure unknown)	DS5443 (1)

[1]Acorga Ltd (ICI); [2]American Cynamid; [3]Daihachi Chemical Co; [4]Farbwerke Hoechst AG; [5]Henkel Corp.; [6]King Industries; [7]Mobil; [8]Rohm and Haas; [9]Shell Chemical Co; [10]Sherex; [11]Tenneco; [12]Chinese Reagents; [13]Various manufacturers;
* Reagent discontinued

metal–extractant bonds are strong, more specific reagents are required. As the stripping process involves the substitution of one ligand (strip reagent) for another (extractant) a study of relative stability or formation constants can provide the desired information for a good strip reagent. The commercial use of organophosphorus esters to date has been largely limited to the use of tri-n-butyl phosphate, and then mainly in the nuclear industry where it is the preferred extractant for fuel reprocessing (Naylor and Wilson 1983; Naylor and Eccles 1988). TBP has also been used for zirconium/hafnium (Nandi *et al.* 1983) niobium/tantalum separations (Ritcey and Ashbrook 1979), where it competes with MIBK and amines (Murthy *et al.* 1986); and in the Falconbridge nickel process where iron in the leachate is extracted as $FeCl_3$ (Thornhill *et al.* 1971). TBP does suffer some degradation problems, being hydrolysed to di-n-butylphosphoric acid, which is of course an acidic extractant. In spite of this potential disadvantage TBP is still widely used and studied, as a recent comprehensive text shows, (Schulz and Navratil 1987). The phosphine oxide, tri-n-octylphosphine oxide, although having the highest extractive power, is more expensive and only currently features in one commercial operation as a component with di-2-ethylhexylphosphoric acid in the synergistic extractant mixture for uranium from wet-process phosphoric acid. However, a number of other processes, including niobium/tantalum separation, where the niobium purity compares very favourably with that from the MIBK process (Hahn and Retelsdorf 1984; Eckert and Bauer 1984), rhenium recovery from catalyst wastes (Bright 1984) and arsenic removal from copper electrolyte solutions (Marr *et al.* 1984), have been patented.

Tri-n-octylphosphine oxide is a solid and some problems can be found with solubility in common hydrocarbon diluents, especially at low temperatures. To overcome these problems American Cyanamid have produced Cyanex 923 which a mixture of four trialkylphosphine oxides where the alkyl groups are combinations of n-hexyl and n-octyl. This reagent is completely miscible with all common diluents at all temperatures. A further development has been the manufacture of a branched chain derivative, Cyanex 925, with 2,4,4,trimethylpentyl substituents. This is reported (American Cyanamid) as being more selective than the straight chain analogue with advantages in extraction of tin, arsenic, and gallium over TOPO .

In addition to these oxygen donor extractants, other solvating reagents are available commercially. For example, sulphur analogues of the ketones and organophosphine oxides have recently been produced for metal extraction. The rationale behind this development is the 'hard and soft acid and base theory' (Pearson 1963) which concludes that some metal ions termed 'soft' would form stronger complexes when associated with 'soft' donors like sulphur, rather than 'hard' donors like oxygen. These 'soft' metal ions include Cu(I), Ag(I), Cd(II), Hg(II), Pd(II), etc. in contrast to ions such as Cu(II) and Pt(IV) which are termed 'hard'. Thus with the reagent Cyanex 471, tri-isobutylphosphine sulphide, silver can be separated from copper(II) (Rickelton and Robinson 1987) and palladium(II) can also be extracted by this reagent (American Cyanamid), as well as the alternative sulphur donor, di-hexylsulphide (Edwards 1979).

Another recent development is the production of a solvating reagent for copper(II) from chloride media. The reagent, Acorga DS5443, is based on a pyridine dicarboxylic acid ester, and extracts copper(II) chloride as the neutral complex, CuL_2Cl_2 (Dalton *et al.* 1983).

Other less common solvating extractants include amides, $RCONR_2'$, largely studied in China (Zhou *et al.* 1982) and also in the Soviet Union (Laskorin *et al.* 1978) and Musikas and Germain (1988) for the extraction of uranium and transuranic elements. These compounds, like the amines, are capable of protonation in acid solution so may function either as a solvating or anionic extractant depending on the aqueous phase conditions. Also bifunctional compounds have been prepared which include di-phosphine oxides, $R_2P(O)(CH_2)_nP(O)R_2'$ (Schulz and Navratil 1981), and reagents with both amide and phosphorus-oxygen groupings, carbamoylmethylphosphine oxides, $R_2NCO(CH_2)P(O)R_2'$ (Schulz and Navratil 1981; Horwitz *et al.* 1983), both of which have potential for actinide purification.

1.2.4 Mixed extractant systems

One of the few examples of the commercial use of mixed extractants is the extraction of uranium from phosphoric acid with DEHPA/TOPO mentioned above. However, with the reagent manufacturers continuing to suggest that it is becoming economically more difficult to produce new reagents, the possibility of modifying extraction performance by using mixed systems is attractive. Thus it has been shown in a laboratory situation that metals can be extracted by mixed reagents under conditions where neither extractant used singly would extract, and cleaner separations have been achieved by favouring the extraction of the chosen metal over the impurities. These modifications are the result of the formation of complexes involving both reagents attached to the metal atom (eqn (1.17)).

$$\overline{MR_n} + \overline{yS} \rightleftharpoons \overline{MR_nS_y}\,. \tag{1.17}$$

Clearly this addition reaction is more likely to occur if the metal atom is coordinately unsaturated, but examples are known where an increase in the normal coordination number does occur. Often, although not always, mixed complex formation is accompanied by synergism, where the performance by the mixed extractants exceeds the combined effects of the individual compounds. This must be associated with an increase in the magnitude of the formation constant for the mixed complex. In all these systems it is important to eliminate any possibility of interaction between the extractants, and to establish coordination of both extractants to the central metal atom. Otherwise variation in extraction behaviour could be due to changes in the polarity of the organic phase or changes in the performance of one of the components by compound formation. Systems which offer the possibility for commercial use will be outlined under the individual extractant classes.

1.2.4.1 *Acid extractants*

Mixed complex formation in acid extraction can occur in several ways. By far the most common is the addition to the metal-acid complex of neutral donors such as organophosphorus esters, amines, or undissociated acids. These can all be represented by the addition reaction (1.17), where S is the neutral donor, or eqn (1.18), where the undissociated acid molecule HR is partially substituted by S.

$$\overline{MR_n(HR)_x} + y\overline{S} \rightleftharpoons \overline{MR_n(HR)_{x-y}S_y} + y\overline{HR}\,. \tag{1.18}$$

Substitution reactions also occur when coordinated water in a partially hydrated chelate metal complex is replaced by a neutral donor. In this case the reason for any enhancement is identical to that found with solvating extractants, that is, decreasing the polarity of the complex by replacement of water increases the compatability with the organic diluent.

Several interesting synergistic systems involving commercial reagents have been investigated. For example, the alpha-hydroxyoxime, LIX63, and organophosphoric (Joe *et al.* 1966) or carboxylic acids (Flett and West 1971), where the undissociated acids seem to function as the donating molecule. Thus it has been shown that the degree of synergism for nickel extraction is directly related to the pK_a of the carboxylic acid and the polarity of the carbonyl oxygen (Castresana *et al.* 1988) and that mixed complexes are only formed over a certain range of acid pK_a values (Schwarz 1978). Other systems where an undissociated acid acts as a donor include the beta-hydroxyaryloximes and carboxylic (Nyman and Hummelstedt 1974), organophosphoric (Moore and Partridge 1972), and organosulphonic (Nyman and Hummelstedt 1974) acids for copper, nickel, and cobalt; alkylated 8-quinolinols with carboxylic acids for cobalt (Lakshmanan and Lawson 1973), nickel (Nyman and Hummelstedt 1974), zinc and cadmium (Harrison *et al.* 1976). In this system cobalt oxidation, which inhibits stripping in the case of the chelating extractant alone, is eliminated by the formation of a six coordinated synergistic complex (Flett *et al.* 1975).

Carboxylic acid extractants have also been studied with amines and organophosphorus esters as donors. However, although mixed complexes were formed, little synergism was observed (Flett and Jaycock 1973). In contrast, when non-chelating aliphatic oximes were used synergism was found for a number of first row transition metals whereas non-transition metals, in particular the alkaline earth elements, showed antagonistic depression of extraction. Thus in several instances improved separations between mixtures of elements can be obtained (Preston 1983a,b,c; 1985). Even so, the mixed extractant system did tend to degrade on recycling, which would tend to limit any commercial use (Rice and Smith 1986). Finally the systems organophosphorus acids/organophosphorus esters, particularly using DEHPA, have been extensively studied with, as expected, the stability of the mixed complex reflecting the increasing polarity of the phosphorus oxygen bond in going from TBP to TOPO.

1.2.4.2 *Anion extractants*

Little is known of the formation of mixed complexes in amine extraction systems. Some evidence is available that synergistic extraction of uranium (Deptula and Minc 1967) and hafnium (Liem and Sinegribova 1971) occurs with mixtures of tri-n-octylamine and di-2-ethylhexylphosphoric acid from high concentrations of sulphate. Here, in addition to forming the amine organophosphate salt, the phosphate species coordinates to the metal and enhances the extraction; but it is more usual for mixtures of these reagents to react to give the amine salt and for the metal extraction to be depressed. No information is available on extraction with mixtures of amine salts and solvating reagents.

1.2.4.3 *Solvating extractants*

Mixtures of these extractants with acids and amines have already been considered. Mixtures of two solvating extractants have not been extensively studied (Kertes 1971) although it is likely that mixed complex formation and synergism could be found, especially where the two extractants differ in dielectric constant or hydrogen bonding capabilities as interactions with the diluent are likely to be important.

1.2.5 Modifiers

Another occasion when mixtures of chemicals are required in the diluent occurs when the metal–extractant complex has a limited solubility in the chosen diluent. The complex then tends to separate as a third phase intermediate in density between the solvent and aqueous phases. As third phase formation is related to the solubility of the extractant complex it depends on factors such as temperature and the nature of the diluent. Thus raising the temperature or changing to a more aromatic diluent often reduces the problem. Also the phenomenon is more common with some reagent systems than others, for example in extraction with amines and di-2-ethylhexylphosphoric acid. The presence of a third phase causes considerable processing difficulties and so, to minimize any incipient formation, chemicals are added as modifiers to the organic phase. These organic compounds tend to consist of oxygen-containing molecules such as alcohols, phenols and ethers. They should be chosen to have a high organic phase solubility, low aqueous solubility and low cost; those most commonly used include nonyl phenol, isodecanol, 2-ethylhexanol, and tri-n-butylphosphate. In addition to increasing the solubility of the metal–extractant complex modifiers have also been shown to affect other aspects of reagent performance, for example reagent strength, selectivity, kinetics, degradation, entrainment, phase separation, and crud control.

One of the most intensive studies of the effect of modifiers on reagent performance has centred around their use with the copper hydroxyoxime extractants, in particular P50. The incorporation of various modifiers has lead to the formulation of the P5000 series (Tumilty *et al.* 1977a,b) and the M-series of reagents (Dalton and Seward 1986). Nonyl phenol was originally added to the 5-nonylsalicylaldoxime (P50) as a means of modifing its stripping performance so that the circuit could interface with normal spent copper tankhouse electrolyte (see Copper). It had little effect on the extraction circuit so the modifier gave better overall transfer efficiency than the unmodified aldoxime. The incorporation of a different modifier, tridecanol, indicated that the choice of modifier can effect subtle changes to other properties, some of which are not yet understood. Thus copper/iron(III) selectivity was shown to be better with alcohols than nonyl phenol and than certain esters were even better (Table 1.6).

The P50 reagent is known to exhibit fast kinetics and these are little changed by addition of modifiers at room temperature. However, at low temperature, (0°C), P5100 has a much poorer kinetics than PT5050, the

TABLE 1.6 *Comparison of modifier performance for copper/iron(III) selectivity of P50 (Dalton and Seward 1986)*

Modifier	Reagant	Cu/Fe Ratio in extract. feed A	feed B
Nonyl phenol	P5100	1859	386
alcohol (unspec.)	M5615	6024	976
ester (unspec.)	M5640	7435	1210

feed A = 3 g/dm^3 Cu(II); 3 g/dm^3 Fe(III) pH 2.0
feed B = 3 g/dm^3 Cu(II); 30 g/dm^3 Fe(III) pH 2.0

tridecanol modified aldoxime P50. Oximes are prone to hydrolysis especially when in contact with the strip acid; for commercial operation such degradation should be minimized. Generally aldoximes are less stable than ketoximes, and it has been suggested that the presence of modifiers can accelerate this degradation (Sudderth and Kordowsky 1986). This generalisation should be viewed with caution because it has been shown (Dalton and Seward 1986) that the precise nature of the modifier can have a significant effect on the level of degradation. Finally in an extraction circuit the problems of entrainment and crud are very important for efficient operation. Once again it has been shown that modifiers can have an effect on both these parameters. Clear differences were found between types of modifier and also between different molecules within the same class of reagents (Dalton and Seward 1986).

Although discussion has been concentrated on this one example of reagent/modifer combination, similar examples can be found in other systems, for example selectivity, rare earth/DEHPA (Ritcey and Lucas 1974), and phase separation, copper/Kelex 100 (Ritcey and Lucas 1974). The mode of action of modifiers is not yet fully understood and a lot more basic knowledge is required; however, it is seems that in the aldoxime/modifier system the modifiers are involved in hydrogen bonding via the oxygen lone pairs to hydroxy-protons on the extractant (Dalton 1979). It is also likely that interfacial properties of the modifiers will be important for aspects such as phase separation, entrainment, and crud control. The above examples indicate that the choice of whether or not to use a modifier is not simple and that benefits other than elimination of third phase development can accrue from their use. Until the role of the modifier in extraction circuits is better understood this choice can only be made by a series of trials which, as shown by the work above, may have to be quite extensive to attain the desired performance.

1.2.6 Diluents

The major component in the organic phase is the diluent, for, except in very few situations like the use of alcohols, ethers, and ketones, the extractant does not possess suitable properties to be used by itself. Therefore a diluent is required to:

(1) reduce the viscosity of the extractant;

(2) provide a suitable concentration of extractant for the metal in the feed;

(3) reduce the tendency of extractants to emulsify with the aqueous feed, and improve dispersion and coalescence.

For use as a diluent a liquid should satisfy the following criteria: a wide range of miscibility with the extractant, and modifier if used; and a high solvency for the extracted metal complex. It should also have a low aqueous solubility together with a low volatility and a high flash point to minimize losses and fire hazard. Finally it should have a low surface tension for good dispersion and coalescence performance, and be easily obtained and cheap. The majority of diluents used in hydrometallurgy are crude oil fractions with varying proportions of aliphatic, naphthenic, and aromatic components. Initially these were produced for other end uses than solvent extraction, but as the latter market has increased a number of specifically designed formulations has become available.

1.3 Kinetics of extraction
[See also Skelland, Vol. 1]

One of the desirable features of a extraction reagent cited above is speed, defined in terms of rates of extraction and stripping. The kinetics of liquid–liquid extraction are complex, involving mass transfer with chemical reaction in a heterogeneous system. However, it is important to elucidate the kinetic behaviour of an extraction system because it influences such important economic parameters as residence time in the mixer and hence the size of plant and volume of reagent. Being a heterogeneous process, one of the major factors governing the overall kinetics is the interfacial area, which depends largely on the degree of agitation. Although other parameters such as interfacial tension are also involved.

Metal extraction with a solvent extraction reagent involves three steps: diffusion of reagents to the reaction zone; chemical reaction; diffusion of products away from the reaction zone. Any one of these can be the slowest step and hence rate determining. This suggests that two limiting cases exist, involving diffusion control and chemical reaction control. The former depends on the interfacial area and the concentration of the slow-diffusing species. In the case of chemical reaction control it is important to determine the site of reaction, that is, within the bulk phase or at the interface, or more likely in the interphase region. For bulk phase reactions the important kinetic parameters are reactant solubilities, their distribution coefficients, ionization constants and phase volumes. For interfacial chemical control the interfacial area will be important, together with the interfacial activity of the species involved, which in turn are influenced by the molecular geometry of the reagent. The study of extraction kinetics has attracted considerable attention over the past few years especially for acid chelating extractants.

As implied, two limiting models have been proposed to explain the observed results. The first (Flett 1977) was concerned with the influence of the liquid–liquid interface and diffusional processes on the extraction rate.

On the other hand, the second model considered that the bulk aqueous phase was the site of the chemical reaction (Freiser 1984). The first model has been widely adopted (Danesi and Chiarizia 1980; Tarasov and Yagodin 1988) for extractants used commercially which normally have low aqueous solubilities. Recently, Aprahamian and Freiser (1987) have shown by a careful study of the extraction of nickel by 8-quinolinol extractants, including Kelex 100, that even when large interfacial areas, up to specific interfacial areas (i.e. interfacial area/phase volume), $>200\ cm^{-1}$, are generated by vigorous stirring, the rate of extraction may be influenced by the interface under some pH conditions. Thus a controversy which has existed for 20 years concerning the mechanism of extraction in these systems has now been largely resolved with the acceptance that in regimes of low agitation diffusion will be important: even with vigorous stirring, which effectively eliminates diffusion layers around droplets, the organic/aqueous interface does have an effect on the reaction under certain conditions.

The interfacial areas generated by these very high stirring speeds are much higher than those normally experienced in commercial contacting equipment so that the majority of experimental laboratory studies still use equipment with defined interfaces (Hanna and Noble 1985). The results of these studies show that depending on the experimental conditions and hydrodynamics of the system, regimes exist where either chemical reaction or diffusion control dominate. The latter becomes important, as expected, in regimes where fast chemical kinetics are found, that is, with high reagent concentrations and, where appropriate, high pH values. Thus it is very important in studies of kinetics of metal transfer to distinguish between chemical reaction and diffusion control. This can be achieved either by using techniques which operate in a kinetic-controlled regime or by allowing for the diffusion component of the overall mass transfer rate.

Several techniques have been developed over the years for this purpose. The stirred quiescent interface cell as originally designed by Lewis has undergone many modifications and provides one of the simplest devices to operate. But as indicated by Hanna and Noble (1985) there are still many limitations which need to be recognized. One of the problems associated with these cells is maintaining a quiescent interface while stirring both bulk phases. This can be overcome in the rotating diffusion cell where a porous membrane is situated at the interface and adapts the well-known hydrodynamics of the rotating disc electrode to a liquid–liquid system (Albery *et al.* 1976). One of the obvious difficulties with this cell is the presence of the membrane with its unknown free pore area for mass transfer and the effect of diffusion of the extracted species through the membrane. This system has been used both for diffusion-controlled measurements (Guy and Fleming 1979) and also for interfacial chemical reactions (Albery and Fisk 1981). Liquid jet contactors (Freeman and Tavlarides 1980) also possess known hydrodynamic parameters but are difficult to set up and operate. Trouve *et al.* (1982) developed a falling-film device where the organic phase flows over a hydrophobic rod immersed in the aqueous phase. A considerable number of earlier studies used single-drop methods as approximating to the conditions within a stirred tank contactor. Although this equipment is easy to operate, detailed analysis of the data is more difficult and its use has decreased in favour of the stirred cell and other methods.

Studies on the kinetics of metal extraction have been published using both Lewis-type cells and single drop methods. Probably the most extensive are those on the copper–hydroxyoxime system with both commercial and pure synthesized reagents (Hanson and Whewell 1981; Danesi and Chiarizia 1980). These have included an extensive study by Szymanowski *et al.* (1984) in which they concluded that for molecules where the total carbon chain substituents on the basic molecular structure consisted of less than three atoms then more than 90 per cent of the reaction occurred in the aqueous phase. However, as the carbon atoms increased above a total of 11 atoms this bulk reaction accounted for less than 10 per cent of the total. This observation was linked to the hydrophilic–lipophilic balance which is a measure of the interfacial activity. Similar conclusions were reached by Miyake *et al.* (1986).

Extraction reagents all exhibit interfacial activity arising from the presence of both polar groups and non-polar alkyl carbon chains. This activity can be estimated by the lowering of the interfacial tension caused by the penetration of the interfacial surface by the hydrophilic groups. The extent of this lowering depends on reagent type and geometry, bulk phase concentration, diluent and water solubility of the reagent. Values of interfacial concentration which feature in the kinetic equations can be obtained from interfacial tension measurements by the application of the Gibbs isotherm. However, because of limitations to the use of the single solute form often quoted (Cox and Flett 1979) and possible variation in the molecular geometry of the reagent between the bulk and adsorbed species, great care must be taken in the use of the isotherm to obtain quantitative data for solvent extraction systems.

Other more direct methods are available to determine reagent concentrations close to the interface (Hanna and Noble 1985) but these have not yet been used for metal extraction systems. For all these interfacial studies extreme purity of the reagents is required which often limits their application with commercial reagents. When considering hydrometallurgical applications the concentration of reagents employed ensures that either the interface is saturated by adsorbed species or the reagent is above its critical micelle concentration, so that unless the reagent impurities are orders of magnitude more interfacially active than the reagent they should not affect the overall kinetics too much. It should be noted that for reagents that are only weakly adsorbed, these can become desorbed from the interface by vigorous agitation. This will result in an apparent decrease in extraction rate with increasing interfacial area.

Because the nature of the diluent will influence the interfacial properties of the extractant and extracted metal complex it follows that it will also affect the rate of extraction. Thus in the system copper-hydroxyoxime the use of an aromatic diluent has been shown by a number of authors to decrease the rate of extraction when compared with an aliphatic hydrocarbon diluent.

The interfacial area can also be increased by the addition of surfactants which reduce the interfacial tension of the diluent/aqueous interface. These compounds are normally more surface active than the metal extractants and so, for extraction systems without chemical reaction, addition of surfactants normally reduces the rate of extraction per unit area. For systems with chemical reaction either an enhancement or reduction of rate is observed, depending on the sign of the charge developed at the interface by the adsorbed surfactant molecules. Thus the adsorption of a positively charged species will reduce the rate of cation extraction, whereas adsorption of a negatively charged species would enhance this extraction rate (Michel and Nitsch 1988; Tarasov and Yagodin 1988).

For anion extraction the opposite conditions would apply. The reduction in rate can be explained by postulating a mechanism of mutual charge repulsion, or a reduction in interfacial area available for the extractant by blocking with adsorbed surfactant molecules. The enhancement is achieved by electrostatic interaction between the surfactant and the transferring species to form a compound or ion-pair which is then rapidly desorbed into the organic phase where it can undergo fast ligand exchange reactions to give the expected product in the absence of surfactant. This process is similar to 'phase transfer catalysis' observed in organic reactions. Several examples of this type of behaviour have been investigated including copper extraction by hydroxyoxime extractants and dinonylnaphthalenesulphonic acid or dioctylsuccinic acid (Morin and Peterson 1975), di-2-ethylhexylphosphoric acid (Joe *et al.* 1966; Hazen and Coltrinari 1975), and carboxylic acids (van der Zeeuw and Kok 1979).

With the addition of diluent modifiers like nonyl phenol and iso-decanol, variation in extraction rate can occur by modification of the extraction chemistry by hydrogen bonding between the modifier and extractant or extracted complex. This usually leads to a reduction in rate presumably by a reduction in the concentration, or activity, of the extractant through the formation of kinetically inactive modifier–extractant complexes, as found in the copper–hydroxyoxime/nonyl phenol system (Hanson *et al.* 1976; Dalton 1979).

A number of compounds have been shown to act as accelerators for the copper–beta-hydroxyaryloxime system; these include alpha-hydroxyoximes, in particular LIX63, and dioximes (van der Zeeuw and Kok 1979). Although they have been investigated for many years there is still no clear understanding of the mechanism of this acceleration process.

So far the discussion has been concerned with factors relating to the extractant, but the mass transfer rate may also depend upon the nature of the metal ion. The extraction reaction involves the replacement of coordinated water by extractant molecules, it might thus be expected that the rate of metal extraction would reflect the corresponding variation in rates of water exchange, which follows the order: Cu(II) > Mn(II) > Fe(II) > Co(II) > Ni(II) > Fe(III) > Cr(III). Although no systematic study has been carried out on metal variation with any one extractant it has been observed that with Kelex 100 copper is extracted more rapidly than iron(III) (Lakshmanan *et al.* 1974) and palladium(II) can be separated from platinum(IV) with Cyanex 471 for the same reason (American Cyanamid).

Other factors which can affect the rate of metal extraction include the presence of hydrolysed or complex aqueous metal species, or the formation of interfacial films, which may block the surface as found in the nickel/DEHPA (Hughes 1978) or zirconium/hafnium extraction with amines (Yagodin and Sinegribova 1983).

From the above it is obvious that the kinetics of metal extraction are complex and involve both diffusion and chemical reaction processes. It is now generally accepted that, except under the most extreme conditions, the chemical reactions are interfacial in nature. This poses the question of a definition for a liquid–liquid interface in this situation. Some workers consider this to be a defined boundary with a thickness of a nanometer, whereas others prefer an interfacial zone of the order of a micrometer. Perhaps, until the picture is clearer, the term 'interphase reactions' might be more suitable, implying reactions in a region in which the properties are different from either of the bulk phases. Certainly the presence of adsorbed extractant molecules will modify this interphase region from that found in the absence of any adsorbed species, and the thickness of any boundary layer will vary according to the adsorbed species. However, as the interfacial/interphase nature of such process is now accepted little is to be gained from just establishing more examples of the same. In the investigation of interfacial reactions more effort is required to determine actual interfacial concentrations and interfacial rate constants. For hydrometallurgical applications the comparison of rate data for different metal systems is required to see if these can be used as a basis for metal separations. Generally it is found that mass transfer rates for 'ionic' systems, for example, simple acid and amine salt extractants, are faster than the chelating acids, but even with these rapid extractions are there kinetic differences which can be exploited? Finally, can kinetic accelerators, like synergistic extractant mixtures, provide an additional means of developing new processes?

1.4 Arsenic, antimony, and bismuth extraction

1.4.1 Introduction

These metalloids tend to accumulate in the tankhouse liquor during copper refining and need to be removed because they, and in particular arsenic, affect the overall quality and efficiency of copper production. Antimony and bismuth tend to precipitate as anode slimes and so their concentration in the electrolyte is self-regulating, but of course all three elements will be present in solution and overall control of concentration may be achieved by purification of a bleed stream from the tankhouse. The use of solvent extraction to treat this bleed stream as an alternative to the traditional methods has advantages and a number of processes have been described.

1.4.2 Tri-n-butylphosphate (TBP) extraction

Giganov *et al.* (1978) showed that tributylphosphate would extract arsenic from sulphuric acid solutions containing copper and this reagent showed good selectivity over antimony and bismuth with copper not being extracted at all. This system has provided the basis for at least two commercial operations. The extraction of arsenic by TBP is favoured by high sulphuric acid concentration and high reagent concentration. However, the extractive power of TBP for arsenic is quite low so that even with undiluted reagent the free sulphuric acid concentration must be greater than 250 g/dm^3 in order to obtain a distribution coefficient greater than unity (De Schepper 1985). Therefore in this MHO process the feed steam was first concentrated to about 500 g/dm^3 sulphuric acid. The effect of variation of TBP concentration on arsenic extraction was also tested and, in addition to lowering of the overall loading on dilution, the formation of a third phase was observed with a 50 per cent TBP/Escaid 110 mixture. This could be eliminated by the addition of a modifier such as tridecanol but because of the increased fire risk associated with the use of a diluent when operating the circuit at 50–55°C it was decided to use the undiluted reagent with the added advantage of an increased loading capacity. The process described by O'Kane *et al.* (1985) operates at a much lower level of acid (175–250 g/dm^3 at ambient temperature) and uses 50 per cent TBP as organic phase in a mixed aliphatic/aromatic diluent, Shellsol 2046. No third phase was reported and this may be due to the choice of a partially aromatic diluent.

In addition to arsenic, the TBP also extracts sulphuric acid and water to give a complex organic species reported to be TBP. $H_2SO_4.H_3AsO_4.xH_2O$ by Tserekova *et al.* (1970). Because of this coextraction, scrubbing of the organic phase is necessary using water (De Schepper) or an aqueous arsenic solution (O'Kane) as suggested by Demikiran (1978) to reduce arsenic losses from the solvent. The arsenic can be easily stripped by water to give a dilute acidic solution of arsenic with low levels of antimony and bismuth. One additional problem noted with the MHO process, namely the precipitation of calcium sulphate and some antimony and bismuth oxides during extraction, required the design of a special mixer–settler incorporating a decanter to remove the solids. To increase the rate of phase separation in this system centrifugal extractors have been proposed by Kalijta *et al.* (1988).

Some advantages on overall performance have been described by Kerfoot (1978), who showed that a mixture of 60 per cent v/o TBP and 5 per cent v/o of Aliquat 336, a quaternary ammonium extractant, in Escaid 100 would extract arsenic at 25°C. Stripping was achieved with sulphuric acid and the process gave an overall removal of arsenic from the feed of 96–98 per cent.

1.4.3 Trialkylphosphine oxide extraction

Trialkylphosphine oxides have also been used for the extraction of these elements. These reagents have the advantage of being stronger extractants than the phosphates, but even so it is still found necessary to use them either undiluted or in a concentrated solution to obtain an adequate extraction of arsenic. This requirement is a disadvantage in the use of the most common commercial extractant, tri-n-octylphosphine oxide (TOPO), which is a solid at room temperature and soluble with difficulty in common hydrocarbon diluents. However, a patent has been published using TOPO in p-xylene solution at ambient temperature and stripping with an acidic sodium sulphate solution (Marr *et al.* 1984).

To provide a reagent that is more compatible with common hydrocarbon diluents American Cyanamid have developed a liquid reagent, Cyanex 923, which is a mixture of four trialkylphosphine oxides with n-hexyl and n-octyl substituents, having an overall average molecular weight of 350 and soluble in aliphatic hydrocarbon diluents. The use of this reagent to extract arsenic from a copper tankhouse liquor has been described by Rickelton and Brown (1985). Extraction performance was compared with undiluted TBP and, as expected, the phosphine oxide had much higher loadings both at 100 per cent and 75 per cent v/o solutions; for example, D(As) 0.86 (100 per cent), 0.42 (75 per cent) under conditions where TBP did not extract arsenic. In addition the separation factors against copper were higher than TBP. Unfortunately the stronger extracting power of the reagent also results in higher extraction of sulphuric acid, $D(H_2SO_4)$ 0.44 (100 per cent Cyanex), 0.03 (100 per cent TBP). The acid can be scrubbed with water to give an aqueous acid solution for recycle to the tankhouse and an organic phase with low acid concentration. Tests indicate that an approximate 15 per cent loss of arsenic from the organic phase would result from this operation but presumably this could be reduced by scrubbing with an aqueous arsenic solution as described above. The loaded organic phase can again be stripped with water. In contrast to TBP, the phosphine oxide does extract significant amounts of antimony and bismuth, the latter quantitatively.

Once again isotherms have shown that the sulphuric acid can be selectively scrubbed from the loaded organic phase and that water can be used as a strip feed for the arsenic and antimony. Bismuth is not completely stripped under these conditions, however, and will recycle in the organic phase. The use of the new reagent Cyanex 925, 2,4,4,trimethylpentylphosphine oxide, for arsenic has been briefly described (American Cyanamid 1988). The alkyl chain branching makes it a weaker extractant for arsenic but, as it also extracts less sulphuric acid, the scrubbed solvent contains a higher As/H_2SO_4 ratio than is obtained with Cyanex 923.

1.4.4 Alcohol extraction

In addition to the organophosphorus solvating extractants processes involving alcohols have recently been described. These have the advantage of being non-toxic and are also capable of extracting As(III) in addition to As(V) (unlike TBP), athough the extraction behaviour for the two oxidation states are different (Baradel *et al.* 1986b). Thus 60–70 per cent of As(III) is removed by the extractant in a single run through a pulsed column compared with three such runs to achieve 50 per cent extraction of As(V). Stripping can be carried out by water or dilute solutions of sodium hydroxide, sulphuric acid, or weakly acidic sodium sulphate (Guerriero *et al.* 1985). These initial studies using 2-ethyl hexanol (Baradel *et al.* 1986a) have been extended to other compounds such as mono- or di-alkyl derivatives of pyrocatechol or pyrogallic acid with increased recovery of arsenic up to >90 per cent (Baradel *et al.* 1986c).

The use of a commercial reagent based on these studies, ENIM 100, has been described by Baradel *et al.* (1986b) on a copper refinery liquor at pilot scale level. As with TBP, extraction yields were found to increase with the acidity of the feed and decrease with increasing temperature. Selectivity of arsenic over the other elements was good and organic additives to the electrolyte did not affect extraction. Stripping was possible with water and no pH control was necessary. However, poor kinetics both for stripping and phase separation made choice of equipment for this operation critical. Thus for extraction either a mixer-settler or pulsed column was found to be satisfactory, giving in excess of 50 per cent extraction at 55°C, 180g/dm^3 sulphuric acid and a phase ratio O/A 2/1. Using the same pulsed column under the same operating conditions but with a temperature of 62°C gave only 13 per cent stripping. Doubling the column size increased this to 44 per cent, but it was found advantageous to increase the residence time still further by using a Graesser contactor when the yields matched those of the theorectical stage. This type of contactor had further advantages in that the gentle mixing characteristics avoided problems of emulsion formation.

1.5 Separation of cobalt and nickel

1.5.1 Introduction

The separation of cobalt and nickel has been a problem for metallurgists for many years. The two elements occur together in ores and concentrates and their adjacent positions in the periodic table results in similar chemical behaviour. Differences in their chemistry do exist, however, and hydrometallurgists have relied on these differences to devise separation schemes, a number of which have achieved commercialization. Before discussing these schemes it will be useful to consider the background chemistry which led to their development.

Although with non-coordinating anions cobalt(II) and nickel(II) form hexahydrated ions in aqueous solution, the presence of coordinating species can effect some very important changes. For example, in the presence of halide and thiocyanate ions the cobalt(II) species exhibit a strong tendency to produce tetrahedrally coordinated anionic species, e.g. $CoCl_3(H_2O)^-$, $CoCl_4^{2-}$, whereas nickel(II) retains its octahedral cationic configuration. The oxidation of cobalt(II) to cobalt(III) is extremely facile especially if preceded by complex formation. Thus although the potential for cobalt oxidation in aqueous solution is very similar to nickel, following complex formation it becomes much easier to oxidize cobalt; indeed this can be achieved by molecular oxygen. Finally, cobalt(II) hexa-aquo species have a higher rate of water exchange than nickel(II), which implies that rates of complex formation by cobalt would be faster than nickel. Conversely, cobalt(III) is a kinetically inert species exhibiting very slow substitution kinetics. Therefore, because of these slow kinetics, if oxidation to cobalt(III) occurs then it will be impossible either to extract or strip the metal easily without a prior reduction step.

The solvent extraction reagents used for cobalt/nickel separation include examples from the acid and anion exchangers. No commercial examples of the use of solvating reagents are known, although a cobalt/nickel separation factor of 10^3 has been reported for tri-n-butylphosphate from chloride media (Specker and Werding 1964). This involves solvation of the $CoCl_4^{2-}$ ion formed under these conditions.

1.5.2 Amine extraction

The species $CoCl_4^{2-}$ also forms the basis for the separation of cobalt using alkylammonium cations. As noted above, nickel does not form anionic

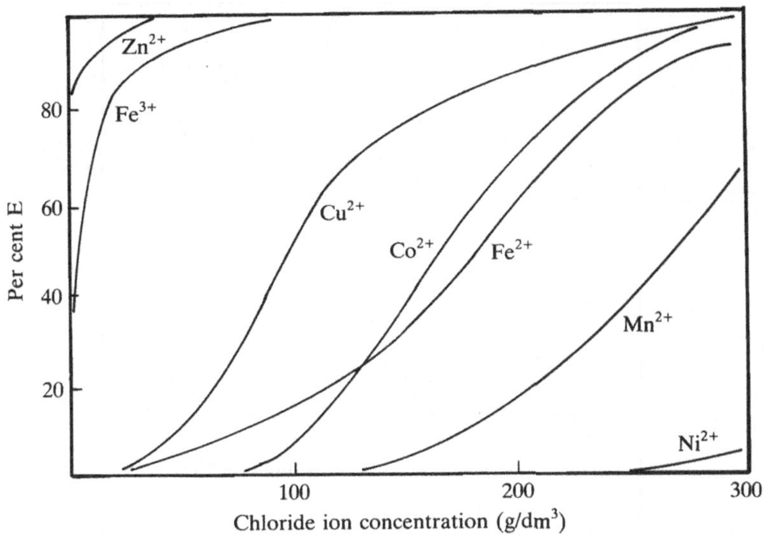

Fig. 1.3. Extraction of metals with amines as a function of chloride ion concentration. (From Ritcey and Ashbrook 1979.)

species in aqueous solution so providing a means of separation. The degree of formation of the tetrahedral $CoCl_4^{2-}$ ion depends on the chloride ion concentration. As shown in Fig. 1.3, variation of the chloride concentration will allow the separation of a number of metals present in a leach liquor. This type of operation is used commercially, for example in the Metallurgie Hoboken nickel/cobalt alloy scrap flowsheet (Fletcher 1973) where zinc is first extracted at low chloride concentration (5–10 g/dm^3 hydrochloric acid), followed by cobalt at 110 g/dm^3 hydrochloric acid, both with a tertiary amine. Also iron(III)/cobalt separation is similarly achieved in the Gullspang (Fletcher 1973) and Societe le Nickel (Bozec *et al.* 1974) processes, with a concentration step following iron removal with a secondary amine, to facilitate the formation of $CoCl_4^{2-}$, which was then extracted with a tertiary amine. The chemistry of this extraction is shown in eqn(1.19):

$$CoCl_4^{2-} + \overline{2R_3HNY} \rightleftharpoons \overline{(R_3HN)_2CoCl_4} + 2Y^- \tag{1.19}$$

As well as depending on the concentration of the extractant the level of extraction also depends on extractant type with tertiary > secondary > primary amine salt (Fig. 1.4). Operating in the hydrochloric acid cycle with the high concentrations required (8–10 mol/dm^3), some competition for the extractant from the presence of HCl_2^- is indicated by the maximum in the extraction curves which is eliminated if metal chloride media is used. Cobalt can be easily recovered from the organic phase by stripping with water and, in multistage operation, complete separation from nickel is possible. This

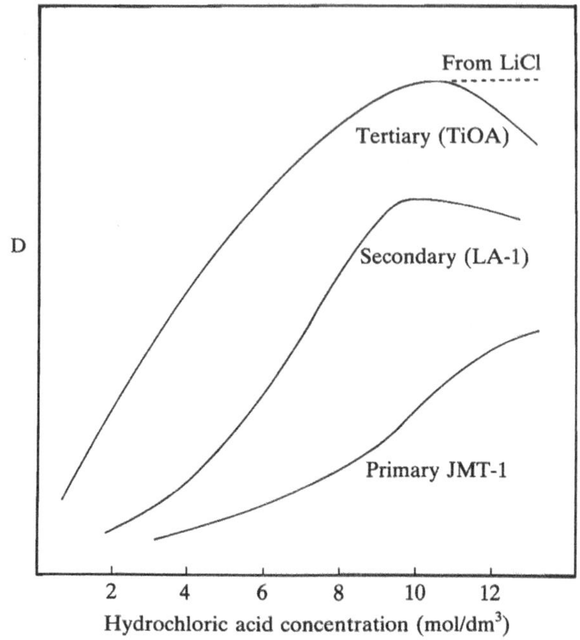

Fig. 1.4. Extraction of metal chlorocomplexes as a function of hydrochloric acid concentration.

anion exchange process has also been used commercially by Falconbridge Nikkelwerk in Norway (Wigstol and Froyland 1972) and Sumitomo in Japan (Suetsuna *et al.* 1980).

In addition to the chloride process, INCO Canada have also studied the extraction of cobalt from sulphate media with quaternary ammonium thiocyanates (O'Neill *et al.* 1976). Zinc, if present, will be coextracted, as will copper but only after prior reduction to copper(I) by the thiocyanate ions. Stripping is achieved by aqueous ammonia. Although extensively studied by INCO in laboratory and pilot plant trials, and giving high degrees of purity in both process streams, the process was never commercialized.

1.5.3 Simple acidic extraction

In the case of acidic extractants, consideration of the extraction versus pH plots for extraction by simple carboxylic (Fig 1.5) and sulphonic acids indicate that the separation factors between nickel and cobalt are not sufficiently high (S ~2) to allow the development of a commercially viable process. However, prior to the studies with organophosphorus acids several papers concerned with carboxylic acid separation of cobalt and nickel were

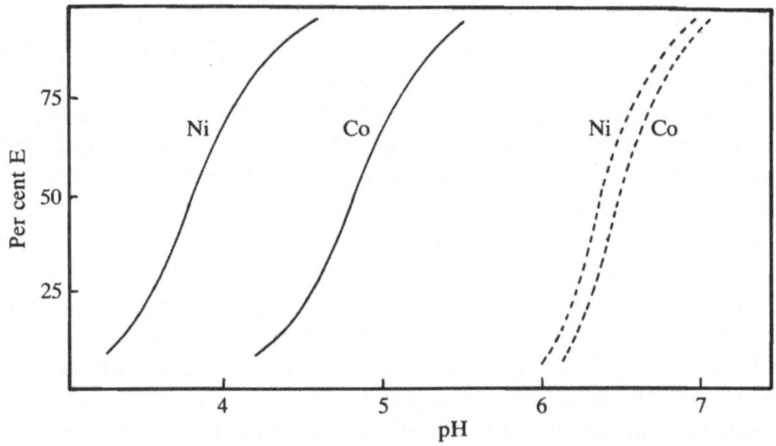

Fig. 1.5. Extraction of cobalt and nickel by 0.50 mol/dm^3 naphthenic acid (dashed lines) and a mixture of 0.50 mol/dm^3 naphthenic acid and 1.0 mol/dm^3 isotridecanal oxime in xylene at 20°C from ammonium sulphate 40 g/dm^3, cobalt/nickel sulphate 3 g/dm^3. (From Preston 1983a.)

published, for example Blumberg and Melzer (1964) and Flett and Fletcher (1970). Also, in the Sherrit Process ammoniacal sulphate solutions are produced containing both nickel and cobalt. A solvent extraction route has been investigated by Ritcey and Lucas (1971) to separate these metals from a nickel precipitate raffinate solution containing 1 g/dm^3 of nickel and cobalt using a carboxylic acid, Versatic 10 or 911. In this case cobalt was favoured over nickel and the separation factor was shown to be dependent on both pH and ammonium sulphate concentration. With the optimum conditions of pH 8.0 and ~500 g/dm^3 $(NH_4)_2SO_4$ a value of S(Co/Ni) = 20 was obtained. The loaded solvent was scrubbed with a cobalt ammine sulphate solution to remove any coextracted nickel before being stripped with, for example, 10 per cent sulphuric acid.

With the organophosphorus acids, cobalt is again extracted in preference to nickel, but this time separation is achieved under acid conditions. The separation factor has been shown to depend on the nature of the extracted cobalt species and this in turn can be related to metal concentration, reagent structure, diluent, modifier, and temperature. Thus with increasing cobalt concentration the colour of the organic phase changes from pink to blue, indicating a change from an octahedrally coordinated $CoR_2(HR)_2$ species to an unsolvated polymeric $(CoR_2)_n$ complex with a consequent increase in distribution coefficient. Similar behaviour is shown by changing the other process variables. Nickel on the other hand remains in the green octahedral form $NiR_2(RH)_x(H_2O)_{4-x}$. This behaviour was first observed for di-2-ethylhexyl phosphoric acid (Ritcey and Ashbrook 1969) and the chemistry of the separation process elucidated by Barnes *et al.* (1976). This process was first suggested for commercial use for cobalt/nickel separation by Ritcey *et al.* (1972) with DEHPA as an extractant and a column contactor. This system was later modified by Flett and West (1978) for use with mixer settlers and the process was operated by Matthey Rustenberg (Clemente *et al.* 1980).

Some problems have arisen with the oxidative degradation of the diluent by cobalt with a concomitant deterioration in plant performance. The oxidation was demonstrated by Flett and West (1986) to depend on the aromatic content of the diluent, temperature, and cobalt loading. The rate of oxidation was also shown to be a function of the nature of the organophosphorus extractant with the phosphonic acid ester PC88A causing faster degradation than DEHPA which, in turn, was faster than Cyanex 272. The reasons for these variations were not determined. The diluent oxidation may be eliminated by the addition of phenolic antioxidants but these must be constantly monitored to ensure long-term stability.

With the commercial introduction of a phosphonic acid ester (PC88A) and a phosphinic acid (Cyanex 272) further investigation has shown that the cobalt/nickel separation factor was also reagent-dependant, increasing in the order phosphoric $<$ phosphonic $<$ phosphinic acid so that at pH 4 and 25°C the separation factor S(Co-Ni) was 14 (di-2-ethylhexylphosphoric acid), 280 (PC88A), and 7000 (Cyanex 272) (Rickelton *et al.* 1983). A change of diluent from aliphatic to aromatic increases the separation factor by increasing the solvation of the polymeric complex but this can be reversed by the presence of a diluent modifier such as nonyl phenol or isodecanol, which tend to depolymerize the system and partially solvate the extracted complex. The factors are summarized in Fig. 1.6, where the separation factor is plotted against percentage tetrahedral cobalt species determined spectroscopically.

Rickelton (1986) demonstrated that bis (2,4,4-trimethylpentyl)phosphinic acid can be used to separate cobalt and nickel from ammoniacal solution where the cobalt is present as Co(III). A diluent modifier is required to prevent third phase formation, and separation factors S(Co/Ni) 30–60, depending on the modifier used, can be obtained. These should be compared with the separation factors obtained with PC88A (14–18) and DEHPA (5–7) under the same conditions. The coextracted nickel can be scrubbed with a cobalt-ammine solution giving cobalt/nickel ratios in the scrubbed solvent of 3000–4000 depending on the organic/aqueous ratios used. The cobalt can be easily stripped by sulphuric acid at 50°C.

1.5.4 Chelating acid extraction

The commercial extractants in this category were designed initially for copper extraction but they are also capable of extracting other metals. Thus

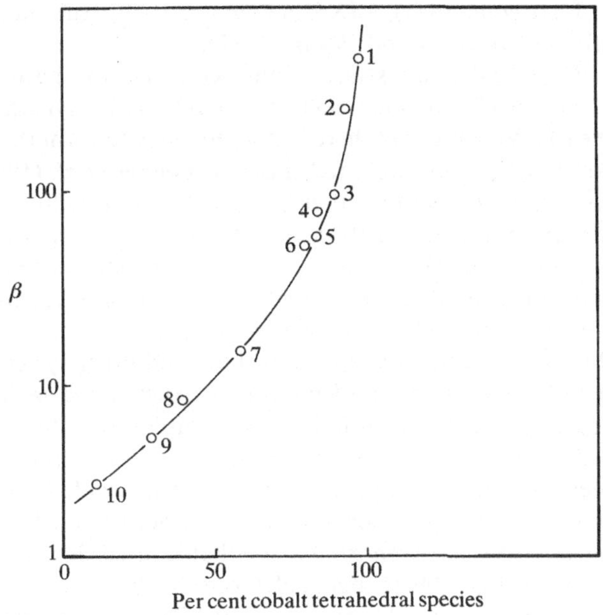

	Extractant	Diluent	Modifier	Temperature °C
1	SME418	xylene	–	25
2	SME418	MSB210	1 per cent TBP	25
3	SME418	MSB210	1 per cent isodecanol	25
4	DEHPA	MSB210	–	75
5	DEHPA	MSB210	–	50
6	DEHPA	xylene	–	25
7	DEHPA	MSB210	–	25
8	DEHPA	MSB210	1 per cent TBP	25
9	DEHPA	MSB210	1 per cent isodecanol	25
10	DEHPA	MSB210	4 per cent isodecanol	25

Fig. 1.6. Cobalt/nickel selectivity (β) for phosphonic acid ester (SME418) and DEHPA versus percentage cobalt tetrahedral complex in the organic phase. (From Flett 1982.)

hydroxyoximes do extract both cobalt and nickel under acid conditions with an increase in the separation factor ranging from α-hydroxyoxime, LIX63, to the aryl-aldoxime P50, for which pH 50 values of nickel(II) and cobalt (II) are respectively 3.5 and 5.0 (Price and Tumilty 1975). Normally this difference in pH would be suitable for a separation process, but this particular system has hidden complications. The rate of extraction of nickel is slow whereas that for cobalt is fast and, in addition, although the cobalt is initially extracted as the divalent species atmosperic oxidation to the inert cobalt(III) complex can readily occur which prevents easy stripping. Such behaviour occurs also with LIX64N (Takahashi *et al.* 1983), Kelex 100 (Ritcey and

Lucas 1975; Flett *et al.* 1975), LIX65N (Tammi 1979; Hubred *et al.* 1978), and DK16 (Przeszlakowski and Wydra 1982).

Flett *et al.* (1975) demonstrated that with the formation of a six-coordinated cobalt(II) complex such as found in the synergistic system cobalt/Kelex 100/Versatic acid there is less tendency to form the cobalt(III) species and stripping is not inhibited. Finally, Penner *et al.* (1986) showed, in a study using pulsed columns, that by completely excluding air from the extraction circuit the extraction efficiency of cobalt by LIX64N from ammoniacal solution exceeded 99 per cent and the stripping efficiency reached 98.5 per cent. Some build-up of non-strippable cobalt still occurred in the circuit, but no loss of extraction efficiency was noticed.

The mechanism of cobalt oxidation in these and other systems probably involves the initial coordination of molecular oxygen, and such a proposal for the oxidation of the cobalt/Kelex 100 complex was demonstrated by Guesnet *et al.* (1980). In the light of the known oxidative properties of the cobalt–oxygen system it is surprising that so little evidence of oxidative degradation of solvent extraction mixtures has been published. However, Takahashi *et al.* (1983) have reported that cobalt caused significant degradation of the oxime LIX64N to benzophenone in the ammoniacal feed to a nickel extraction circuit. As noted above, Flett and West (1986) also observed degradation of aromatic diluents by cobalt in the presence of organophosphorus acids.

Other developments regarding the production of a nickel extractant from acid solution have centred on the vicinal dioximes, following the well-known analytical nickel reagent, dimethylglyoxime. Burkin and Preston (1975) showed that several symmetrical dioximes showed promise but suffered from low rates of extraction and non-strippability of cobalt. Using asymetric vicinal dioximes, which had been tested as potential accelerators for the copper–hydroxyoxime system, Van der Zeeuw and Kok (1979) were able to reduce the equilibration time for extraction from days to hours at ambient temperature. These reagents also extracted nickel in a favourable pH range of 1–3 and virtually excluded iron(III). However, there were a number of disadvantages including poor solubility in suitable diluents, and a pH range which made stripping a little difficult. Also only the anti-isomer was active for extraction so that the synthetic route had to be optimized to produce this isomer. In spite of optimism shown at the time no commercial reagent based on these structures has emerged.

A number of processes have considered extraction by hydroxyoximes from ammoniacal solution. Here prior oxidation of the cobalt is often recommended as a first step followed by nickel extraction with either LIX63 (Drobnick and Millsap 1966), LIX64N (Merigold and Jensen 1974), or SME529.

In addition to hydroxyoximes other chelating species have been studied, for example β-diketones (Przeszlakowski and Wydra 1982), and the sulphonamidoquinoline, LIX34 (Cox and van Bronswijk 1980), but in neither case are the separation factors large enough in acid media for economic commercial operation. One interesting feature has been reported with the β-diketone extractant LIX51 which was shown to extract cobalt(III) from ammoniacal solution (Flett and West, unpub.). No data have been published but it is suggested that the extracted species could be a cobalt-ammine derivative, as similar complexes have been demonstrated in the case of nickel extraction from ammoniacal solution (Flett and Melling 1979).

The 8-quinolinol reagent, Kelex 100, has also been shown to extract nickel and cobalt from ammoniacal solution with the order of extraction cobalt before nickel, but the nickel loading is considerably less than LIX64N. It has also been reported (Ritcey 1983) as extracting cobalt(III) from ammoniacal solution with the level of extraction increasing from pH 7–9 and with increasing ammonium sulphate concentration from 50–500 g/dm^3. Extracted cobalt(II) can be easily stripped from the organic phase with 10 per cent sulphuric acid, but if the loaded solution is exposed to air the expected oxidation occurs, resulting in a reduction of stripped cobalt.

1.5.5 Synergistic mixtures

A number of studies of cobalt/nickel separation using synergistic mixtures have been published. These include mixtures of LIX63 and di-2-ethylhexylphosphoric acid (Joe *et al.* 1966), carboxylic acids (Flett and West 1971), and di-nonylnaphthalenesulphonic acid (Osseo-Asare *et al.* 1981). With the organophosphorus and carboxylic acids nickel is extracted before cobalt and with LIX64N separation factors of between 150 and 300 have been reported by Nyman and Hummelstedt (1974) depending on the nature of the carboxylic acid. Recently a correlation has been shown between the degree of synergism and the pK_a of the carboxylic acid (Castresana *et al.* 1988). However, for all these mixtures the extraction of nickel was anomalously slow which precluded any industrial development (Flett and West 1971). These slow kinetics could be reversed by the addition of dinonylnaphthalenesulphonic acid, which presumably acted as a phase transfer catalyst. Because of the suggestion by Oliver and Ettel (1976) that sulphonic acids may cause hydrolytic degradation of oximes, the commercial use of such systems as LIX70/Versatic acid/dinonylnaphthalenesulphonic acid as suggested by Nyman and Hummelstedt (1974) seems unlikely.

Kelex 100 was found to participate in similar synergistic mixtures with carboxylic acids (Hummelstedt *et al.* 1974). A large degree of synergism for nickel over cobalt has been shown for mixtures of non-chelating oximes and carboxylic acids (Preston 1983a), with values of S (Ni–Co) ~ 200 being obtained together with a lowering of the pH of extraction (Fig. 1.5). In contrast to the LIX63/carboxylic acid system, no reduction in the rate of nickel extraction occurs and no oxidation of cobalt is observed. The structure of the extracted complex for both metals is $MR_2(oxime)_4$ and the mechanism for synergism has been proposed as arising from the different contributions from ligand field stabilization energies shown by these elements. One problem that has been found in continuous processing is that the oxime has a tendency to be hydrolysed by the acid strip solution (Rice and Smith 1986). This problem would have to be solved before any commercialization. In the system di-2-ethylhexylphosphoric acid/2-ethylhexyloxime, a selectivity reversal for nickel over cobalt occurs on addition of the oxime (Preston 1983b). Although the separation factor is not large enough to allow nickel/cobalt separation, the system may have potential as a nickel extractant as extraction occurs at a pH low enough to virtually eliminate coextraction of ammonia from ammoniacal feed solutions (Preston and Fleming 1983). An alternative system has been proposed by Grinstead and Tsang (1980), a mixture of an N-alkylated bispicolylamine and dinonylnaphthlenesulphonic acid, which forms an ionic salt $H_2L^+(DNNS)_2^-$. This will extract divalent metals according to eqn (1.20):

$$\overline{H_2L^+\ (DNNS)_2^-} + M^{2+} \rightleftharpoons \overline{ML(DNNS)_2} + 2H^+ \qquad (1.20)$$

With this mixture both nickel and cobalt can be recovered selectively from iron(III) and separation of nickel from cobalt is also possible.

1.6 Copper extraction

1.6.1 Introduction

One of the most successful applications of liquid–liquid extraction in hydrometallurgy is for the recovery of copper. The development of reagents selective for copper over iron(III) based on the hydroxyoxime and 8-quinolinol structures in the mid 1960s allowed the introduction of liquid extraction into the leach electrowinning circuit with the advantage of being able to operate on the impure low tenor copper solutions derived from many sulphuric acid leaching operations. Currently a significant proportion of the world's copper supply is obtained via this route and a selection of the operations up to 1988 is given in the Table 1.7. Ammoniacal leaching of copper can also be used as an alternative to sulphuric acid with examples in scrap metal recovery (Flett 1974) and lead refinery drosses (Hopkin 1982). Liquid extraction is also used in these circuits with reagents based on β-diketones. Considerable interest has recently been shown in chloride leaching of complex sulphide ores with iron(III) chloride and copper(II) chloride favoured as leachants. The need to extract copper from a leachate containing significant concentrations of iron, zinc, and lead in a high chloride medium has led

TABLE 1.7 *World copper solvent extraction plants: September 1987*—continued overleaf

Company and location	Start-up date	Method of leaching	Production rate (mt/annum)	Feed composition Cu tenor (g/dm^3)	pH	Extractant	Diluent	No. of streams	Flow rate per stream (m^3/hr)	Staging Extract	Staging Strip
Bluebird Mine Miami, Arizona (closed)	1968	Dump leach	7 500	~2	1.8	LIX 64N	Napoleon 470B	1	460	3	2
Cyprus, Baghdad, Arizona	1970	Dump leach	6 500	1.1	2.0	LIX 64NHS	Kermac 470B Chevron IES	4	190	4	3
Zambia Consolidated Copper Mines, Zambia	1974	Agitated tank leach of tailings	100 000 120 000	5	1.9	PT-5050 LIX 864/984	Escaid 100/ LSGO	4	750	3	2
ANAMAX Twin Buttes, Arizona (closed)	1974	Agitated tank	33 000	2.0	1.9	LIX 64N (LIX & (ACORGA)	Chevron	2	700	4	2
Cyprus Johnson, Arizona (closed)	1976	Dump leach	5 000	1.3	2.1	(LIX) LIX 64N	Napoleon 470B	5	120	3	2
Pinto Valley Copper Co. (Miami), Arizona	1976	In place leach	4 200	0.8	2.0	LIX 865	Phillips SX-7	2	320	3	2
Cerro Verde, Peru	1977	Dump leach	33 000	2.7	2.0	LIX 64NHS LIX 860	Local Kerosene	4	280	3	2
Duval Corp., Battle Mountain, Arizona (closed)	1979	Heap leach	4 000	1.2	2.0	LIX 64N	Chevron	1	400	3	2

TABLE 1.7 — continued

Company and location	Start-up date	Method of leaching	Production rate (mt/annum)	Feed composition Cu tenor (g/dm^3)	pH	Extractant	Diluent	No. of streams	Flow rate per stream (m^3/hr)	Staging Extract	Staging Strip
Inspiration CC Co. Arizona	1979	Heap ferric cure leach	45 000	5.5	1.8	M.5615 P.5100	Chevron IES	1	1020	2	2
Asarco Ray Mine (formerly Kennecott) Arizona	1980	Dump leach Heap leach	39 000	5.0 0.96	1.8 2.2	M.5397	Exxon 200	1 2	680 680	2 1	2 1
Soc. Min Pudahuel, Chile	1980	Thin layer leaching	15 000	3.0	1.9	LIX 864	Escaid 100	3	200	2	1
Cia Minera de Cananea, Mexico	1980	Dump leach	12 700	2.5	2.0	M.5615 P-5100 LIX 622	Hydrosol 1000	1	730	2	2
Centromin, Cerro de Pasco, Peru	1980	Mine water	6 000	1.0	2.0	P-5100 M.5615	Local Kerosene	2	360	3	2
Pinto Valley Copper Co (Pinto), Arizona	1981	Dump leach	7 000	0.6	2.1	LIX 864 LIX 984	Phillips SX-7	2	700	2	2
Cyprus Casa Grande (was Noranda Lakeshore), Arizona	1981	In-situ leach Roast leach	6 000 40 000	1.5 9.0	1.6 1.9	P-5100	Phillips SX-1	1 1	450 500	1 2	2 2
BHAS, Port Pirie S. Australia	1984	$Cl/SO_4/O_2$ Agitated tank leach	3 500	35.0	1.5	P-5100	Shellsol 2046	1	20	2	2

Phelps Dodge, Burro Chief, Tyrone, New Mexico	1984	Dump leach	39 000	2.6	2.0	LIX 984	Phillips SX-7	2	910	2	1
CODELCO- El Teniente Chile	1985	Mine water	5 000	1.2	3.0	PT-5050 LIX 864	Escaid 100	1	340	2	2
Sunshine Mining, Idaho	1985	Agitated tank	2 800	8.0	2.0	PT-5050	Chevron IES	1	46	3	2
Magma, San Manuel Arizona	1985	Dump/in-situ	23 000	1.5	2.0	LIX 622	Chevron IES	1	1820	2	1
Gibraltar Mines, Canada	1986	Dump leach	6 000	1.0	2.0	PT-5050 P-5100	Shellsol 160	1 1	690 690	2 1	1 1
Phelps Dodge Morenci, Arizona	1987	Dump leach	17 000 3 000 55 000	1.5 0.3 2.5	2.0 2.0 2.0	LIX 84	Phillips SX-7	1 1 1	1360 1360 2720	2 1 2	1 1 1
Minera Tocopilla, Chile	1987	Heap leach	15 000	8.0	1.5	PT-5050	Escaid 100	1	24	2	1
CODELCO Chuquicamata Chile	1988	In-situ	80 000	6.0	2.0	PT-5050	Shellsol 2046	1	1600	2	2
Roxby Downs/ Olympic Dam. S. Australia	1988	Agitated tank	35 000	1.0	1.1	LIX 622	Shellsol 2046	1	450	2	1

to the development of a new type of solvating extractant based on a pyridine dicarboxylic ester with excellent extraction performance. Thus the choice of copper extractant depends upon the nature of the leaching agent and, because of the extensive range of these reagents, they will be considered under their chemical structure. Also because of the considerable amount of literature on these systems only the current situation will be covered in detail. Earlier processes are contained in Ritcey and Ashbrook (1979, 1984) and Fisher and Notebaart (1983).

1.6.2 Hydroxyoxime extraction

The first commercial hydroxyoximes specifically designed for copper extraction were produced by General Mills (now Henkel) under the trade names LIX63 (Swanson 1965) and LIX 64 (Swanson 1969). These were respectively an α-hydroxyoxime and a substituted benzophenone oxime (Table 1.3). Rapid developments by General Mills gave the reagent LIX64N which was a mixture of a benzophenone oxime and a small amount of LIX63, added to improve the kinetic performance. This reagent was for many years the recognized work-horse for copper extraction from acid leach liquors and a number of commercial plants were designed around it. Because of the commercial interest in the process many studies were carried out on the chemical and chemical engineering aspects of the extraction in both academic and industrial laboratories. Other reagent manufacturers produced their own extractants based on acetophenone oxime (Shell Chemical Co) and salicylaldehyde oxime (ICI/Acorga). A summary of these first generation reagents and their performance is provided in a number of texts including Ritcey and Ashbrook (1984), Fisher and Notebaart (1983), Whewell and Hanson (1981), Kordowsky *et al.* (1985), and references contained therein.

Recently a series of second generation reagents have become available designed to match the requirements of particular feedstocks and the changing perception of the industry. It is now generally agreed that aldoximes, based on salicylaldehyde oxime, offer a number of advantages in terms of very fast transfer kinetics, rapid phase separation, good copper over iron selectivity and high reagent strength, giving the ability to extract copper from highly acidic leach solutions. However, a limitation to their general commercial acceptance was their poor stripping performance with a typical copper tankhouse electrolyte, resulting in poor overall copper transfer performance when compared to the ketoximes. Two different approaches by leading reagent manufacturers have been developed to overcome this limitation. ICI found that the addition of modifiers such as nonyl phenol or tridecanol markedly improved the stripping performance and transfer efficiency with little adverse effect on the extraction characteristics (Dalton and Seward

1984). These modified reagants are available as the Acorga P5000 series and PT5050. More recently Dalton and Seward (1986) have reported that other unexpected benefits such as reduced entrainment and crud formation and increased hydrolytic stability can result from the addition of chemical modifiers and have added long chain alkyl esters to the previously reported alcohols. These reagents are incorporated with the aldoxime P50 and marketed as the Acorga M series.

An alternative approach to improve the aldoxime stripping performance, developed by Henkel, is to use a 1 : 1 mixture of aldoxime and ketoxime. The logic behind this is to combine the extractive strength, fast kinetics, and selectivity of the aldoximes with the proven good physical performance of the ketoximes in terms of hydrolytic stability and stripping. The overall copper transfer ability is greater than expected, which is considered to result from the ketoxime acting as a modifier for the aldoxime on stripping, whereas in the extraction circuit the extractive strength of the aldoxime is almost fully retained (Kordowsky *et al.* 1987). Initially such mixtures incorporated LIX64N or LIX65N as the ketoxime but with the discontinued manufacture of these benzophenone oximes in 1987 they have been replaced by an acetophenone oxime LIX84. The aldoxime used has a 5-dodecyl ring substituent which is reported as being more stable than the corresponding 5-nonyl derivative (Kordowsky *et al.* 1987).

The majority of the pubished studies to date of both series of second generation reagents have been from the respective manufacturers and it will be interesting to see if their claims are substantiated by independent studies.

In their report on commercial extraction plants Fisher and Notebaart (1983) show that prior to 1980 the general circuit configuration was 3 extraction and 2 strip stages. However, a gradual change has taken place, with a trend to a 2 extraction–1 strip configuration for feeds containing between 0.5–3 g/dm^3 copper and a 2–2 circuit with feeds $>3\ g/dm^3$ copper as the operators have gained in experience of running large extraction plants and have become more concerned over capital rather than running costs (see Table 1.7). Both of the above reagent mixtures are suitable for these configurations and only extensive operational use will show whether the modified aldoxime or the aldoxime/ketoxime mixture have a significant advantage under commercial conditions.

Hydroxyoximes, in particular LIX63 and LIX65N, were also considered for extraction of copper from ammoniacal leach liquors with typical processes being described by Ritcey and Ashbrook (1979). The kinetics of extraction were very fast but because these reagent also tended to extract ammonia they have been superseded by the β-diketones which do not have this disadvantage.

1.6.3 Quinoline-based reagent extraction

Kelex 100 (Schering), a 7-alkyl derivative of 8-quinolinol, has also been designed for copper extraction from acidic leach liquors. The separation of copper from iron(III) was based on the slow chemical reaction between iron and Kelex and is therefore based on kinetic rather than thermodynamic principles (Fleming *et al.* 1980). With the short contact times and low pH values normally encountered in copper extraction circuits, separation factors >1000 are obtained (Ritcey and Ashbrook 1979). Kelex 100 also has the advantage of good extraction performance from high copper tenor, high acid liquor and does not require high acid concentration for stripping. Unfortunately, because of the ready protonation of the quinolinic nitrogen atom, Kelex 100 readily takes up acid on stripping and this would either be transferred to the last extraction stage in a counter-current operation or would have to be removed by a water wash stage which would complicate the circuit. Thus it is unlikely, with the dominance of hydroxyoximes, that this reagent would be used commercially for copper extraction.

Henkel also have an 8-quinolinol derivative (LIX26) in their product range but the published data do not suggest that this would perform in a different manner to Kelex 100 for copper extraction.

Another reagent based on quinoline is LIX34 (Henkel), having a chemical structure based on 8-alkarylsulphonamidoquinoline (Virnig 1979). This reagent has similar extraction performance to LIX64N but with excellent selectivity against iron and may be stripped with lower strength acids. It also does not have the disadvantage of protonation in the strip solutions so no carry over of acid is found. However, the major disadvantage was the cost of manufacture and its production has therefore been discontinued.

1.6.4 β-Diketone extraction

Although it is possible to extract copper from ammoniacal solution by hydroxyoximes, as noted above, the coextraction of ammonia can cause problems. No such problem occurs with the β-diketone extractants which are now commonly used in this application. Some early studies by Flett and Melling (1979) which did indicate some ammonia carry over was subsequently shown to be due to the presence of nonyl phenol in the samples used (Flett and Melling 1980).

The reagents extract copper from its ammine complex by the following reaction,

$$Cu(NH_3)^{n}_{2+} + \overline{2RH} \rightleftharpoons \overline{CuR_2} + (n-2)NH_3 + 2NH_4^+ \qquad (1.21)$$

The extent of extraction will depend on the concentration of ammonia and ammonium salts; separation from both nickel and zinc can be achieved by varying the ammonium content of the aqueous phase. The kinetics of extraction are fast and stripping can be readily achieved with aqueous solutions of low acidity and high copper concentration, as found in typical tankhouse electrolytes, giving a much higher copper transfer capacity than the hydroxyoximes. The use of LIX54 (Henkel) been described for copper extraction from printed circuit board etch liquors (the MECER process) by Andersson and Reinhardt (1983), and also extraction from lead smelter drosses by Hopkin (1976). In the latter case, tests showed a 10-fold reduction in plant size by using LIX54 in place of a the hydroxyoximes then available.

The manufacture of two other β-diketone extractants Hostarex DK16 (Hoechst) and LIX51 (Henkel) has been discontinued.

1.6.5 Pyridine di-carboxylic acid ester extraction

Chloride leaching of ores produces a leachate in which the metals are present in high concentrations as chlorocomplexes. In addition, in concentrated chloride media the proton has a very high activity. Thus in these systems there is a requirement for a copper extractant which can operate without the need for pH control and with a high selectivity over a wide range of metals, especially iron and the metalloids. This would preclude the use of the amine salt ion-pair extractants because copper has a low tendency to form chloroanions whereas these are favoured by, for example, iron(III), zinc, and lead. A solvating reagent which operates on a chloride ion concentration cycle has been developed by ICI and marketed under the Acorga series. The reagent, DS5443, is a pyridine dicarboxylic acid ester, the precise structure of which has not been released by ICI. This is marketed as Acorga CLX50, a solution containing 50 per cent of the active ingredient, a earlier formulation CLX20 having been discontinued. The reagent extracts copper at high chloride concentrations as a neutral complex, CuL_2Cl_2, which can be stripped by low chloride media.

$$Cu^{2+} + 2Cl^- + \overline{2L} \rightleftharpoons \overline{CuL_2Cl_2} \tag{1.22}$$

Unlike the amine extractants, DS5443 does not readily undergo protonation of the nitrogen, therefore excellent selectivity is shown for copper from many metals and metalloids which exist in such media as chloroanions. Extraction of the latter can be shown to follow the extent of protonation of the reagent (Dalton *et al.* 1983) which only becomes significant above 8 mol/dm^3 chloride. The integration of this extractant into the Cuprex process for the recovery of copper has been described by Dalton *et al.* (1987).

Reagents therefore exist for copper extraction from all types of leach liquors and with the general maturity of this particular technology, especially in the case of the acidic leach liquors, it is unlikely that there will be significant advances made in the design of reagents for copper in the future. The introduction of DS5443 as a solvating reagent for copper from chloride media introduces a new type of system which may be applicable in other situations.

1.7 Gallium extraction

1.7.1 Introduction

Gallium is present as a constituent of a number of ores including aluminium and zinc. In the processing of these ores the gallium may report in either acid or alkaline solution which leads to alternative solvent extraction routes for separation and purification.

1.7.2 Extraction from acid solution

In the processing of zinc ores, residues are obtained which contain a number of elements including gallium. These residues are usually leached with acid from which gallium may be extracted in a number of ways. For example, in high concentration of hydrochloric acid the species $HGaCl_4$ is formed which can be extracted either by amines (for example, Haskett *et al.* 1979) or solvating reagents like methyl isobutyl ketone or tri-n-butylphosphate. De Schepper (1979) has reported a process where gallium is extracted from a mixed sulphuric/hydrochloric acid solution with 10 per cent tri-n-butylphosphate in Escaid 100 with 25 per cent isodecanol as modifier. The critical part of these flowsheets is the gallium/iron(III) separation, which is often achieved by first reducing the iron(III) to iron(II) with iron powder thereby largely eliminating any coextraction of iron with gallium. In the TBP flowsheet good separation factors from iron are obtained because, first, iron(II) is not extracted by TBP and, second, the iron(II) acts as a salting-out agent enhancing the gallium extraction (Silaev *et al.* 1980). As expected, the distribution coefficient also varies with hydrochloric acid concentration so that any coextracted iron can be scrubbed with 3–4 mol/dm^3 hydrochloric acid before stripping the gallium with dilute hydrochloric acid. This process was tested in a continuous countercurrent operation which gave a gallium recovery of 99.5 per cent with a gallium : iron ratio of 50 : 1. Another process devised by Helgorsky and Leveque (1982) also used hydrochloric acid solutions to produce high purity metal. Here the gallium solution is extracted countercurrently with 2-ethyl-1-hexanol in a column. The extract is scrubbed with dilute hydrochloric acid before stripping with water to give a >99.99 per cent product.

Tri-n-octylphosphine oxide (TOPO) will also extract gallium from hydrochloric acid media with a maximum in extraction occurring at lower acid concentration than TBP. Loading tests support the formulae of the extracted complexes as $GaCl_3$. TOPO and $GaCl_3.2TBP.H_2O$ (Sato *et al.* 1984). A different flowsheet, operated by Dowa Mining Company in Japan, has been described by Maeshiro (1978), in which, following the leaching of the zinc residues with acid and removal of copper, arsenic, and aluminium, both gallium and indium are coextracted with Versatic acid at pH 2.5. Air has to be excluded to prevent precipitation of Ga_2O_3. The two metals are stripped with hydrochloric acid and the gallium separated by extraction into isopropyl ether, presumably as $HGaCl_4$. Stripping by water and recovery by electrowinning follows.

1.7.3 Extraction from alkaline solution

Another major source of gallium are the strongly alkaline liquors from the Bayer aluminium process. Although some flowsheets have been devised which extract gallium as chlorocomplexes, following neutralization of the strongly alkaline media by hydrochloric acid with TBP (Bhat and Sundararajan 1967), the economics must be marginal and a reagent which can extract directly from the Bayer solutions must be favoured. Alkyl substituted 8-quinolinols are such reagents and the use of Kelex 100 has been extensively studied and seems to be the reagent of choice. A number of other substituted derivatives have also been examined either as alternatives or as additives with Kelex 100 (for example Unitika Ltd. 1984; Sumitomo Chemical Co. 1984; Bauer *et al.* 1986; Bauer and Pescher-Cluzeau 1987). The process is one of ligand exchange according to eqn (1.23):

$$Ga(OH)_4^- + \overline{3HL} \rightleftharpoons \overline{GaL_3} + 3H_2O + OH^- \qquad (1.23)$$

Aluminium is also extracted in the same manner but the distribution coefficients for this element and sodium with Kelex 100 are much lower than gallium (Leveque and Helgorsky 1979). The rate of extraction is slow, taking several hours to reach equilibrium. Several approaches have been tried to increase this extraction rate; increasing the temperature to 50°C and above reduces the equilibration time to less than an hour (Leveque and Helgorsky 1979). Recently it has been shown that formation of microemulsions can significantly reduce contact times (Fourre and Bauer 1983), and the addition of phosphate esters and other molecules can also achieve similar results. For example, on addition of di-2-ethylhexylphosphoric acid (6 per cent) 80 per cent gallium is extracted in 30 minutes compared with 23 per cent in the absence of the acid (Bauer *et al.* 1985). Similar results were found with the addition of organosulphates, organosulphonates (Bauer *et al.* 1984), or organophosphinic acids (Wynn and Zabelka 1985), and carboxylic acids (Helgorsky and Leveque 1979). The mode of action of these additives has not been deduced for this system but it is likely that they react as phase-transfer catalysts, transferring the element across the interface and allowing reaction with the hydroxyquinoline to occur in the organic phase. High loadings of the organic phase do tend to give problems with third phase formation during stripping and so it is usual to add modifiers such as isodecanol (Leveque and Helgorsky 1979) or 2-ethylhexyl alcohol (Sato 1986) to prevent this. The loaded organic phase can be easily stripped by contact with acid, releasing the metal as hydrated cations and regenerating the Kelex for recycle.

Although the distribution coefficents of aluminium and sodium are lower than gallium the high concentration of these elements in the Bayer liquors means that significant amounts are extracted into the organic phase. Two different procedures may be used for their selective stripping. Figure 1.7 shows that with hydrochloric acid the gallium tends to be re-extracted at high acid concentration due to the formation of $HGaCl_4$, which is extracted by the protonated form of Kelex 100 via an ion exchange mechanism:

$$HGaCl_4 + \overline{H[Kelex]Cl} \rightleftharpoons \overline{H[Kelex]GaCl_4} + HCl \qquad (1.24)$$

This re-extraction may be avoided by stripping with sulphuric acid, removing the aluminium and sodium with 0.5 mol/dm^3 acid before stripping the gallium with >1.5 mol/dm^3 acid. Some loss of product is to be expected from gallium reporting in the dilute acid solution. Indeed Leveque and Helgorsky have shown that a better recovery of higher purity material can be achieved by initially stripping the impurities with 6 mol/dm^3 hydrochloric acid (these elements do not re-extract with the gallium), and then

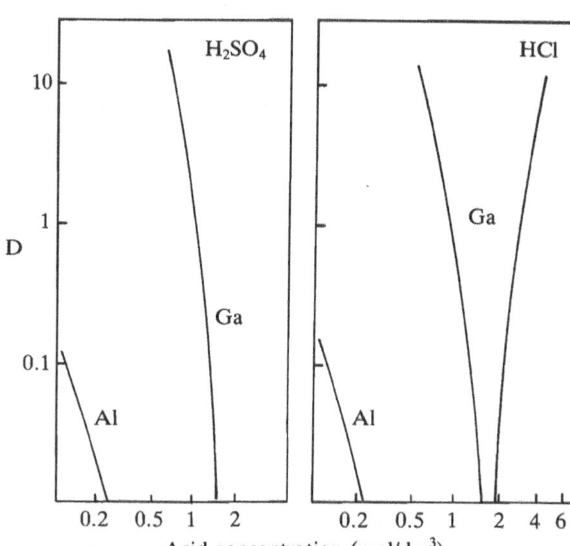

Fig. 1.7. Stripping of aluminium and gallium by acid. (From Leveque and Helgorsky 1979.)

using a 2 mol/dm^3 acid solution to recover the gallium. In this way loss of gallium with the impurities was avoided (Table 1.8). Because the reagent has been protonated in the stripping circuit, it must be water washed to remove acid before recycling to extraction. An alternative reagent for the re-extraction of gallium, following stripping by hydrochloric acid, could be TBP or butyl acetate as described by Wynn (1985).

TABLE 1.8 *Stripping tests of loaded 8 per cent Kelex 100/10 per cent decanol/82 per cent kerosene with different concentrations of hydrochloric acid (see Figure 1.7) (Leveque and Helgorsky 1979)*

	Loaded solvent	Strip solutions	
		0.67 mol/dm^3 HCl	2.0 mol/dm^3 HCl
Gallium (g/dm^3)	0.175	0.017	0.78
Aluminium (g/dm^3)	1.43	1.31	0.30
Al/Ga ratio	12.3		0.385
		6.4 mol/dm^3 HCl	2.0 mol/dm^3 HCl
Gallium	0.175	not detected	0.60
Aluminium	1.43	3.07	0.03
Al/Ga ratio	12.3		0.05

Recently two patents have described the use of substituted hydroxamic acids as extractants for gallium. In the first of these Iwaya (1985) used compounds with the general formula RCONHOH for the extraction from strongly basic sodium aluminate solutions. It has been reported however by Coleman *et al.* (1988) that these compounds readily decompose in the extraction/strip cycle and also that any coextracted iron(III) is difficult to strip. These latter authors prefer the use of N-substituted hydroxamic acids which not only have a greater stability towards hydrolysis over a wide pH range but also exhibit good selectivity for gallium over aluminium, iron(III), and zinc. Extraction was demonstrated from both acid, pH 1–2, and alkaline solutions, pH 8–11, although above pH 1.7 extraction of iron becomes more significant.

Equilibrium was reached in about 30 minutes at 60°C, and stripping could be achieved by either 20 per cent sulphuric acid or alkaline solutions above pH 12. The effect of a range of modifiers was examined, as was the addition of a second substituted hydroxamic acid. The use of a long chain aliphatic alcohol at 10 per cent w/v level was beneficial to the overall process and, using mixed hydroxamic acids, the extraction of gallium was slightly increased over a single reagent. A flowsheet for the extraction from a phosphate treater dust leachate was demonstrated using a 30 per cent solution of N-methyl nonahydroxamic acid in Kermac 470-B with four stages of countercurrent extraction, three-stage stripping with 30 per cent sulphuric acid followed by two-stage scrub of the organic phase with 1 : 1 mixture of 20 per cent sulphuric acid and 20 per cent phosphoric acid before recycle of the organic phase to extraction. The whole circuit was operated at 40–50°C at O/A ratio of 1 : 1 and a pH range of 1.03–1.36.

1.8 Extraction and separation of precious metals

1.8.1 Introduction

In recent years one of the successes in the application of liquid–liquid extraction in the field of hydrometallurgy has been its acceptance into the processing of precious metals. This group of elements, comprising the platinum group metals, gold and silver, have traditionally been separated and recovered by laborious precipitation and dissolution routes which have the disadvantage of many recycle streams in which considerable metal values are locked up. The sources of these elements often contain all the metals in varying proportions which offer a challenge to solvent extraction technology. In addition, their application as catalysts and in the electronics industry provides an important secondary source which again is often a complex mixture of elements. This challenge has been met by the industry and currently three refinery flowsheets have been published. Not surprisingly, because of the complexity of the system, the three companies involved, INCO, Lonrho, and Matthey Rustenberg Refiners, have produced flowsheets (Figs 1.8, 1.9, 1.10) which are different in a number of aspects and which will be discussed in this section, together with other recent studies on individual separation processes. However, it is first necessary to consider the nature of the feed solution to the extraction process.

The precious group elements are noble metals, which means that they can only be dissolved with great difficulty. The choice of lixiviant is hydrochloric acid, usually with added chlorine to increase the oxidation potential. This strong chloride medium ensures that the aqueous chemistry of these leach solutions is concerned with anionic and occasionally neutral chlorospecies. Only under special circumstances are cationic species formed. However, for the precious group metals there are a wide range of available oxidation states and also many anionic species which may be in equilibria in these solutions, which, taken together, leads to a complex chemistry dependent upon the halide concentration and the redox potential of the leach solution. A summary of the most common chloro-species found in these leach solutions is given in Table 1.9, from which mixed aquo-chloro- and polynuclear species have been omitted. The latter are found especially with the heavier elements, rhodium, iridium, ruthenium, and osmium. The halo-species also differ in their kinetics of reaction, which in general terms are slower than those found for the base metals. This means that kinetic factors tend to be more important for the precious group metals, as separations which should be thermodynamically possible are sometimes not as successful as expected because of the presence of small amounts of labile species in equilibrium in the solution. This lability also depends on oxidation state, the order of reactivity for substitution reactions being: Ag(I),Au(I) >> Pd(II) > Au(III) > Pt(II) > Ru(III) >> Rh(III) > Ir(III) > Os(III) >> Ir(IV),Pt(IV) with the states from rhodium(III) being termed 'inert'.

TABLE 1.9 *Common chlorospecies of precious group metals*

	Oxidation state	Major chlorospecies	Comments
Silver (Ag)	I	AgCl	Insoluble
		$AgCl_2^-$	High HCl conc.
Gold (Au)	I	$AuCl_2^-$	
	III	$AuCl_4^-$	Very stable
Platinum (Pt)	II	$PtCl_4^{2-}$	Conversion IV–II, slow
	IV	$PtCl_6^{2-}$	Most common species, kinetically inert
Palladium (Pd)	II	$PdCl_4^{2-}$	Most common
	IV	$PdCl_6^{2-}$	Conversion II–IV difficult
Iridium (Ir)	III	$IrCl_6^{3-}$	Both species stable and conversion IV–III easy
	IV	$IrCl_6^{2-}$	
Rhodium (Rh)	III	$RhCl_6^{3-}$	
Ruthenium (Ru)	III	$RuCl_6^{3-}$	Complex equilibria between III and IV depending on
	IV	$RuCl_6^{2-}$	redox potential and chloride concentration
Osmium (Os)	III	$OsCl_6^{3-}$	
	IV	$OsCl_6^{4-}$	Os(IV) more stable than Ru(IV)

Rhodium, ruthenium, and osmium also form a number of chloroaquospecies.

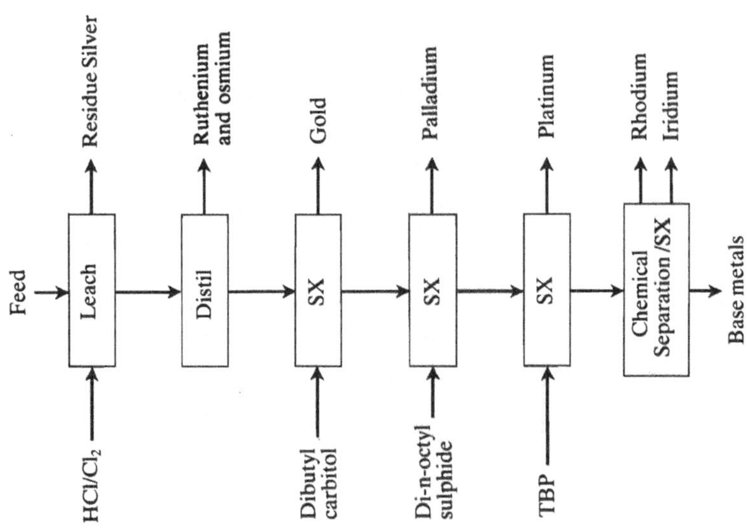

Fig. 1.8. INCO precious metal refining flowsheet

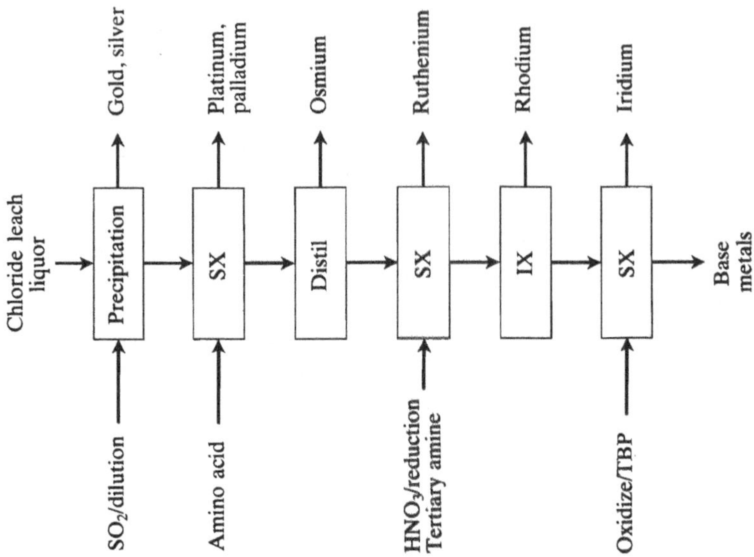

Fig. 1.9. Lonrho precious metals refinery flowsheet.

Thus a solvent extraction process for the precious metals will need to cope with a wide variety of complexes and oxidation states which vary both in kinetic and thermodynamic stability. With this range of parameters it is not surprising that many approaches have been tried and different flowsheets produced.

1.8.2 Gold and silver extraction

These two elements have significant arisings in which they are not associated with the platinum group metals and also, because they can be easily separated from the latter by precipitation or extraction, they tend to be removed early in a flowsheet.

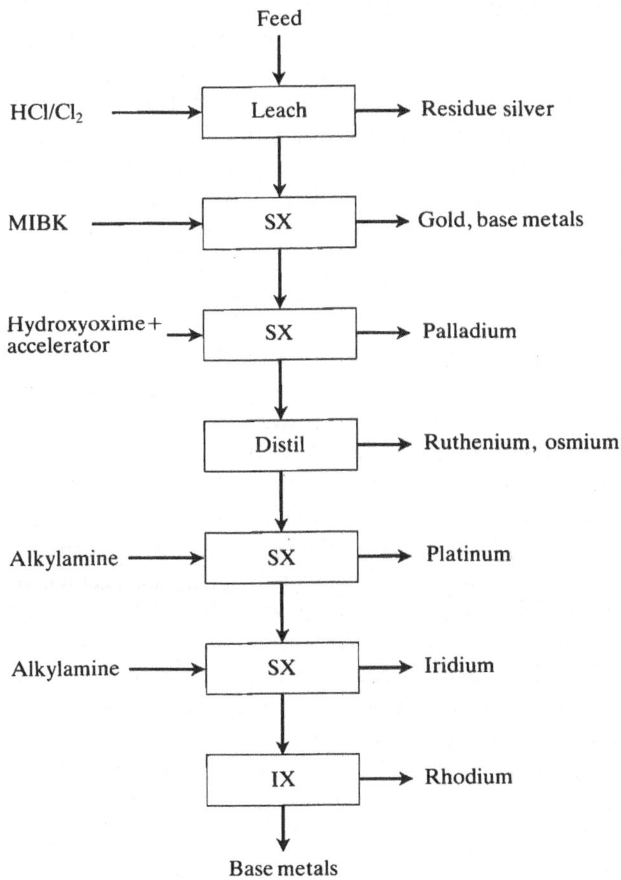

Fig. 1.10. Matthey Rustenberg Refiners precious metal refinery flowsheet.

1.8.2.1 *Gold*

Gold can be extracted, probably as the $HAuCl_4$ species, by most oxygen-containing solvating reagents, presumably by solvation of the proton. The early studies by Morris and Khan (1968) led to the adoption of dibutylcarbitol (diethyleneglycol dibutylether) by INCO in their gold refinery several years before it was incorporated into their complete flowsheet (Rimmer 1974). The extractant has a high selectivity for gold, but any base metals which are coextracted, for example iron(III), antimony(III) and (V), and tin(IV) (Fig. 1.11), can easily be scrubbed by dilute hydrochloric acid. The gold distribution coefficient is also high so that stripping is best carried out by chemical reduction of the loaded organic phase. The use of oxalic acid solution at 90°C produces an easily filtered gold grain, but as the kinetics are slow the plant has to operate as a batch rather than a continuous operation. The Matthey Rustenberg (MRR) flowsheet uses one of the alternative solvating extractants, methyl isobutylketone, which does not seem to have the selectivity of dibutylcarbitol (Fig. 1.12), so a simple scrub is incapable of reducing the level of impurities to an acceptable level. However, it seems that the gold circuit is operated at high acidities to remove some of the base metal impurities at an early stage to minimize potential problems later in the process (Flett 1982). Again a batch reductive chemical step is used to recover the gold, but this time iron is used to give an impure product.

Isodecanol has also been studied and here, as well as having good selectivity over the platinum group metals, the gold can easily be stripped with very weak hydrochloric acid. This has been suggested as a basis for a continuous operation by Fieberg and Edwards (1978).

Other extractants that have been investigated include the amine salts. But because they also tend to extract other precious metal anions they do not offer the selectivity or operating advantages of the solvating extractants. However, by using an aqua regia leachate it has been shown by Baroncelli *et al.* (1988) that tri-octyl amine can selectively extract gold from a range of other metals including iron, copper, zinc, and tin present in the leachate as chloranions. They attribute the selectivity towards the gold complexes to the presence of $TOA.HNO_3$ in the organic phase. Any coextracted metals can be scrubbed with dilute nitric acid and the gold stripped with a hydrochloric acid/thiourea solution. Extraction of gold from aqua regia solutions with N,N′-di-n-butyloctanamide is also described in this paper. Good selectivity is shown over other species in the leachate. In this case the loaded organic phase is scrubbed with dilute hydrochloric acid before stripping with the same solution as used in the amine flowsheet.

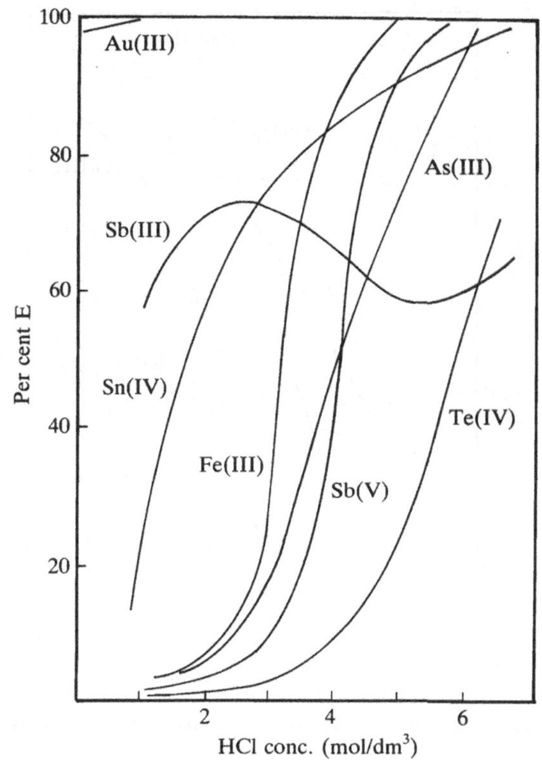

Fig. 1.11. Extraction of metals from chloride media by dibutylcarbitol. (From Rimmer 1974.)

Recently Demopoulos *et al.* (1986) suggested the use of the 8-quinolinol reagents, Kelex 100 or LIX 26, operating in the protonated form for gold extraction via an ion-pair reaction. Recovery as high purity gold grain was achieved by hydrolytic stripping.

1.8.2.2 *Silver*

Silver is not treated by solvent extraction in any of the three flowsheets under consideration, but is either left as a leach residue, or precipitated from solution by reduction as in the Lonrho scheme. Silver may also be a component of other hydrometallurgical streams based on chloride media which may contain base metals such as copper, lead, and zinc. To cope with these and other aqueous solutions American Cyanamid Co have developed a reagent Cyanex 471 based on tri-isobutylphosphine sulphide. It is reported (American Cyanamid Co) as extracting silver from chloride, nitrate, or

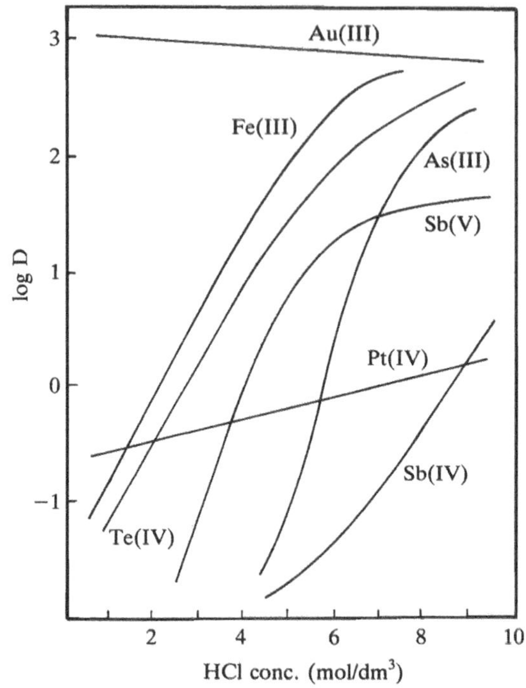

Fig. 1.12. Extraction of metals from chloride media by MIBK. (From Fuwa 1987.)

sulphate media, although with sulphate it is necessary to add di-2-ethylhexylphosphoric acid or nonyl phenol as diluent modifiers, presumably to aid in replacement of residual coordinated water molecules (Rickelton and Robinson 1987).

Stripping is achieved with aqueous sodium thiosulphate solution from which the metal may be recovered by cementation or electrolysis.

Dithiophosphinic acids have also recently been cited as extractants for silver from sulphate media with excellent separation from the alkaline earth metals (Brown *et al.* 1987).

1.8.3 Extraction of the platinum group metals

Having removed gold and silver and some base metal impurities, the process solution now only contains the platinum group metals. To simplify the subsequent separation procedures all the flowsheets retain some of the earlier established technology. Thus ruthenium and osmium are the only elements which can be oxidized to produce a tetroxide, MO_4, which is volatile. Both elements are removed in the INCO and MRR processes, but Lonrho only remove osmium in this way before separating ruthenium as the pentachloronitrosyl complex $Ru(NO)Cl_5^{2-}$ with a tertiary amine. The reason for this variation is not known.

1.8.3.1 *Palladium and platinum*

Like silver, alternative sources of these elements are available from recycle materials such as catalyst wastes, and a number of alternative processes have been suggested for their separation. Pouskouleli *et al.* (1987) discussed these, outlining two distinct routes: (a) selective extraction with different reagents and (b) coextraction followed by selective stripping. The former is operated by INCO and MRR, whereas Lonrho practise coextraction but without selective stripping.

Before discussing these and other processes it is essential to understand some of the relevant features of the solution chemistry of these elements. The two common oxidation states under the process conditions will be palladium(II) and platinum(IV), favouring the formation of square planar and octahedral complexes respectively. Of these, palladium will have the more rapid substitution kinetics, with platinum being virtually inert. However, even in the case of palladium, substitution will generally be a lot slower than for base metals, requiring long contact times for extraction equilibrium to be reached.

The INCO process uses di-n-octylsulphide to separate palladium, but because gold can also be extracted, care must be taken to ensure an essentially gold-free feed to this extraction step. Extraction is slow, with 2–3 hours required to reach equilibrium, hence this must be a batch operation. A recent observation by Daamach *et al.* (1987) that tertiary amines can act as a kinetic accelerator in extraction with di-hexylsulphide, reducing contact times to minutes, may allow the flowsheet to be redesigned, but the effect on the palladium–platinum separation has not yet been investigated. Loading with these sulphides is high (30 g/dm^3), and stripping is easy with aqueous ammonia. The raffinate from palladium extraction is acidified to 5–6 mol/dm^3 to optimize platinum extraction by tri-n-butylphosphate. As this is a solvating extractant kinetics are rapid, allowing continuous operation with 4-stage extraction, 2-stage scrub with 5–6 mol/dm^3 hydrochloric acid, followed by 2-stage water strip.

The second selective extraction route, MRR, again removes palladium first, this time using a hydroxyoxime. The actual compound used has not been specified but published data refer to an aliphatic hydroxyoxime LIX63 (Cleare *et al.* 1979) and a β-hydroxyaryloxime LIX70 (Reavill and Charlesworth 1980). The former has the faster extraction kinetics but suffers from problems with stripping, whereas the latter requires the use of a kinetic accelerator in the form of an unspecified amine to provide a contact time of ~45 minutes. As with other kinetic accelerators, the mode of operation of this amine is not known but it is interesting to speculate on the formation of interfacially active species such as protonated amines participating in the rapid extraction of ion-pairs involving $[PdCl_4]^{2-}$ followed by organic phase reactions with the hydroxyoximes as suggested for the sulphide extraction (Daamach *et al.* 1987; Rong and Freiser 1987). Again good levels of palladium loading are found (10–15 g/dm^3), and palladium–platinum separation factors of 10^3–10^4 are possible, with easy stripping by hydrochloric acid.

Having removed the palladium, platinum can be recovered by an amine extractant. Stripping is difficult because of the high distribution coefficient and can only be achieved by using strong alkali. Another problem concerns the presence of iridium. Because iridium(IV) is highly extractible by amines it must be reduced to iridium(III) before the platinum extraction step, or it may be possible to coextract the iridium(IV) and platinum(IV) and remove the iridium by a selective reductive strip. The published flowsheet seems to suggest the former route has been adopted by MRR.

The possibility of using the phosphine–sulphide, Cyanex 471, as an alternative to the alkyl sulphide or hydroxyoxime is worth considering. The separation of palladium and platinum is kinetically based so that separation factors will depend on contact time as well as reagent concentration. Thus with a relatively short contact time of 10 minutes, albeit at a temperature of 50°C, separation factors S(Pd/Pt) of about 10 : 1 were achieved with a 333 g/dm^3 reagent in a diluent modified with nonyl phenol (American Cyanamid). Good loading of palladium (20 g/dm^3) is possible, but stripping is slow with the recommended sodium thiosulphate solution. There does not seem to be any reason why aqueous ammonia cannot be used as in the alkyl sulphide system. Various other separations have been investigated, for example the use of tri-n-butylphosphate to preferentially extract platinum(IV) chloroanions from palladium(II) has been suggested by Knothe (1980) and Casey *et al.* (1967). In this case iridium(IV) has the highest distribution coefficient and, if present, would first have to be reduced to iridium(III).

Coextraction of palladium and platinum only features commercially in the Lonrho process, which has been based on development work at NIM (Edwards 1979). The original proposals employed two extraction steps with di-n-hexylsulphide for palladium followed by amine extraction for platinum, very similar to the INCO process. The possibility of amine coextraction for palladium and platinum was investigated but abandoned because of irreversible palladium extraction via the formation of neutral chloroammine complexes with primary and secondary amines (eqn 1.25), and platinum stripping problems when tertiary amines were used.

$$\overline{(R_3HNH)_2[PdCl_4]} \longrightarrow \overline{Pd(R_3HN)_2Cl_2} + 2HCl \qquad (1.25)$$

The current process uses a novel amino acid to coextract palladium and platinum. The reagent, made by reaction between chloracetic acid and a secondary amine (Amberlite LA-2, Rohm and Haas), has the general formula R_2NCH_2COOH. It possesses excellent loading performance, 20 g/dm^3 palladium and platinum, good kinetics, and easy stripping with hydrochloric acid (Edwards 1979). However, in the Lonrho flow-sheet no selective stripping is employed and so no palladium–platinum separation is achieved by solvent extraction.

Alternative coextraction routes which use amine extractants have been proposed by Baltz and Coltinari (1977) using a secondary amine for extraction with selective stripping by hydrazine which forms a complex with palladium, followed by bicarbonate to strip the platinum by deprotonation of the amine salt. Another process developed by Warshawsky (1983) uses a tertiary amine for extraction, followed by palladium stripping with thiourea, and platinum with thiocyanate. However, both these schemes suffer from metal lock-up in the organic phase, problems with stripping, and metal recovery from the strip solutions. Amines have also been proposed by Dhara (1984) for the separation of platinum group metals. The flowsheet (Fig. 1.13), shows similarities to the Warshawsky scheme for platinum and palladium extraction.

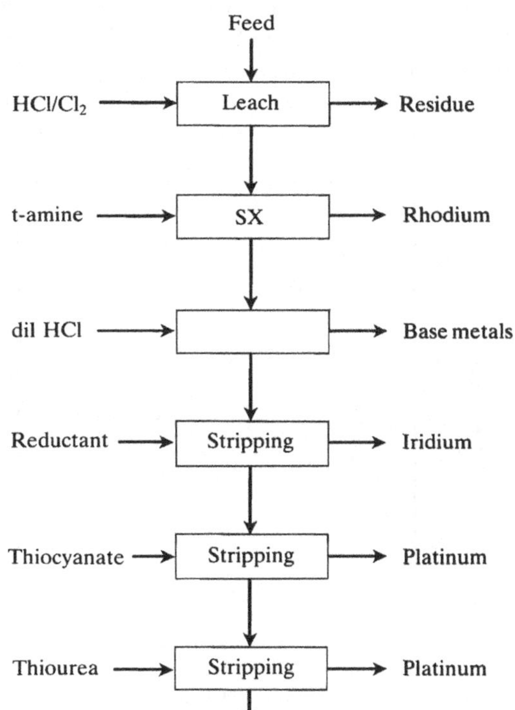

Fig. 1.13. Flowsheet for platinum metal separation. (From Dhara 1984.)

As mentioned earlier, Pouskouleli *et al.* (1987) have suggested another coextraction route using Kelex 100 or LIX 26. These reagents can operate both as a chelating acid or, in the protonated form, as an anion extractant. It was observed that palladium extraction was favoured at low acidity whereas platinum extraction was favoured at high acidity and at about 2 mol/dm^3 the level of extraction was about equal (Fig. 1.14). This suggested two different extraction schemes with the extractant behaving as a chelating acid for palladium and an anion extractant for platinum (eqns 1.26, 1.27).

$$2PdCl_4^{2-} + \overline{4H_2L^+} \rightleftharpoons \overline{(H_2L)_2PdCl_4} + \overline{PdL_2} + 4H^+ + 4Cl^- \qquad (1.26)$$

$$PtCl_6^{2-} + \overline{2H_2LCl} \rightleftharpoons \overline{(H_2L)_2PtCl_6} + 2Cl^- \qquad (1.27)$$

The reaction producing the palladium chelate is similar to that previously observed with amine salts, although in this case, being a chelating molecule, no chlorine atoms are retained in the molecule and the system is easy to strip. As the postulated extraction schemes both involve initial ion-pair extraction, followed in the case of the palladium by an organic phase rearrangement, fast kinetics are observed with 3–5 minutes contact time for complete extraction at ambient temperature. High loading capacities are achieved (20–25 g/dm^3 total metals), and, because the two metals are extracted by different schemes, the extraction conditions can be modified to suit the relative metal concentrations in the feed. These quinolinol reagents are known to extract base metals via a chelating mechanism, for example copper and iron(III), but

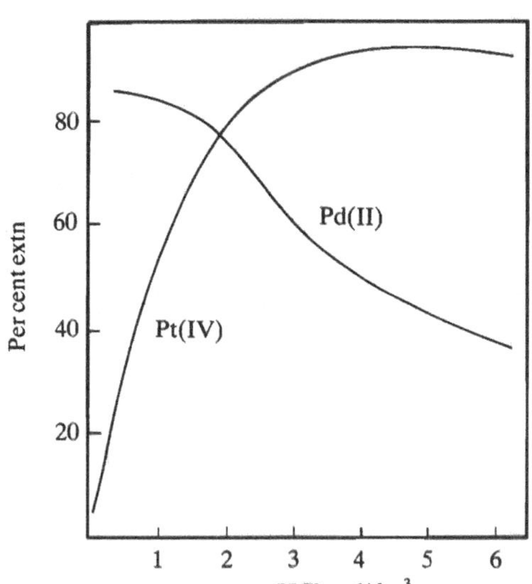

Fig. 1.14. Extraction of platinum (IV) and palladium (II) by LIX26 (5 per cent) in 5 per cent isodecanol and 90 per cent Solvesso 150. (From Pouskouleli *et al.* 1987.)

because the protonated form is used in this process, selectivity over these metals is good (Demopoulos *et al.* 1986), the only problem being gold, which must be removed first. Differential stripping of the organic phase is achieved by varying concentrations of hydrochloric acid. An initial dilute solution (0.02 mol/dm^3) released the platinum, followed by a more concentrated solution, 8 mol/dm^3, to free the palladium. Using these solution a separation factor of S(Pt-Pd) > 2000 is obtained. This process has a number of desirable features and it will be interesting to see if it is introduced into any commercial flowsheet.

Another type of reagent has recently been patented for platinum–palladium separation (Grant and Murrer 1986). These N-substituted alkylamides extract platinum before palladium which has both commercial and operating advantages. Extraction occurs from strong acid media with the mode of extraction believed to involve the protonated form of the amide forming an ion-pair with the chloro-anions of the platinum group metals (eqns (1.28) and (1.29))

$$\overline{\mathrm{RCONHR'}} + \mathrm{H^+} \rightleftharpoons \overline{\mathrm{RCONH_2R'^+}} \tag{1.28}$$

$$\overline{2\mathrm{RCONH_2R'^+}} + \mathrm{MCl}_x^{2-} \rightleftharpoons \overline{[\mathrm{RCONH_2R'}]_2[\mathrm{MCl}_x]} \tag{1.29}$$

$$x = 4 \text{ or } 6.$$

Stripping is easily carried out by contact with dilute acid or preferably water, which causes deprotonation of the reagent and liberation of the metal. The process seems to favour metals in high oxidation states (for example Pt(IV) and Ir(IV)), with distribution coefficients and platinum–palladium separation factors depending on amide structure. The diluent can be either a hydrocarbon, with the optional addition of tri-n-butylphosphate as modifier, or a ketone. As the separation factors are fairly small, (S(Pt/Pd) 25–55), scrubbing of the loaded organic phase is required before stripping to give a good separation. Separation of both platinum and palladium from the other platinum group metals has been demonstrated by reduction of the feed solution prior to extraction. Following removal of the platinum and palladium, the raffinate can be reoxidized for removal of iridium.

1.8.3.2 *Iridium and rhodium*

These two metals are left in the raffinate after palladium and platinum separation. Iridium can be readily extracted in the + IV oxidation state either by amines, as used by MRR, or tri-n-butylphosphate, in the INCO, and Lonrho flowsheets. Although these separations are easy, recovery may not be quantitative because of the presence of other non-extractable iridium halo-species present in varying amounts in the feed solution. In the flowsheet proposed by Dhara (Fig. 1.13) iridium is coextracted with platinum and palladium by a tertiary amine and subjected to selective stripping by reduction to iridium (III). Grant and Murrer (1986) have shown that very good separation factors for iridium from rhodium are possible using the substituted amides (S Ir/Rh = 5×10^5).

Finally, rhodium can be recovered by resin ion exchange either following iridium recovery, INCO, or coextracted with iridium(IV) followed by selective reductive elution of the latter from the resin, MRR and Lonrho. Dhara (1984), on the other hand, favours liquid–liquid extraction with a primary amine followed by stripping with thiourea.

1.8.4 Equipment

The separations discussed above vary considerably in their kinetics and so it is impossible to provide an easy overall picture of the equipment used. In addition the nature of the industry means that relatively small quantities are processed and equipment can therefore be produced in-house or assembled as required. This is also an advantage in that the construction can be of glass to accommodate the highly corrosive chloride solutions used. A number of processes require batch operation with contact times of up to several hours which may be conveniently carried out in stirrer reactors; for example, gold extraction. On the other hand, the coextraction of palladium and platinum with the amino acid in the Lonrho flowsheet has been reported as using a pulsed packed column, 75 mm × 6 m, treating 20 dm^3/h (Anon 1979). This contactor was chosen as the acidity of the feed and so the number of stages required for extraction was variable. By suitable choice of column height any expected variation could be accommodated. Stripping was also carried out in pulsed columns. However, in general little information is available on the type and size of equipment used commercially.

1.9 Extraction and separation of rare earths

1.9.1 Introduction

The rare earths, which are generally defined as the lanthanides with the addition of yttrium, present one of the most difficult separation problems encountered by the hydrometallurgist. The elements (Table 1.10) possess the same outer shell configuration so the chemical properties of adjacent members are very similar. Both cerium and europium possess oxidation states other than III, so they can be selectively precipitated after first oxidation in the case of cerium to Ce(IV), and reduction of europium to Eu(II). Before the introduction of ion exchange and latterly solvent extraction, separation was achieved by lengthy and laborious processes such as fractional precipitation and crystallization requiring extensive recycle streams. Liquid–liquid

TABLE 1.10 *The rare earth elements*

Element		Atomic number	Atomic masses	
Lanthanum	La	57	138.9	Light earths (La–Eu)
Cerium	Ce	58	140.1	
Praesodymium	Pr	59	140.9	
Neodymium	Nd	60	144.2	
Promethium	Pm	61	*	
Samarium	Sm	62	150.4	Middle earths (Sm–Ho)
Europium	Eu	63	152.0	
Gadolinium	Gd	64	157.3	Heavy earths (Gd–Y)
Terbium	Tb	65	158.9	
Dysprosium	Dy	66	162.5	
Holmium	Ho	67	164.9	
Erbium	Er	68	167.3	
Thulium	Tm	69	168.9	
Ytterbium	Yb	70	173.0	
Lutetium	Lu	71	175.0	
(Yttrium	Y	39	88.9)	

extraction has provided a much more flexible approach which allows for the variability in distribution of individual elements in the various minerals and other sources as well as providing processes for the production of individual rare earths at purities and quantities previously unattainable.

The commercial separations of the rare earths are largely based on the use of four reagents representing the three classes of extractants: tri-n-butylphosphate; quaternary amines; Versatic acid and di-2-ethylhexyl-phosphoric acid. The main problem with processing is that separation factors for adjacent pairs of elements are typically in the range 1.5–2.5, which therefore requires a large number of stages whether to achieve high-purity products or for the separation of adjacent pairs. Thus in a number of flowsheets extraction is carried out under a high degree of reflux, enabling two or three products to be obtained in the same plant (Sherrington and Kemp 1966; Brown 1972). Reflux ratios are defined as the amount of solute returned to the process divided by the amount removed as product. As this concept is rather uncommon in other systems but essential in rare earth processing it is appropriate to consider it in some detail here. A multistage countercurrent system under partial reflux continuously fed with a mixture of components A and B is illustrated in Fig. 1.15. The two components (A and B) leave the extraction stages in the organic phase and aqueous raffinate respectively. The two exit streams are split into two fractions, one of which

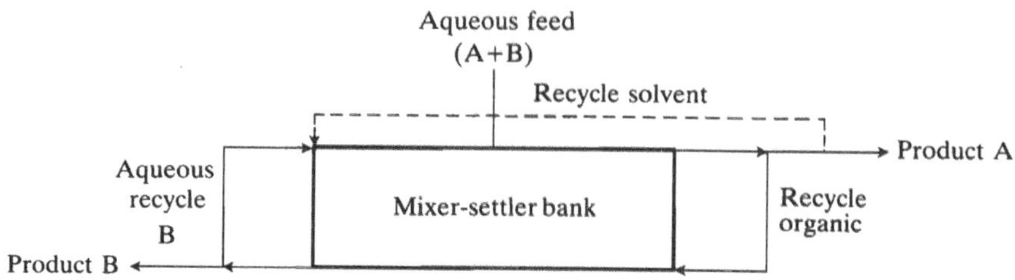

Fig. 1.15. Mixer-settler bank under reflux configuration.

returns to the extraction bank and the other is removed as product. By returning the whole of the exit streams back to extraction no product is withdrawn and the process experiences 100 per cent reflux. Under this configuration, following a period during which the extraction stages are forced-fed to build up the required concentration of components, the feed is stopped and the process operated under total reflux until the desired separations have been achieved.

Results of a typical run (Brown and Sherrington 1979) for the separation of lanthanum, praesodymium, and neodymium using tri-n-butylphosphate are shown in Fig. 1.16. The number of extraction stages can be adjusted to contain the lanthanum–praesodymium crossover, pure praesodymium, and the praesodymium–neodymium crossover sections within the extraction bank with pure lanthanum concentrating in the aqueous raffinate and pure neodymium in the loaded organic phase, both accumulating in holding tanks external to the mixer-settler. The various flowsheets proposed for possible commercial operation will be considered according to the reagents employed.

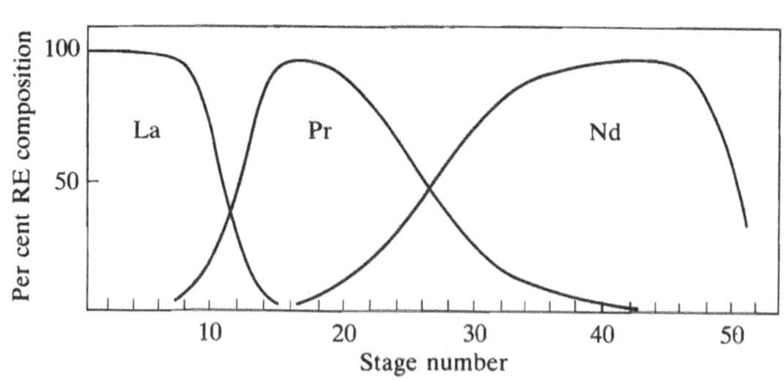

Fig. 1.16. Separation of rare earths by total reflux method using TBP, stage-wise composition after 285 hours. (From Brown and Sherrington 1979.)

1.9.2 Tri-n-butylphosphate extraction

Tri-n-butylphosphate is a solvating reagent and extracts the rare earths from nitrate media by the following process:

$$Ln(NO_3)_3 . xH_2O + \overline{3TBP} \rightleftharpoons \overline{Ln(NO_3)_3 . 3TBP} + xH_2O \quad (1.30)$$

Separation factors between adjacent elements are given in Table 1.11 and indicate that the process is best suited to separation of the light rare earths and little separation would be possible for the heavy elements; in practice Thorium Ltd used a minimum separation factor between adjacent elements of 1.4 (Brown and Sherrington 1979).

The process depends on a number of parameters, the most important of which are:

(1) acidity of the aqueous phase;

(2) ionic strength of the aqueous phase;

(3) concentration of rare earths in the aqueous phase;

(4) concentration of extractant;

(5) concentration of nitric acid in the organic phase.

The presence of nitric acid in the organic phase arises from the ability of tri-n-butylphosphate to extract the acid as a solvate. Thus the acidity of the aqueous phase and concentration of acid in the organic phase are related. It has been shown (Judin and Sund 1981) for representative elements of the light and heavy rare earths that the distribution coefficient (D) depends on both these factors (Fig. 1.17). At low nitric acid concentration in the organic phase the value of D increases rapidly but also coextraction of the acid

TABLE 1.11 *Separation factors from nitrate media, extraction with 50 per cent v/v TBP in Shellsol A (Brown and Sherrington 1979)*

Aqueous phase conc. (Ln_2O_3 g/dm^3)	Separation factors					
	Sm/Nd	Gd/Sm	Dy/Gd	Ho/Dy	Er/Ho	Yb/Er
460	2.26	1.01	1.45	0.92	0. 96	0.81
430	–	–	1.20	0.96	0.65	–
310	2.04	1.07	1.17	0.94	0.82	–
220	1.55	0.99	1.08	0.89	0.78	–
125	1.58	0.82	0.92	0.83	0.72	–
60	1.40	0.18	0.89	0.77	0.70	0.63

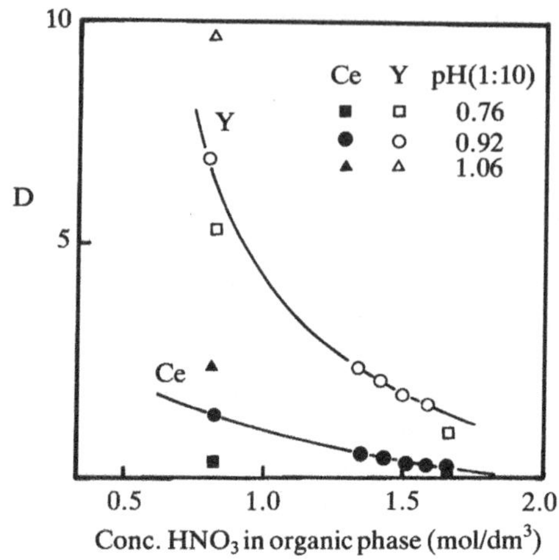

Fig. 1.17. Dependance of distribution coefficient of cerium and yttrium on organic phase acid concentration and pH, 50 per cent TBP/kerosene. (From Judin and Sund 1981.)

occurs, causing the pH of the feed to rise. The pH of the feed will also affect the value of D especially with respect to the light elements. Thus optimization of the process has to be between the highest aqueous feed pH and the lowest possible concentration of nitric acid in the organic phase.

As mentioned in section 1.2.3, ionic strength has a noticeable effect with solvating reagents 'salting out' the complexes into the organic phase. Judin and Sund have shown that the presence of added cations arising from neutralization of free nitric acid is beneficial to extraction and this salting out can enhance the effect of the lower pH of the aqueous phase (Fig. 1.18).

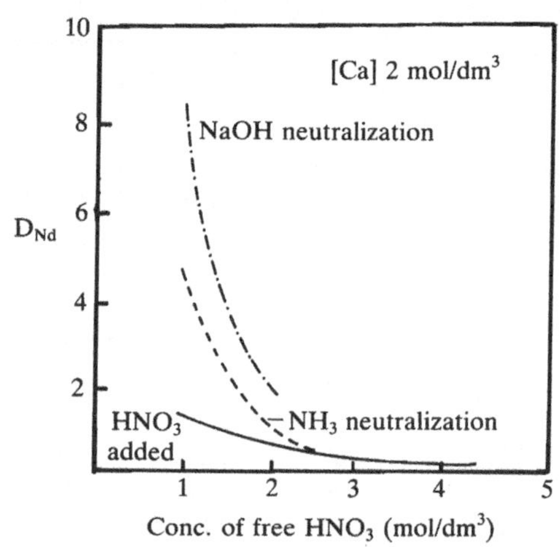

Fig. 1.18. Synergic effect of neutralization and salting-out on distribution coefficient of neodymium with 50 per cent TBP in kerosene. (From Judin and Sund 1981.)

Table 1.11 shows that separation factors depend on the rare earth aqueous feed concentration and that at high concentration the maximum occurs at dysprosium, but changes to samarium on dilution. The separation factors between the lanthanides and yttrium also depend on the concentration of rare earths in the feed, (Table 1.12); in this case the factors increase with dilution, so that a process using TBP/Shellsol A with a dilute aqueous rare earth feed could be used to produce purified yttrium. However, the extraction isotherms indicate a lower limit to the concentration of rare earths of approximately 50 g/dm^3, below which little extraction occurs (Brown and Sherrington 1979). A consequence of these isotherms is that stripping is easily achieved by water.

Commercially, 50 per cent TBP/Shellsol A has been used by Thorium Ltd to separate pure lanthanum from a feed containing 98 per cent La, 1 per cent Pr and 1 per cent Nd as nitrates at 450 g/dm^3 rare earth oxides (REO). Pure lanthanum separates in the raffinate with praesodymium and neodymium reporting in the extract. The extract is stripped with water, the product solution concentrated and returned to a scrub section as a reflux of about 70 per cent REO. Lowering the feed flow and increasing the reflux and number of stages allows the separation of praesodymium and neodymium. Finally, by operating under continuous reflux for 5–6 days with 65 separation stages a complex rare earth feed containing 38 per cent La, 10 per cent Pr, 35 per cent Nd, and 17 per cent heavy rare earths can be separated by 50 per cent TBP/Shellsol A to give 99.6 per cent lanthanum in the raffinate; 95 per cent praesodymium, and 99 per cent neodymium in two bands in the mixer settler train, with the heavy rare earths concentrating in the evaporated strip holding tank (Sherrington and Brown Patents). The profile of the extraction mixer–settler bank would be similar to Fig. 1.17.

Preston and du Preez (1986) have studied the extraction of rare earths with dialkylphosphonates and shown that the structure of the extracted metal

TABLE 1.12 *Variation of yttrium/rare earth separation factor with rare earth concentration, nitrate media, with 50 per cent v/v TBP in Shellsol A (Brown and Sherrington 1979)*

Aqueous phase conc. (Ln_2O_3 g/dm^3)	Separation factors							
	Ce/Y	Nd/Y	Sm/Y	Gd/Y	Dy/Y	Ho/Y	Er/Y	Yb/Y
460	–	0.39	0.88	0.89	1.30	1.20	1.15	0.93
430	–	0.50	–	1.14	1.37	1.31	0.85	–
310	–	0.69	1.41	1.51	1.76	1.65	1.35	–
220	0.60	1.30	2.02	1.99	2.15	1.91	1.48	0.83
125	0.75	1.77	2.79	2.29	2.10	1.75	1.25	0.83
60	–	4.08	5.71	4.44	3.96	3.03	2.13	–

complexes from nitrate media are of the type $Ln(NO_3)_3\ L_3$ in which the nitrate group may be mono- or bi-dentate according to the size of the central metal ion. The actual number of carbon atoms in the alkyl groups does not seem to affect the overall extraction performance but steric effects are important with, for example, extraction by di-n-butyl butylphosphonate exceeding that of the cyclohexyl analogue. Also the butyl compound shows maximum extraction for elements near the centre of the lanthanide series, europium, whereas for the cyclohexyl species the variation in extraction performance is less marked but the maximum appears to be shifted towards the lighter element, praesodymium. Further studies from thiocyanate media (Preston *et al.* 1988) with dibutyl butylphosphonate have shown no such maximum in extraction across the lanthanide series, (Fig. 1.19) reflecting similar behaviour of Aliquat 336. Note also the different positions of yttrium in these two media, showing similar behaviour with amine extraction.

The use of a different organophosphorus solvating extractant, di-1-methylheptyl ester of methylphosphonic acid, has been reported and results indicate that it is more effective than TBP for the separation of lanthanum from nitric acid (Chen Chia-yung 1983).

1.9.3 Tertiary and quaternary amine extraction

Rare earths can be extracted from nitric acid media by tertiary amines and quaternary ammonium compounds. In either case the metals are extracted as ion-pairs involving the amine salt and a complex anionic metal complex.

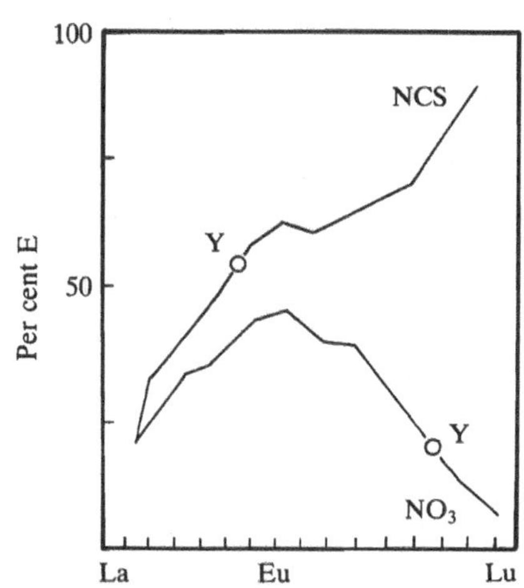

Fig. 1.19. Extraction of rare earths from nitrate and thiocyanate media by di-n-butylester of n-butylphosphonic acid. (From Preston *et al.* 1988.)

Extraction does not occur in acidic media because of competition for the extractant by the nitric acid, but the rare earths can be extracted above pH 2 in the presence of metal nitrates as salting-out agents. The distribution coefficients increase with concentration of the salting-out agent and decreasing acid concentration. The extraction also depends on the structure of the amine with the highest distribution coefficients occurring with asymmetric compounds, especially where one of the substituents is a benzyl group (Jedinakova 1986). With a system based on asymmetric benzyldialkylamines it should be possible to separate the lighter lanthanides from the medium/heavy elements and separation of the lanthanides from yttrium will proceed readily as S(Ln/Y) range from 5–20 (Jedinakova 1986).

These ion-pair extractants have been used commercially in a yttrium process operated by MCI/Megon (Gaudernauk *et al.* 1974), the raw material being the mineral xenotime having the composition: *c.* 60 per cent Y; *c.* 10 per cent light and 30 per cent heavy rare earths. Separation is based on the different extraction behaviour of the rare earths in nitrate and thiocyanate systems and the fact that yttrium behaves as a heavy rare earth in nitrate but as a light rare earth in thiocyanate media (Fig. 1.20). Thus it is possible to extract the light rare earths from yttrium by using nitrate media. Then by using a thiocyanate system with a quaternary amine the heavies can also be extracted from yttrium. Finally yttrium itself is extracted to separate it from other impurities, stripped with pure water and processed to Y_2O_3.

1.9.4 Versatic acid 911 extraction

This reagent is a mixture of synthetic tertiary carboxylic acids containing 9–11 carbon atoms and extracts the rare earths by a cation exchange mechanism (eqn (1.31)).

$$\overline{3R_3CCOOH} + Ln^{3+} \rightleftharpoons \overline{(R_3CCOO)_3Ln} + 3H^+ \qquad (1.31)$$

Maximum extraction occurs close to the pH of hydrolysis of the element, thus requiring pH adjustment to approximately 6.0. Unlike the other systems where extraction is favoured by the presence of particular anion (for example TBP requires nitrate media and DEHPA favours chloride over nitrate), Versatic acid exhibits similar separation behaviour for both chloride and nitrate systems (Table 1.13). Separation factors of adjacent rare earths also have a much lower dependance on total rare earth concentration than the tri-n-butylphosphate system.

The heavy rare earths are more strongly extracted by this reagent, and with the high cerium/lanthanum separation factor the use of Versatic acid would be more suitable for production of pure lanthanum than tri-n-butylphosphate, for which the separation factor Ce/La is 1.7.

Yttrium is seen to behave as an element lying between gadolinium and

Table 1.13 *Separation factors for extraction of rare earths with Versatic 911 in Shellsol A (Brown and Sherrington 1979)*

	Ce/La	Pr/Ce	Nd/Pr	Sm/Nd	Gd/Sm	Dy/Gd	Ho/Dy	Er/Ho	Yb/Er
	3.00	1.60	1.32	2.20	1.96	1.94	1.17	1.28	1.49
La/Y	Ce/Y	Pr/Y	Nd/Y	Sm/Y	Gd/Y	Dy/Y	Ho/Y	Er/Y	Yb/Y
0.028	0.085	0.14	0.18	0.39	0.77	1.50	1.75	2.25	3.35

Table 1.14 *Separation factors for extraction of rare earths with DEHPA in Shellsol A from chloride media (Brown and Sherrington 1979)*

Pr/La	Nd/Pr	Sm/Nd	Eu/Sm	Gd/Eu	Tb/Gd	Dy/Tb	Ho/Dy	Er/Ho	Tm/Er
1.9	1.4	4.0	2.3	1.5	5.7	2.0	1.8	2.4	3.4

dysprosium in contrast to tri-n-butylphosphate where in concentrated solution it lies between gadolinium/dysprosium and erbium/ytterbium, but in dilute solutions behaves as a light rare earth. This allows the development of a yttrium extraction process initially using Versatic acid under total reflux to remove light rare earths in the raffinate and producing 95 per cent yttrium in the middle section of the mixer train. Purification of this product from neodymium and europium is achieved in a 56-stage plant, again operating under total reflux using tri-n-butylphosphate/Shellsol A as solvent (Sherrington 1983).

1.9.5 Organophosphorus acid extractants

Like Versatic acid, these reagents extract by a cationic exchange mechanism, and of the three types of available acids, the phosphoric acids and, in particular, di-2-ethylhexylphosphoric acid have received most attention. However, some evidence seems to indicate that the phosphonate esters (Daihachi Chem. Ind. 1981) and phosphinic acids (Sabot and Rollat 1985) may have some processing advantages. Inoue *et al.* (1986) have determined extraction coefficients for praesodymium and neodymium and found the order phosphoric > phosphonic > phosphinic. The little data that are currently available for these other organophosphorus acids is given later.

1.9.5.1 *Di-2-ethylhexylphosphoric acid*

This is a more powerful extractant than Versatic acid with higher separation factors for adjacent elements in the middle rare earths, mean value about 2.8 (Table 1.14). Yttrium is shown to behave as an element between holmium and erbium with this reagent. Variation of the pH of the feed allows control of the degree of extraction, thus a high pH favours extraction of light earths whereas the heavy elements can be extracted even at low pH (Fig. 1.21). The corollary of this is that the heavy rare earths are difficult to strip with ytterbium and lutetium, requiring 20 per cent hydrogen fluoride as a strip solution. At high loadings a third phase tends to be formed which can be minimized by the presence of a modifier, for example tri-n-butylphosphate. Unlike Versatic acid, extraction from chloride media produces better separation factors than nitrate or sulphate. It has also been shown that increasing the acidity of the chloride solution generally decreases the separation factors of adjacent elements except erbium/yttrium which increases in this case, which provides an additional parameter to aid separation. Similar effects have been found to a lesser extent in sulphate media where extraction of cerium(IV) increases up to 0.8 mol/dm^3 sulphate and then decreases above 2.0 mol/dm^3, and maximum extraction is also found at a pH of 1.3 in this medium.

Di-2-ethylhexylphosphoric acid has been used commercially in a number

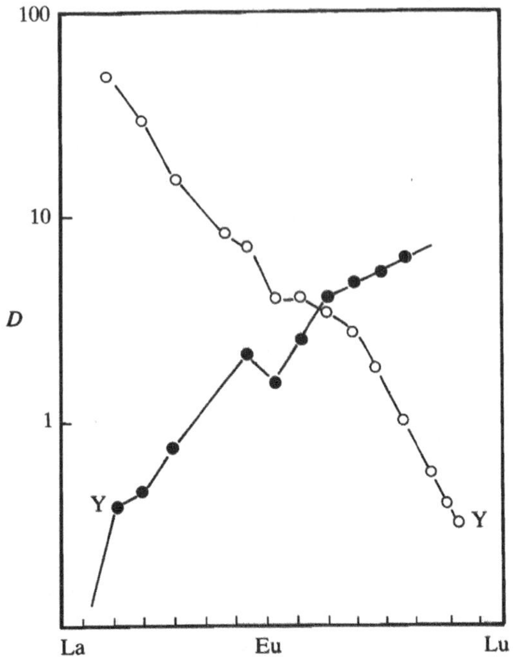

Fig. 1.20. Distribution coefficients for rare earths in Aliquat 336/Solvesso/nitrate (○) and Aliquat 336/xylene/thiocyanate (●) media. (From Sherrington 1983.)

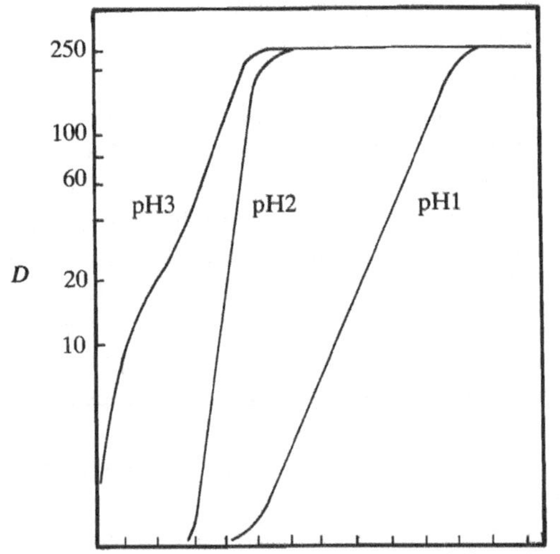

Fig. 1.21. Effect of pH on the extraction of rare earths from sulphate solution with 0.1 mol/dm^3 DEHPA. (From Ritcey and Ashbrook 1979.)

of flowsheets for rare earth separation. As the composition of the feed mixtures vary, the individual flowsheets take advantage of the variation in extraction which can be achieved by change of pH of extraction, concentration of extractant, and concentration of strip solution. In addition, the ability to separate groups of elements allows the recovery and storage of intermediate products until such time as their separation and further processing becomes economic.

The Molybdenum Corporation flowsheet takes a chloride leach solution from bastnasite and extracts with 10 per cent DEHPA/kerosene at pH 1.0 in five stages of mixer-settlers. The light rare earths, lanthanum, praesodymium, and neodymium, report in the raffinate. Stripping the loaded extractant with 4 mol/dm^3 hydrochloric acid in four stages is followed by precipitation and removal of coextracted iron, before re-extraction of the europium with di-2-ethylhexylphosphoric acid. The remaining light rare earths are returned to the primary extraction circuit. The loaded organic phase is stripped with 5 mol/dm^3 hydrochloric acid and the europium precipitated after reduction to Eu(II) with zinc. The filtrate contains gadolinium and samarium which can be recovered as a secondary product and separated by another process with this extractant using a ten extraction/five strip circuit (Kruesi and Schiff 1968).

A similar flowsheet was designed by Gaudernauk and Braaten (1976) to process xenotime concentrates as an alternative to the Megon amine process described earlier (page 75). In this process the nitrate leach liquor is extracted in four stages by 30 per cent DEHPA/Shellsol T to provide an intermediate split of heavy rare earths in the extract and light in the raffinate. The extract is subjected to differential stripping, first with 1.5 mol/dm^3 nitric acid, four stages, to remove the neodymium–terbium fraction, and then with 6 mol/dm^3 nitric acid in eight stages to remove yttrium and most of the heavy rare earths. Ytterbium and lutetium are only partially recovered so a final strip with 10 per cent NaOH, or 20 per cent HF, or 50 per cent H_2SO_4 is used before recycle of the extractant phase. The main product of this process was yttrium and so the yttrium nitrate intermediate solution was further processed using a quaternary ammonium extractant to give a 99.999 per cent grade product.

By changing the concentration of the extractant a different intermediate split of the rare earths can be obtained (Fig. 1.22). Thus the USBM were able to use a dilute DEHPA solution with an aqueous sulphate feed to separate yttrium, dysprosium, erbium, and ytterbium as an extract from gadolinium, europium, samarium, neodymium, and cerium in the aqueous raffinate. Here again the main product was yttrium and after ten stages of extraction and five stages of stripping with 6 mol/dm^3 hydrochloric acid a product was obtained analysing at 76.8 per cent Y_2O_3 (Borrowman and Bridges 1963).

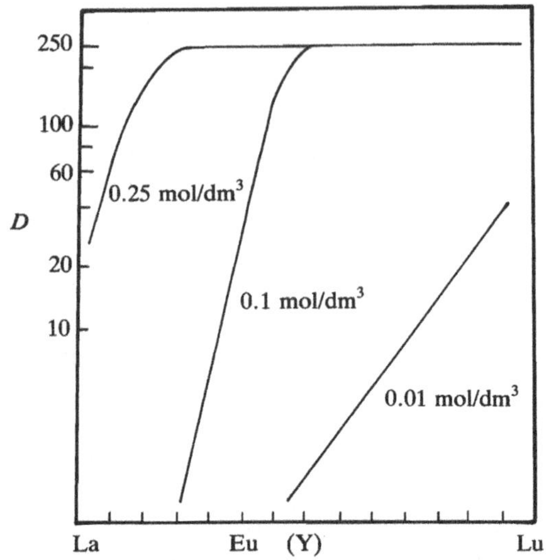

Fig. 1.22. Effect of DEHPA concentration on extraction of rare earths at pH 2.0. (From Ritcey and Ashbrook 1979.)

A number of flowsheets have been devised to extract rare earths from the barren liquors resulting from uranium extraction. The uranium ores contain some rare earths, a proportion of which are coleached with uranium by sulphuric acid. An extensive study by Ritcey and Lucas (1969) showed that variation of the pH and extractant concentration (Figs 1.21 and 1.22) allowed separation of some adjacent rare earth pairs, but with others, for example neodymium/praesodymium, no separation was possible. However, it was feasible to separate yttrium and heavy rare earths from the lights by variation of pH, DEHPA molarity, and a number of extraction stages. A plant for the production of yttrium using such barren uranium liquors employing counter-current centrifugal contactors for both extraction and stripping was constructed at Denison Mines. With a contact time of 10 seconds, feed pH 2.38, 10–12 v/v DEHPA, and a phase ratio A/O of 4, virtually complete extraction of the rare earths was obtained. Stripping was carried out using nitric acid but only achieved 60–80 per cent efficiency; various flow rates and acid concentrations were tried to improve the stripping but with no conclusive results (Lucas and Ritcey 1975). Once again a yttrium concentrate was produced at 30–35 per cent Y_2O_3.

In addition to yttrium, di-2-ethylhexylphosphoric acid has featured in flowsheets for production of other individual rare earths. For example, Brown and Sherrington (1979) describe processes for the production of europium, lanthanum, praesodymium, neodymium, gadolinium, and samarium. All these rely on operating under a high degree of reflux during extraction and force-feeding of the circuit to achieve sufficient loading of the organic phase. The feed was typically a liquor derived from chloride leaching of bastnesite. The rare earth composition of this mineral suggests the separation of the minor elements by a split between neodymium and samarium, allowing an approximate 50-fold concentration of the minor elements by adjusting the phase ratio of the extract phase. Reference to Tables 1.13 and 1.14 indicates that in chloride media DEHPA/Shellsol has a higher neodymium/samarium split than Versatic/Shellsol and so the former was chosen. Initial concentration of the heavy rare earths (> Sm) was achieved in a bank of 17 mixer-settlers which were fed at stage 7 with the chloride leach liquor at 200 g/dm^3 rare earth oxide. The heavy earths were extracted and stripped by 3.7 mol/dm^3 hydrochloric acid operating under a total reflux. The aqueous raffinate was cycled via a batch holding tank to load the recycle extract phase to 100 g/dm^3 REO before return to the extraction circuit. After continuous operation for an appropriate time the heavy rare earth concentrate was removed from the strip circuit. The europium could be removed by chemical reduction to Eu(II) and precipitation, leaving a filtrate containing largely gadolinium and samarium. This solution contained, typically 70–160 g/dm^3 REO with Gd 5.5 per cent, Sm 23 per cent, Nd 45 per cent, Pr 4 per cent with the balance Ce, Y, and heavy earths. Once again DEHPA/Shellsol was selected for the gadolinium/samarium separation and, using a 60 mixer-settler extraction train, either gadolinium or samarium could be produced using the techniques of force-feeding and total reflux and varying the flow rates to trap the desired element in the product section of the mixer-settler bank. This process readily produced products of 99.99 per cent purity. The aqueous raffinate containing the light earths was treated chemically to precipitate cerium as $Ce(OH)_4$ by oxidation with hypochlorite and then, after filtration, the remaining elements could be precipitated as carbonates. This allowed the production of a rare earth nitrate solution which was suitable for treatment with tri-n-butylphosphate. This system possesses the highest separation factors for the light rare earths, and has already been described on page 73.

1.9.5.2 *Other organophosphorus acids*

The use of the 2-ethylhexyl ester of 2-ethylhexylphosphonic acid for rare earth extraction has been described in a patent filed by Daihachi Chemicals Co. (1979). This compares the performance of this extractant with DEHPA for the extraction of neodymium and samarium from hydrochloric acid solution. Although the rate of extraction for these elements is slower than for DEHPA the respective distribution coefficients are approximately 100 times larger for equivalent acid concentration. This not only allows extraction from much more dilute solutions but also, because the percentage extraction decreases with increasing acidity, the efficiency of acid stripping

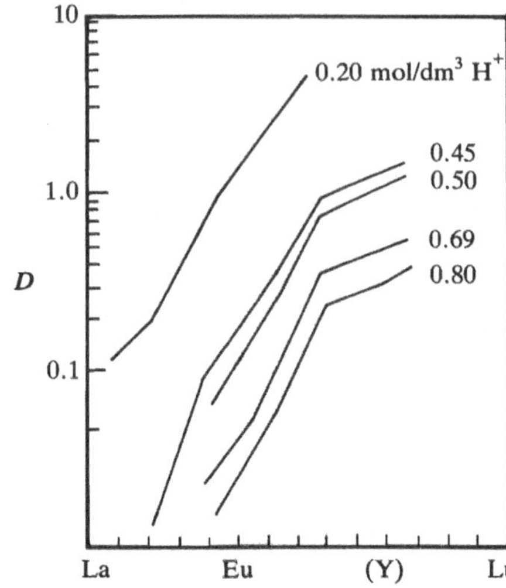

Fig. 1.23. Effect of acidity on extraction of rare earths by 2-ethylhexylphosphoric acid 2-ethylhexylester. (From Daihachi 1978.)

is increased (Fig. 1.23). A diluent effect on extraction was also noted with the extraction of samarium following the order kerosene > n-heptane > toluene. This reagent has also been used by the Chinese to separate lutetium from terbium and ytterbium in 99.95 per cent purity (Chen Chia-yung 1983) and is currently (1990) used commercially in several plants in Japan and China for rare earth separation.

Preston and du Preez (1988) compared the extraction of the rare earths with the three types of organophosphorus acids, all containing the same alkyl group, namely 2-ethylhexyl. As expected from their respective pK_a values, the pH_{50} increased phosphoric < phosphonic < phosphinic (Table 1.15) and extraction increased with the atomic number of the rare earth. This was attributed to increased electrostatic interactions between the acid anion and the rare earth cation as the size of the latter decreased. Also the position of yttrium in the series agrees with its cationic radius. Extraction by these three acids follows eqn 1.32 but the

$$Ln^{3+} + \overline{3H_2A_2} \rightleftharpoons \overline{Ln(HA_2)_3} + 3H^+. \quad (1.32)$$

n-octylphosphinic showed a dependency of extraction of 1.5 with respect to the reagent dimer concentration, implying the stoichiometry of the metal complex as $Ln(A)_3$.

Plotting the value of pH_{50} for these elements demonstrated the tetrad effect observed by other systems (Fig. 1.25). Here the experimental points can be joined by smooth curves without inflections. The values for lanthanum (f^0), gadolinium (f^7), and lutetium (f^{14}) fall on a straight line. The

TABLE 1.15 *Values of pH_{50} for extraction by 0.20 mol/dm^3 solutions of 2-ethylhexyl derivatives in toluene at 20°C (Preston and Du Preez 1988)*

	Phosphoric acid	Phosphonic acid	Phosphinic acid
pK_a (75% Ethanol)	3.51	4.77	6.03
La	2.17	3.08	4.00
Ce	1.93	2.70	3.59
Pr	1.84	2.52	3.50
Nd	1.80	2.46	3.46
Sm	1.47	2.08	3.08
Eu	1.36	1.95	2.97
Gd	1.26	1.86	2.93
Tb	–	1.58	2.67
Dy	–	1.58	2.67
Ho	–	1.30	2.44
Y	–	1.24	2.36
Er	–	1.15	2.30
Tm	–	1.01	2.12
Yb	–	0.82	1.98
Lu	–	0.74	1.92

tetrad effect, which is also shown by the actinides, has been explained in terms of interelectronic repulsion energies of the 4f electrons. Inspection of Fig. 1.24 shows that this effect causes the separation of certain pairs of elements to be particularly difficult (for example Pr-Nd, Eu-Gd, Dy-Ho, and Yb-Lu), although Eu/Gd can be easily separated by access to the divalent europium state.

Sabot and Rollet (1985) have described the application of the organophosphinic acid Cyanex 272 to the extraction and separation of rare earths from nitric acid feed solution. They carried out a number of separations based on a general flowsheet of extraction with pH control by ammonia, followed by stripping with nitric acid. Thus for the separation of ytterbium and lutetium from 0.1 mol/dm^3 nitric acid a 16/10/4 stage extraction/wash/strip circuit was used with 1 mol/dm^3 Cyanex 272 in kerosene at a phase ratio of 1.1 : 1. The organic phase was scrubbed with 0.45 mol/dm^3 nitric acid at an O/A ratio 0.37 : 1 with the aqueous solution after scrubbing fed into the last mixer-settler of the extraction circuit. The aqueous raffinate from extraction/scrub units contained the majority of the ytterbium at a purity of 99.3 per cent. The loaded organic phase was stripped with 1 mol/dm^3 nitric acid at O/A 16.6 : 1 and gave a product solution containing 74 per cent pure lutetium.

By incorporating a second extraction/wash/strip circuit to either the aqueous raffinate or the strip product solution the separation of the rare

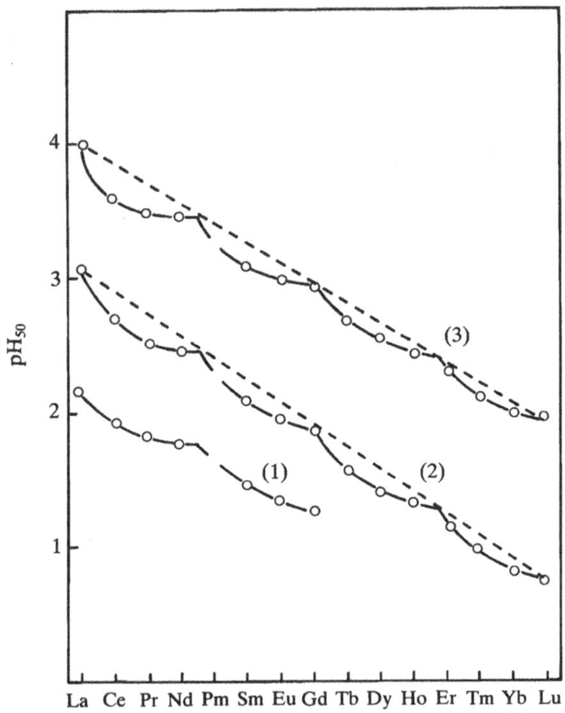

Fig. 1.24. pH$_{50}$ values for extraction of trivalent lanthanide ions by organophosphorus acids (0.20 mol/dm^3 in toluene at 20°C. 1. di-2-ethylhexylphosphoric acid; 2. 2-ethylhexyl mono-ester of 2-ethylhexylphosphonic acid; 3. di-2-ethylhexylphosphinic acid. (From Preston and du Preez 1988.)

earths could be enhanced. Thus after an initial separation of terbium, ytterbium, lutetium, and yttrium in a circuit of 19/16/10 stage configuration, the aqueous raffinate from the extraction/scrub stages was concentrated and fed to a second circuit. This improved the product from 99 per cent to 99.9 per cent Tm_2O_3 with an overall recovery of 99 per cent.

Separation factors of neighbouring elements were determined from a dilute nitric acid feed solution and the results are given in Table 1.16. Li and Freiser (1986) have also studied the extraction of these elements with Cyanex 272.

1.10 Uranium extraction from ores

[See also Batey, Chap. 2; Weatherley, Chap. 5]

1.10.1 Introduction

One of the earliest uses of liquid–liquid extraction in hydrometallurgy was for the extraction of uranium from leach solutions.

TABLE 1.16 *Separation factors for rare earth extraction with Cyanex 272 from dilute nitric acid freed (Sabot and Rollat 1985)*

	Nd/Sm	Sm/Eu	Eu/Gd	Gd/Tb	Tb/Dy	Dy/Ho	Ho/Y	Y/Er	Er/Tm	Tm/Yb	Yb/Lu
Separation factor	14.9	2.21	1.28	4.78	2.38	1.68	1.44	1.53	2.79	2.87	1.47
Organic phase loading of REO (g/dm^3)	←	8.15	→	←	10.8	→	←	11.7	→ ←	15.7	→

The most common lixiviant for uranium is sulphuric acid in the presence of an oxidant and at slightly elevated temperatures. Following leaching and clarification, the leach liquor contains at most a few grams per litre of uranium present as uranium(VI) sulphate so that either ion exchange or liquid–liquid extraction is required for concentration and separation. Both processess have been successfully used, with solvent extraction becoming latterly more popular unless the leach liquors are very dilute when ion exchange is favoured. In addition, combinations of the two operations have been developed for particular feeds. Two types of reagent have been used commercially for uranium extraction from sulphate media, amines and organophosphoric acids, and these will be discussed in turn.

1.10.2 Amine extraction

The mode of extraction follows the formation of ion-pairs between the amine salt and uranium(VI) sulphate as follows:

$$UO_2(SO_4)_3^{4-} + 2\overline{(R_4N)_2SO_4} \rightleftharpoons \overline{(R_4N)_4UO_2(SO_4)_3} + 2SO_4^{2-} \quad (1.33)$$

Difficulties can arise by the coextraction of other ions capable of forming anionic complexes, in particular vanadium and molybdenum, and high concentrations of anions such as bisulphate, chloride or nitrate, which can replace the sulphate in the amine salt. The problems associated with these ions influence the choice of strip solution and the necessity of introducing scrubbing of the loaded organic phase.

A number of alternative strip solutions have been proposed including neutral sulphate or chloride, acidic chloride or nitrate, and alkaline carbonate solutions. Although there are some processing advantages for some of these the major factor determining choice is economic. It should be noted that if an anion other than sulphate is involved in the strip circuit the reagent will require a regeneration step before recycle to extraction.

The overall process using amine extractants has been termed the Amex process in North America and the Purlex process in South Africa. The various operating conditions for these processes will be discussed together with typical flowsheets; additional details of a number of circuits have been discussed by Ritcey and Ashbrook (1979).

Although all three types of amine will extract uranium (Ishimori *et al.* 1963, 1964, 1966) reports for primary and secondary amines are limited (Ritcey and Ashbrook 1979). The reagents used commercially are tertiary amines which have been shown to have a higher loading capacity than the secondary amines (Coleman *et al.* 1958). These workers also showed that the distribution coefficient increased with increasing pH for secondary amines, although for tertiary amines the converse was found with the distribution coefficient following a negative half-order dependance on the hydrogen ion concentration over the pH range 0.5–2.5. Also the extraction decreases with increasing sulphate concentration, in this case following a negative second order dependency over the range 30–150 g/dm^3 sulphate. These linear relationships are sufficiently accurate to allow extrapolation from a single distribution measurement to cover other conditions (Lloyd 1983).

Both Alamine 336 and Adogen 364 have been used in different flowsheets in the concentration range of 5–10 per cent amine in aliphatic diluents. Modifiers in the form of long chain alcohols are used as third phase inhibitors. The concentration of modifier is related to the amine concentration, typically in the range 2–5 per cent. Too high a concentration can cause problems with loss of modifier by solubility in the aqueous phase. Thus in the Highland Uranium plant circuit, using a solvent phase containing 2.5 per cent Alamine 336 in kerosene, levels of isodecanol greater than 1.8 per cent caused an increase in modifier solubility in the aqueous phase (Ritcey and Ashbrook 1979). In another circuit based on 5 per cent tri-isooctylamine/kerosene/4 per cent nonanol the reagent losses were 4 ppm amine and 3 ppm nonanol (Bellingham 1961). The need for a modifier can be reduced or totally eliminated by incorporating an aromatic diluent into the organic phase. This can have other advantages as in certain situations modifiers have caused problems by supporting fungal growths.

Extraction is generally carried out under organic continuous conditions to minimize solvent losses by entrainment. However, because of the wide disparity between the aqueous and organic flows required to achieve satisfactory loading of the organic phase, recycle of the organic phase into the mixer is necessary to maintain the organic continuity.

In addition to uranium, other elements contained in the leach liquor may be extracted under these operating conditions, including iron, silica, magnesium, calcium and molybdenum. The nature and level of impurities in the leach liquor depends to a large extent on the leach operation and subsequent treatment such as partial neutralization or adsorption. Thus vanadium can be largely eliminated by controlling the redox potential of leaching to maintain the vanadium in the +IV oxidation state when it is non-extractable. In situations where high purity products are required the loaded organic phase can be scrubbed in a single stage with water to remove most of the impurities with the exception of molybdenum. Stripping, as indicated above, can be carried out using a number of different reagents, under acidic, neutral, or alkaline conditions.

1.10.2.1 *Acidic stripping*

In this case the uranium is stripped by reversal of the extraction process, regenerating the amine salt and releasing uranium(VI) sulphate to the aqueous phase. Both hydrochloric and nitric may be used which then converts the amine to the corresponding salt (eqn (1.34)).

$$\overline{(R_4N)_4UO_2(SO_4)_3} + 4HX \rightleftharpoons \overline{4R_4NX} + UO_2(SO_4) + 2H_2SO_4 \quad (1.34)$$

Stripping with nitric acid produces the amine nitrate which is a poor extractant for uranium from sulphate media; in addition it is difficult to replace the nitrate ion by sulphate in the amine salt. Therefore to recycle the amine to extraction the free base must first be prepared. This places additional costs on the process, hence nitric acid stripping has not been widely used.

Hydrochloric acid does not produce such problems because chloride is easily displaced from the amine salt by sulphate, and the amine chloride will also extract uranium from sulphate media. Contact with the chloride solution at pH 2 does not release any coextracted molybdenum which therefore tends to build up in the circuit. This can easily be controlled via a bleed stream or scrubbing with an alkaline chloride solution before recycle of the organic phase to extraction.

1.10.2.2 *Neutral stripping*

Most commercial operations use a concentrated ammonium sulphate solution for stripping, with pH adjustment with ammonia. The uranyl sulphate complex is hydrolysed and thus transferred to the aqueous phase where it is kept in solution by the high sulphate concentration. Four stages of stripping are normally used with the pH ranging from 3.5–5 through the bank of mixer-settlers. The addition of ammonia also tends to hydrolyse some of the amine salt, increasing the concentration of ammonium sulphate in the strip solution. Uranium is precipitated from the strip as ammonium diuranate by the addition of ammonia at pH 7. This operation produces ammonium sulphate as a by-product for recycle to the stripping circuit, with a bleed stream to maintain the appropriate concentration.

The stripped organic phase can be periodically scrubbed by an alkaline carbonate solution at pH 8.5 to minimize build-up of impurities, especially molybdenum, which is not stripped by the neutral solution. This treatment also removes chloride ions which have a detrimental effect at concentrations $> 2\,g/dm^3$. Ammonium or sodium chlorides may also be used as alternative strip solutions.

1.10.2.3 *Alkaline stripping*

By employing an alkaline sodium or ammonium carbonate solution the amine salts are hydrolysed, releasing the uranium, which forms uranyl carbonate complexes in the aqueous phase. Once again four stages of stripping are generally used with a mixer residence time of about 1 minute. The uranium is precipitated as ammonium or sodium diuranate by the appropriate hydroxide. The barren strip solution can be recycled after pH adjustment and recarbonation. Molybdenum is stripped by this process but does not precipitate with the uranium, and can therefore be recovered from the uranium barren solution. The recycle organic phase does not usually require scrubbing, but as the amine is in the free base form care must be taken to ensure that sufficient free sulphuric acid is available in the feed to convert this into the amine sulphate without significantly raising the pH. This is also true of the recycle amine extractants which have been scrubbed.

Combination of ion exchange and solvent extraction unit operations have also been developed for uranium extraction. These have been called the Eluex or Bufflex processes and involve initial loading of the ion exchange resin with the leach solution followed by elution with sulphuric acid. Concentration of the uranium occurs as a result of the ion exchange process giving a feed to the solvent extraction plant of approximately 12 g/dm^3 U_3O_8. This is then treated by a typical amine extraction/ammonium sulphate strip with inclusion of a water/dilute ammonia scrub of the loaded organic phase prior to stripping. The stripped solvent phase is regenerated by sodium carbonate/hydroxide which removes any coextracted molybdenum.

1.10.3 Di-2-ethylhexylphosphoric acid extraction

Organophosphoric acids provide an alternative to amines for uranium extraction from sulphate solutions. This Dapex process has a number of disadvantages, however, in that the extraction kinetics are slower than with the amines (eqn (1.35)):

$$UO_2(SO_4) + \overline{2HDEHPA} \rightleftharpoons \overline{UO_2(DEHPA)_2} + H_2SO_4 \qquad (1.35)$$

In addition the reagent is not as selective for uranium as the amines, especially in the presence of impurities such as iron(III), thorium(IV), rare earths, molybdenum, and titanium(IV). Even so the system does tend to be more tolerant of the presence of suspended solids, and, as before, the process will be considered in terms of typical flowsheets.

Although a number of organophosphorus extractants have been examined at different times, including alkyl and aryl phosphoric acids, and alkylphosphinic acids (Sekine and Hasagawa 1977), the choice of reagent for commercial operation is di-2-ethylhexylphosphoric acid. The solvent phase consists of 3–10 per cent DEHPA in kerosene with the addition of a modifier, usually tri-n-butylphosphate or a long chain alcohol. A modifier is required to prevent third phase development during the stripping cycle, and the concentration employed is usually close to equimolar with the reagent. The extraction of uranium increases with increasing pH as normal with acidic extractants; however, because impurities such as iron, aluminium, thorium, etc. are extracted more strongly at the higher pH values, the pH of extraction is usually kept below pH 1. In addition, the presence of high concentrations of sulphate ions reduce the extraction due to the formation of uranyl(VI) sulphate complexes. Thus sulphate and other coordinating ions like phosphate and nitrate should be kept as low as possible. The use of tri-n-butylphosphate as a modifier can be beneficial for extraction because the mixture with di-2-ethylhexylphosphoric acid can behave as a synergist, enhancing the overall extraction. Other reagents have also been shown to behave as synergists in this context, for example tri-n-octylphosphine oxide, but have not been used commercially mainly because of the increased costs and/or inferior performance as modifiers at the optimum synergistic concentration.

The coextraction of impurity elements can be minimized by solution pretreatment prior to extraction. Typically this involves a reduction step mainly to reduce the iron to the less extractable iron(II). This may also reduce vanadium to a more extractable species and so scrubbing of the loaded solvent phase may be required. Dilute sulphuric acid is normally used to remove any vanadium(IV).

Stripping can be carried out either with a strong acid solution, for example 10 mol/dm^3 hydrochloric acid, as in the Vitro Uranium circuit, which reverses the extraction process, or with a sodium carbonate solution. The latter is normally chosen and here the strip reaction involves the formation of soluble uranyl carbonato complexes (eqn (1.36)).

$$\overline{(\mathrm{DEHPA})_2\mathrm{UO}_2} + 3\mathrm{Na_2CO_3} \rightleftharpoons \overline{2\mathrm{NaDEHPA}} + \mathrm{Na_4UO_2(CO_3)_3} \tag{1.36}$$

The sodium di-2-ethylhexylphosphate is largely responsible for the third phase formation as it has a limited solubility in the hydrocarbon diluent. The solvent phase now containing some sodium salt is returned to extraction where sufficient free acid must be available to prevent an undue rise in pH. Coextracted impurities such as titanium or iron are hydrolysed and precipitated from the aqueous solution by alkali and so a filtration step is introduced to remove these before precipitation of the uranium.

Di-2-ethylhexylphosphoric acid has also been used in Eluex type operations. In one example the strip solution is regenerated by carbonation of the barren uranium filtrate after precipitation as sodium diuranate with sodium hydroxide (Ritcey and Ashbrook 1979).

1.10.4 Tri-n-butylphosphate extraction

Nitric acid has also been proposed for the leaching of some ores as well as in the refining of some products from the primary extraction circuits considered above. The uranium dissolves as complex uranyl nitrates and a number of extractants have been used including ethers, ketones, and carboxylic esters, but the majority of studies have centred on tri-n-butylphosphate. Similar processes are used for the reprocessing of irradiated nuclear fuel and are discussed elsewhere (Chap. 2).

The extraction of uranium from nitrate solution by tri-n-butylphosphate depends both on nitrate and extractant concentration, as indicated by the eqn (1.37).

$$UO_2^{2+} + 2NO_3^- + \overline{2TBP} \rightleftharpoons \overline{UO_2(NO_3)_2(TBP)_2}\,. \qquad (1.37)$$

However, with high concentrations of nitric acid an alternative process is favoured with the extraction of an anionic nitrato complex by protonated tri-n-butyl phosphate, namely $(HTBP)_2(UO_2(NO_3)_3)$. This has a lower solubility in the kerosene diluent and leads to third phase formation. The presence of coordinating anions like sulphate, fluoride, phosphate, etc. will suppress the extraction but such effects can be reduced or eliminated by the addition of cations to preferentially complex the interfering species.

In addition to uranium, thorium, which also may be present in the feed solution, will also be extracted under these conditions. This can be removed by a scrubbing process, using uranyl nitrate solution. Stripping of the uranium by water is easy, and the stripped solvent is regenerated by washing with sodium carbonate to remove any acidic hydrolysis products of the extractant, followed by acid pre-equilibration before being returned to extraction (Tunley and Nel 1974).

Refining of the uranium products to produce nuclear grade material from any of these primary extraction routes is usually carried out by tri-n-butylphosphate extraction from nitric acid. The overall process is similar to that described above. The concentration of impurities in the organic phase can be minimized by adjusting the flowrates to maintain virtual saturation of the extractant by uranium. The loaded phase may be scrubbed with either water or a uranyl nitrate solution if required.

1.10.5 Extraction of uranium from phosphoric acid

Phosphate rock may contain up to 0.02 per cent U_3O_8 which is released into solution by acidulation of the rock to produce phosphoric acid. A number of processes have been developed to extract uranium from these solutions using both ion exchange or solvent extraction. Wet-process phosphoric acid may contain a number of metallic impurities, in particular iron, aluminium, calcium, as well as the anions fluoride and sulphate. Where acid leaching of the ore follows a roasting step the levels of humic material are low and the solution is called 'green acid'. If this preroast is omitted the liquor is very dark, and contains significant amounts of organic matter and carbon and is referred to as 'black acid'. The high concentration of humic acids and the organic matter gives rise to the formation of emulsion with extractants, etc. Thus the satisfactory extraction of uranium requires pretreatment of black acid to remove this organic matter normally by wet oxidation followed by carbon adsorption.

Three extractants have been proposed for uranium extraction, all based on organophosphorus compounds. The first extractant to be used was an dialkylpyrophosphoric acid which extracts uranium(IV) (Long *et al.* 1955). These reagents with C8–C10 alkyl groups have a high distribution coefficient for uranium and so can operate with a single extraction/strip cycle. However, stripping is difficult and the flowsheet devised by Kouloheris (1980) uses a mixture of 25 per cent sulphuric acid/15–20 per cent hydrofluoric acid which both strips and precipitates the uranium as the tetrafluoride. The strip circuit is preceded by a sulphuric acid/water scrub to remove calcium and is followed by a further scrub with the same solution before reagent recycle. Disadvantages of the process are that the extractants are not very selective so that a further purification is necessary using nitric acid/tri-n-butylphosphate; and the extractants are also not very stable to hydrolysis so reagent consumption is high at a third of the inventory per cycle. To offset this, the reagent is cheap, easily produced on site, and its instability may be an advantage in reducing entrainment contamination of the phosphoric acid product.

The other two flowsheets which have been developed both involve two-cycle processes because both the uranium distribution coefficients and the feed concentration are too low to provide sufficient loading in one circuit to permit recovery by precipitation of the uranium from the final strip solution. The first of these processes involves mixtures of the mono- and di-octylphenylphosphoric acids (Hurst and Crouse 1974) which extracts uranium(IV) with a value of the distribution coefficient of 6–7 at 0.3 mol/dm^3 extractant. Stripping is by oxidation to uranium(VI) by, for example, sodium chlorate in concentrated phosphoric acid and the cycle repeated using the same reagent or the di-2-ethylhexylphosphoric acid/tri-n-octylphosphine oxide system. Difficulties with the flowsheet include the need for prior reduction of the feed if oxidative pretreatment of the phosphoric acid has been necessary, and the extractants have a small but significant aqueous solubility and are susceptible to hydrolysis. Thus although the flowsheet has been piloted the following system has received more commercial interest.

The use of di-2-ethylhexylphosphoric acid/tri-n-octylphosphine oxide mixture was also developed by Hurst and Crouse (1973) and has been used commercially in several plants world wide. The process involves the synergistic extraction of uranium(VI) by the extractants in the molar ratio DEHPA:TOPO $\sim$ 4:1. However, the distribution coefficient is still not high enough for a single circuit operation. So two extraction/strip circuits are required and as the second circuit involves much lower volumes it has been proposed, on economic grounds, that the first cycle strip solutions from several plants be transported to a central second cycle plant for final treatment (Ross 1975). The overall flowsheet includes a pretreatment step to oxidize uranium, the choice of oxidant being dictated by the need to minimize any contamination of the product phosphoric acid. Thus in one plant (Berry and Dubreucq 1980) hydrogen peroxide is used. This oxidized feed is contacted with the extractant mixture operating in an organic continuous regime to minimize entrainment. Due to the stability of the aqueous uranium phosphate complexes, the distribution coefficient (< 10) is lower than that found in sulphate solution, and decreases as the phosphate concentration increases. Extraction also decreases with increasing temperature and so cooling of the feed is included is the pretreatment. Stripping is carried out using a sidestream of the product phosphoric acid contacted with scrap iron to produce an iron(II) solution. This reduces uranium to the (IV) oxidation state which is not complexed by the extractant mixture and therefore reports in the aqueous phase. The process is very effective and produces a strip product containing 10–12 g/dm^3 uranium. This strip acid is now reoxidized with sodium chlorate and subjected to a second extraction circuit. The loaded organic phase is then scrubbed by water to remove entrained acid, but not iron, before being stripped with ammonium carbonate solution to produce ammonium diuranate. The barren organic phase is treated with mineral acid to regenerate the extractant.

Acknowledgement

The author would like to thank friends and colleagues for their help in the preparation of this chapter by the provision of papers and data prior to publication.

References

Albery, W. J., Burke, J. F., Leffler, E, B. and Hadgraft, J. (1976). *J. Chem. Soc. Faraday Trans. 1* **72**, 1618.

Albery, W. J. and Fisk, P. R. (1981). *Proc. Hydrometallurgy 81*, Paper F1. Soc. Chem. Ind., London.

American Cyanamid Literature: Cyanex 301 Extractant; Cyanex 471 Extractant; Cyanex 923 Extractant; Cyanex 925 Extractant. 1988

Anderson, S. O. S. and Reinhardt, H. (1983). In *Handbook of Solvent Extraction* (ed. T. C. Lo, M. H. I. Baird and C. Hanson) Chapter 25.10, p. 751. Wiley, New York.

Anon (1979). *Chem. Eng.* **83**, 90.

Aprahamian, E. A. Jr and Freiser, H. (1987). *Separ. Sci. Technol.* **22**, 233.

Baltz, J. and Coltinari, E. (1977). US Patent 4041126.

Baradel, A., Guerriero, R., Merrigalli, L. and Vittadini, I. (1986*a*). *J. Metal* **38**, 32.

Baradel, A., Guerriero, R., Merrigalli, L. and Vittadini, I. (1986*b*). *Proc. ISEC86, Munich*, Vol. II, p. 401. DECHEMA, Frankfurt am Main.

Baradel, A., Guerrireo, R. and Veronese, G. (1986*c*). European Patent Application EP 180110.

BARNES, J. E., STETCHFIELD, J. H. and WILLIAMS, G. O. R. (1976). *J. Inorg. Nucl. Chem.* **38**, 1065.

BARONCELLI, F., CARLINI, D., GASPARINI, G. M. and SIMONETTI, E. (1988). *Proc. ISEC88, Moscow*, Vol. IV, p. 236. Vernadsky Inst. Geochem. and Anal. Chem., Academy of Sciences USSR, Moscow.

BAUER, D., COTE, G. and PESCHER-CLUZEAU (1986). *Chem. Sep. Dev., Select. Pap. 1st Int. Symp. Sep. Sci. Technol., New York* (ed. C. J. KING and J. NAVRATIL) Vol. 1, p. 903. Litarvan Press, Colorado.

BAUER, D., FOURRE, P. and SABOT, J.-L. (1984). US Patent 4.485.076.

BAUER, D., FOURRE, P. and SABOT, J.-L. (1985). US Patent 4.559.203.

BAUER, D. and PESCHER-CLUZEAU, Y. (1987). *Hydromet.* **18**, 243.

BELLINGHAM, A. (1961). *Proc. Aust. I. M. M.* **198**, 85.

BERRY, W. W. and DUBREUCQ, A. (1980). *Proc. ISEC80, Liege*, Vol. 3, Paper 80–235. Ass. des Ingenieurs sortis de l'Univ. de Liege.

BHAT T. R. and SUNDARARAJAN, S. (1967). *J. Less-Common Metals* **12**, 231.

BIERMAN, L. W., POLINSKY, S. M., HUMPEL, D. A., and HUMBERGER, R. B. (1985). US Patent 4511541.

BLUMBERG, R. and MELZER, P. (1964). *Proc. Int. Miner. Process. Conf.*, Vol. 1, p. 139. Gordon and Breach, New York.

BORROWMAN, S. R. and BRIDGES, D. W. (1963). USBM RI 6457.

BOZEC, C., DEMARTHE, J. M. and GANDON, L. (1974). *Proc. ISEC74, Lyon*, Vol. 2, p. 1201. Soc. Chem. Ind., London.

BRIGHT, J. H. (1984). European Patent 113912-A.

BROWN, C. G. (1972). British Patent 1262469.

BROWN, C. and SHERRINGTON, L. G. (1979). *J. Chem. Technol. Biotechnol.* **29**, 193.

BROWN, K. B. (1959). *USAEC Report CF 59-9-85.*

BROWN, T. J., RICKLETON, W. A. and BOYLE, R. J. (1987). European Patent Application EP 210 387.

BURKIN, A. R. and PRESTON, J. S. (1975). *J. Inorg. Nucl. Chem.* **37**, 2187.

CASEY, A. T., DAVIES, E., MEEK, T. L. and WAGNER, E. S. (1967). *Solvent Extraction Chemistry (Proc. Int. Solv. Extn. Conf., Goteborg 1966)* (ed. D. DYRSSEN, J. O. LILJENSEN and J. RYDBERG), p. 327. North-Holland, Amsterdam.

CASTRESANA, J.-M., ELIZALDE, M. P., AQUILAR, M. and COX, M. (1988). *Solv. Extn. Ion Exch.* **6**. 265.

CATTRALL, R. W. and SLATER, S. J. E. (1973). *Coord. Chem. Rev.* **11**, 227.

CHEN, CHIA-YUNG (1983). *Hydromet. Res. Dev. and Plant Pract.* (ed. K. OSSEO-ASARE and J. D. MILLER), p. 65. AIME/TMS, New York.

CLEARE, M. J., CHARLESWORTH, P. and BRYSON, D. J. (1979). *J. Chem. Tech. Biotechnol.* **29**, 210.

CLEMENTE, D. DEJ., DEWAR, B. I. and HILL, J. (1980). *CIM 10th Annual Hydrometallurgy Meeting*. Edmonton, Alberta.

COLEMAN, C. F., BROWN, K. B., MOORE, J. G. and CROUSE, D. J. (1958). *Ind. Eng. Chem.* **50**, 1756.

COLEMAN, J. P., GRAHAM, C. R. and MONZYK, B. F. (1988). US Patent 4741887.

COTE. G. and BAUER, D. (1986). *Chem. Ind.* p. 780

COX, M. and FLETT, D. S. (1979). *Proc. ISEC77*, Toronto, CIM Spec. Pub. 21, Vol. 1, p. 63.

COX, M. and FLETT, D. S. (1987). *Chem. Ind.* p. 188.

COX, M. and VAN BRONSWIJK, W. J. (1980). *Proc. ISEC80, Liege*, Vol. 1, paper 80–15. Ass. des Ingenieurs sortis de l'Univ. de Liege.

DAAMACH, S., COTE, G. and BAUER, D. (1987). *Separation Processes in Hydrometallurgy* (ed. G. A. DAVIES), p. 221. Ellis Horwood, Chichester.
Daihachi Chemical Industry Co Ltd (1981). French Patent 2460275.
DALTON, R. F. (1979). *Proc. ISEC77, Toronto*, CIM Spec. Pub. 21, Vol. 1, p. 40.
DALTON, R. F. and PRICE, R. (1982). *Proc. Oslo Symp. 1982*, paper II–23. Soc. Chem. Ind. London.
DALTON, R. F., PRICE, R., HERMANA, E. and HOFFMAN, B. (1987). *Separation Processes in Hydrometallurgy* (ed. G. A. DAVIES), p. 446. Ellis Horwood, Chichester.
DALTON, R. F., PRICE, R. and QUAN, P. M. (1983). *Proc. ISEC83, Denver*, p. 189. AIChE, New York.
DALTON, R. F. and SEWARD G. W. (1984). *Reagents in the Minerals Industry* (ed. M. J. JONES and R. OBLATT), p. 106. Inst. Min. Metal. London,
DALTON, R. F. and SEWARD G. W. (1986). *Proc. ISEC86, Munich*, Vol. II, p. 11. DECHEMA, Frankfurt-am-Main.
DANESI, P. R. and CHIARIZIA, R. (1980). *Crit. Rev. Anal. Chem.* **10**, 1.
DEMIKIRAN, A. (1978). MSc Thesis, University of Leeds.
DEMOPOULOS, G., POUSKOULELI, G. and RITCEY, G. M. (1986). *Proc. ISEC86, Munich*, Vol. II, p. 581. DECHEMA, Frankfurt-am-Main.
DEPTULA, C. and MINC, S. (1967). *J. Inorg. Nucl. Chem.* **29**, 159.
DE SCHEPPER, A. (1979). *Hydromet.* **4**, 285.
DE SCHEPPER, A. (1985). *Extn Metal 85*, p. 167. Inst. Min. Metal, London.
DHARA, S. C. (1984). *Precious Metals, Mining Extraction and Processing* (ed. V. KUDRYK, D. A. CORRIGAN and W. W. LIANG), p. 199. Met. Soc. AIME, New York.
DROBNICK, J. L. and MILLSAP, W. A. (1966). US Patent 3 276 863.
ECKERT, J. and BAUER, J. (1984). Ger. Offn. 3241832.
EDWARDS, R. I. (1979). *Proc. ISEC77, Toronto*, CIM Spec. Pub. No. 21, Vol. 1, p. 52.
FIEBERG, M. and EDWARDS, R. I. (1978). *Report No. 1996*. National Institute for Metallurgy, Randburg, South Africa.
FISHER, J. F. C. and NOTEBAART, C. W. (1983). *Handbook of Solvent Extraction* (ed. T. C. LO, M. H. I. BAIRD and C. HANSON), Chapter 25.1, p. 649. Wiley, New York.
FLEMING, C., GREEN, B. R. and ASHURST, K. G. (1980). *Proc. ISEC80, Liege*, Vol. 2, paper 80–224. Ass. des Ingenieurs sortis de l'Univ. Liege.
FLETCHER, A. W. (1973). *Chem. Ind.* p. 414.
FLETT, D. S. (1974). *Trans. Inst. Min. Met. Sec. C* 83, C30.
FLETT, D. S. (1977). *Acc. Chem. Res.* **10**, 99.
FLETT, D. S. (1982). *Hydromet. Res., Dev. and Plant Pract.* (ed. K. OSSEO-ASARE and J. D. MILLER), p. 39. AIME-TMS, New York.
FLETT, D. S., COX, M. and HEELS, J. D. G. (1975). *J. Inorg. Nucl. Chem.* **37**, 2197.
FLETT, D. S. and FLETCHER A. W. (1970). *SME-AIME* 247, 294.
FLETT, D. S. and MELLING J. (1979). *Hydromet.* **4**, 135.
FLETT, D. S. and MELLING J. (1980). *Hydromet.* **5**, 283.
FLETT, D. S., MELLING J. and COX M. (1983). *Handbook of Solvent Extraction* (ed. T. C. LO, M. H. I. BAIRD and C. HANSON), Chapter 24, p. 629. Wiley, New York.
FLETT, D. S. and JAYCOCK, M. J. (1973). *Solvent Extraction* (ed. Y. MARCUS), Vol. 3, p. 1. Marcel Dekker, New York.
FLETT, D. S. and WEST, D. W. (1971). *Proc. ISEC71, The Hague*, Vol. 1, p. 214. Soc. Chem. Ind., London.

FLETT, D. S. and WEST, D. W. (1978). *Complex Metallurgy '78*, p. 49. Inst. Min. Metal., London.
FLETT, D. S. and WEST, D. W. (1986). *Proc. ISEC86, Munich*, Vol. II, p. 2. DECHEMA, Frankfurt-am-Main.
FLETT, D. S. and WEST, D. W. unpublished.
FOURRE, P. and BAUER, D. (1983). *Solv. Extn. Ion Exch.* **1**, 465.
FREEMAN, R. W. and TAVLARIDES, L. L. (1980). *Chem. Eng. Sci.* **35**, 559.
FREISER, H. (1984). *Acc. Chem. Res.* **17**, 126.
FUWA, A. (1987). *Metal. Rev. M.M.I.J.* **4**, 98.
GATRONE, R. C., KAPLAN, L. and HORWITZ, E. P. (1987). *Solv. Extn. Ion Exch.* **5**, 1075.
GAUDERNAUK, B. and BRAATEN, O. (1976). *9th Rare Earth Research Conference, Virginia, Oct. 1976.*
GAUDERNAUK, B., HANNESTAD, G. and HUNDERE, I. (1974). US Patent 3 821 352.
GIGANOV, G. P., TRAVKIN, V. F. and ANIKINA N. P. (1978). *Tsvet. Metl.* 27, CA **89**,166662e.
GOKCEN, N. A. (1982). *Hydromet., Res., Dev. and Plant Prac.* (ed. K. OSSEO-ASARE and J. D. MILLER), p. 329. AIME/TMS, New York.
GRANT, R. A. and MURRER, B. A. (1986). European Patent. 0210004.
GRINSTEAD, R. R. and TSANG L. (1983). *Proc. ISEC83, Denver, Colorado*, p. 230. AIChE, New York.
GUERRIERO, R., VERONESE, G., BARADEL, A. and RIVOLA, L. (1985). Ger. Offn. DE 3 423 713.
GUESNET, P., SABOT, J. L. and BAUER, D. (1980). *J. Inorg. Nucl. Chem.* **42**, 1459.
GUY, R. H. and FLEMING, R. (1979). *Int. J. Pharm.* **3**, 143.
HAHN, R. and RETELSDORF, H. (1984). *Erzmetall.* **37**, 444.
HANNA, G. J. and NOBLE, R. D. (1985). *Chem. Rev.* **85**, 583.
HANSON, C., HUGHES, M. A., PRESTON, J. S. and WHEWELL, R. J. (1976). *J. Inorg. Nucl. Chem.* **38**, 2306.
HANSON, C. and WHEWELL, R. (1981). *Solvent Extraction and Ion Exchange* (ed. J. A. MARINSKY and Y. MARCUS), Vol. 8, p. 1. Marcel Dekker, New York.
HARRISON, G., LAKSHMANAN, V. I. and LAWSON, G. J. (1976). *Hydromet.* **1**, 339.
HASKETT, P. R., FLYNN, C. M. and CARNAHAN, T. G. (1979). US Patent Application 8298.
HAZEN, W. C. and COLTRINARI, E. L. (1975). US Patent 3872209.
HELGORSKY, J. and LEVEQUE, A. (1979). French Patent 77/37785.
HELGORSKY, J. and LEVEQUE, A. (1982). European Patent Application EP 53971.
HOPKIN, W. (1976). Paper presented at *105th AIME Annual Meeting, Las Vegas, Nevada.*
HOPKIN, W. (1982). *Erzmetall.* **35**, 192.
HORWITZ, E. P., DIAMOND, H., and KALINA, D. G. (1983). *Am. Chem. Soc. Symp. Ser* 216 (Plutonium Chemistry), p. 433.
HUBRED, G. L., KUST, R. N., NATWIG, D. L. and PEMSLER J. P. (1978). US Patent 4083915.
HUGHES, M. A. (1978). *Hydromet.* **3**, 85.
HUMMELSTEDT, L., SUND, H. E., KARAJALUOTO, J., BERTS, L. O. and NYMAN, B. G (1974). *Proc. ISEC74, Lyon*, Vol. 1, p. 829. Soc. Chem. Ind. London.
HURST, F. J. and CROUSE, D. J. (1973). US Patent 3 711 591.
HURST, F. J. and CROUSE, D. J. (1974). US Patent 3 835 214.

INOUE, K., FIJINO, O. and ARITA, H. (1986). *Nippon Koygo Kaishi* **102**, 491. CA **106** 126806t.

IWAYA (1985). Japanese Patent SHO 60 245736.

ISHIMORI, T. and NAKAMURA, E. (1963). *JAERI Report 1047*. Japanese Atomic Research Institute.

ISHIMORI, T., AKATSU, E., CHENG, W-P., TSUKUECHI, K.I. and OSAKABE, T. (1964). *JAERI Report 1062*, Japanese Atomic Research Institute.

ISHIMORI, T., AKATSU, E., TSUKUECHI, K., KOBUNE, T., USUBA, Y., KIMURA, K., ONAWA, G. and UCHIYAMA, H. (1966). *JAERI Report 1106*, Japanese Atomic Research Institute.

JEDINAKOVA, V. (1986). *Recent Developments in Separation Science*, (ed. N. N. LI and J. D. NAVRATIL), Vol. VIII, Chapter 2, p. 45. CRC Press, Boca Raton, Florida.

JENKINS, I. L. and MCKAY, H. A. C. (1954). *Trans. Faraday Soc.* **50**, 107.

JOE, E. G., RITCEY, G. M. and ASHBROOK, A. W. (1966) *J. Metal*, July, p. 13.

JUDIN, V. P. and SUND, H. E. (1981). *Proc. Hydrometallurgy '81*, paper F4. Soc. Chem. Ind., London.

KALIJTA, V. V., KUZNERSOV, G. I., KRAVCHENKO, A. N., LANIN, V. P., MIRAEVSKY, G. P., PUSHKOV, A. A., TRAVKIN, V. F. and SHKL'YAR, L. I. (1988). *Proc. ISEC88, Moscow*, Vol. IV, p. 252. Vernadsky Inst. Geochem. Anal. Chem., Academy of Sciences, Moscow.

KERFOOT, D. G. E. (1978). US Patent 4115512.

KERTES, A. S. (1971). In *Recent Advances in Liquid-liquid Extraction* (ed. C. HANSON), p. 15. Pergamon Press, London.

KNOTHE, M. (1980). *Z. anorg. allgem. Chem.* **470**, 216.

KOLARIK, Z. and GRIMM, R. (1976). *J. Inorg. Nucl. Chem.* **38**, 1721.

KORDOWSKY, G. A., SUDDERTH, R. B. and HOUSE, J. E. (1985). *Extn. Metallurgy '85*, Inst. Min. Metal. London.

KORDOWSKY, G. A., OLAFSON, S. M., LEWIS, R. G. and DEFFNER, V. L. (1987). *Separ. Sci. Technol.* **22**, 215.

KOULOHERIS, A. P. (1980) *Chem. Eng.*, p. 82.

KRUESI, F. R. and SCHIFF, N. N. (1968). Paper at AIME Meeting New York, Feb. 1968.

LAKSHMANAN, V. I. and LAWSON, G. J. (1973). *J. Inorg. Nucl. Chem.* **35**, 4285.

LAKSHMANAN, V. I., LAWSON, G. J. and NYHOLM, P. S. (1974). *Proc. ISEC74, Lyon*, Vol. 1, p. 699. Soc. Chem. Ind., London.

LASKORIN, B. N., YAKSHIN, V. V., FILIPPOV, E. A., CHYMAKOVA, S. M., BELOV, V. A. and ARKHIPOVA, G. G. (1978). *Radiokhimiya* **20**, 511.

LEVEQUE, A. and HELGORSKY, J. (1979). *Proc. ISEC77, Toronto*, CIM Spec Pub No. 21, Vol. 2, p. 439.

LI, K. and FREISER, H. (1986). *Solv. Extn. Ion Exch.* **4**, 739.

LIEM, D. H. and SINEGRIBOVA, O. A (1971). *Acta Chem. Scand.* **25**, 301.

LLOYD, P. J. D. (1983). *Handbook of Solvent Extraction* (ed. T. C. Lo., M. H. I. BAIRD and C. HANSON), Chapter 25.11. Wiley, New York.

LLOYD, P. J. and MASON, E. A. (1964). *J. Phys. Chem.* **68**, 3120.

LONG, R. S., ELLIS, D. A. and BAILES, R. H. (1955). *Proc. Int. Conf. Atomic Energy, Geneva*, **8**, 77.

LUCAS, B. H. and RITCEY, G. M. (1975). *CIM Bull.*, Jan 1975.

MARCUS, Y. and KERTES, A. S. (1969). *Ion Exchange and Solvent Extraction of*

Metal Complexes. Wiley-Interscience, New York.
MARKOVITS, G. Y. and CHOPPIN, G. R. (1973). *Solvent Extraction* (ed. Y. MARCUS), Vol. 3, p. 51. Marcel Dekker, New York.
MARR, R., BART, H. J. and WACHTER, R. (1984). European Patent Application EP 106118 A1.
MAESHIRO, I. (1978). *Proceedings 11th Commonwealth Mining and Mettallurgy Congress, Hong Kong*, p. 315. Inst. Min. Metal., London.
MERIGOLD, C. R. and JENSEN, W. H. (1974). *Proc. ISEC74, Lyon*, Vol. 2, p. 1231. Soc. Chem. Ind. London.
MICHEL, TH. and NITSCH, W. (1988). *Proc. ISEC88, Moscow*, Vol. 11, p. 20. Vernadsky Inst. Geochem. and Anal. Chem., Academy of Sciences, Moscow.
MIYAKE, Y., IMANISHI, Y., KATAYAMA, Y., HAMATANI, T. and TERAMOTO, M. (1986). *J. Chem. Eng. (Japan)* **19**, 117.
MOORE, R. H. and PARTRIDGE. J. A. (1972). Batelle Pacific North West Labs. *Rep No. BNWL-SA 4476.*
MORIN, E. A. and PETERSON, H. D. (1975). US Patent 3878286.
MORRIS, D. F. C. and KHAN, M. A. (1968). *Talanta* **15**, 1301.
MUIR, D. M. and PARKER, A. J. (1982). *Hydromet. Res. Dev. Plant Prac.* (ed. K. OSSEO-ASARE and J. D. MILLER), p. 341. AIME, New York.
MURTHY, T. K. S., KOPPIKER, K. S. and GUPTA, C. K. (1986). *Recent Developments in Separation Science* (ed. N. N. LI and J. D. NAVRATIL), Vol. VIII, Chapter 1, p. 1. CRC Press, Boca Raton, Florida.
MUSIKAS, C. and GERMAIN M. (1988). *Proc. ISEC88, Moscow*, Vol. IV, p. 124. Vernadsky Inst. of Geochem. and Anal. Chem., Academy of Sciences, USSR, Moscow.
NANDI, B., DAS, N. R. and BHATTACHARYA, S. N. (1983). *Solv. Extn. Ion Exch.* **1**, 141.
NAYLOR, A. and ECCLES, H. (1988). *Proc. ISEC88, Moscow*, Vol. 1, p. 31. Vernadsky Inst. of Geochem. and Anal. Chem., Academy of Sciences, USSR, Moscow.
NAYLOR, A. and WILSON, P. D. (1983). *Handbook of Solvent Extraction* (ed. T. C. Lo, M. H. I. BAIRD, and C. HANSON), Chapter 25.12, p. 783. Wiley, New York.
NYMAN, B. G. and HUMMELSTEDT, L. (1974). *Proc. ISEC74, Lyon*, Vol. 1, p. 669. Soc. Chem. Ind., London.
KANE, J. O'., MIDDLIN, B. and ROYSTON, D. (1985). *Extn. Metal 85*, p. 951. Inst. Min. Metal. London.
OLIVER, A. J. and ETTEL, V. A. (1976). *Can. Met. Quart.* **15**, 383.
O'NEILL, C. E., ETTEL, V. A., OLIVER, A. J. and ITZKOVITCH, I. (1976). *CIM Bull.* **69**, 86.
OSSEO-ASARE, K., LEAVER, H. S. and LAFERTY, J. M. (1981). *Process Fundamental Considerations of Selected Hydrometallurgical Systems*, p. 195. AIME.
PEARSON, R. G. (1963). *J. Am. Chem. Soc.* **85**, 3533.
PENNER, L. R., SIEMENS, R. E. and GALVAN, G. L. (1986). *Solv. Extn. Ion Exch.* **4**, 345.
POUSKOULELI, G., KELEBEK, S. and DEMOPOULOS, G. P. (1987). *Separation Processes in Hydrometallurgy* (ed. G. A. DAVIES), p. 175. Ellis Horwood, Chichester.
PRESTON, J. S. (1983a). *J. S. A. Inst. Min. Met.* **83**, 126.
PRESTON, J. S. (1983b). *Hydromet.* **10**, 187.

PRESTON, J. S. (1983c). *Proc. ISEC80*, p. 357. Denver, Colorado, AIChE New York.
PRESTON, J. S. and FLEMING, C. A. (1983). *Hydromet. Res., Dev. Plant Prac.* (ed. J. D. MILLER and K. OSSEO-ASARE), p. 475. AIME/TMS, New York.
PRESTON, J. S. (1985). *Mintek Report M234*.
PRESTON, J. S. and DU PREEZ, A. C. (1986). *Proc. ISEC86, Munich*, Vol. II, p. 83. DECHEMA, Frankfurt-am-Main.
PRESTON, J. S., DU PREEZ, A. C. and JACOBS, A. J. (1988). *Proc. ISEC88, Moscow*, Vol. III, p. 137. Vernadsky Inst. Geochem. and Anal. Chem., Academy of Science, Moscow.
PRESTON, J. S. and DU PREEZ, A. C. (1988). *Mintek Report M378*.
PRICE, R. and TUMILTY, J. A. (1975). *Symp. Ser. No 42*, p. 19. Inst. Chem. Eng., London.
PRZESZLAKOWSKI, S. and WYDRA, H. (1982). *Hydromet.* **8**, 49.
REAVILL, L. R. P. and CHARLESWORTH, P. (1980). *Proc. ISEC80, Liege*, Vol. 3, paper 80–93. Soc. Chem. Ind., London.
RICE, N. M. (1978). *Hydromet.* **3**, 111.
RICE, N. M. and SMITH, D. C. (1986). *Proc. ISEC86, Munich*, Vol. II, p. 633. DECHEMA, Frankfurt-am-Main.
RICKELTON, W. A. (1986). US Patent 4 619 816.
RICKELTON, W. A. and BROWN, T. J. (1985). *Proc. CIM Ann. Hydromet. Meeting*, Impurity Control and Disposal, Vancouver, paper 19.
RICKELTON, W. A., FLETT, D. S. and WEST, D. W. (1983). *Proc. ISEC83, Denver, Colorado*, p. 195. AIChE, New York.
RICKELTON, W. A., FLETT, D. S. and WEST, D. W. (1984). *Solv. Extn. Ion Exch.* **2**, 815.
RICKELTON, W. A. and ROBINSON, A. J. (1987). *Miner. Metall. Proc.* **4**, 7.
RIMMER, B. F. (1974). *Chem. Ind.*, p. 63.
RITCEY, G. M. (1983). unpublished data, cited in *Handbook of Solvent Extraction* (ed. T. C. LO, M. H. I. BAIRD and C. HANSON), Chapter 25.2. Wiley, New York.
RITCEY, G. M. and ASHBROOK, A. W. (1969). *Trans. Inst. Min. Metall.* C78, 57.
RITCEY, G. M. and ASHBROOK, A. W. (1984). *Solvent Extraction*, part I. Elsevier, Amsterdam.
RITCEY, G. M. and ASHBROOK, A. W. (1979). *Solvent Extraction*, part II. Elsevier, Amsterdam.
RITCEY, G. M., ASHBROOK, A. W. and LUCAS, B. H. (1972). Annual Meeting, AIME, San Franscisco.
RITCEY, G. M. and LUCAS, B. H. (1969). AIME Meeting, Washington; *Mines Branch Technical Bulletin TB113*, CanMet 1969.
RITCEY, G. M. and LUCAS, B. H. (1971). *Proc. ISEC71, Hague*, Vol. 1, p. 463. Soc. Chem. Ind., London.
RITCEY, G. M. and LUCAS, B. H. (1974). *Proc. ISEC74, Lyon*, Vol. 3, p. 2437. Soc. Chem. Ind., London.
RITCEY, G. M. and LUCAS, B. H. (1975). *CMI Bull* **68**, 105.
RONG, Q. and FREISER, H. (1987). *Solv. Extn. Ion Exch.* **5**, 923.
ROSS, R. C. (1975). *Eng. Min. J.* **176**, 80.
SABOT, J -L. and ROLLAT, A. (1985). European Patent 0156735.
SATO, T. (1986). *Chem. Sep. Dev. Select Paper*. 1st Int. Symp. Sep. Sci. Technol,

New York (ed. C. J. KING and J. D. NAVRATIL), Vol. 1, p. 285. Litarvan Press, Colorado.
SATO, T., NAKAMURA, T. and KUWAHARA M. (1984). *Solv. Extn. Ion Exch.* **2**, 201.
SCHULZ, W. W. and NAVRATIL, J. D. (1981). *Recent Developments in Separation Science*, Vol. 7, Chapter 2, p. 31. CRC Press, Boca Raton, Florida.
SCHULZ, W. W. and NAVRATIL, J. D. (ed) (1987). *The Science and Technology of Tributyl Phosphate*. CRC Press, Boca Raton, Florida.
SCHWARZ, R. R. (1978). PhD Thesis, Hatfield Polytechnic.
SCOTT, T. R. (1949). *Analyst* **74**, 486.
SEKINE, T. and HASAGAWA, Y. (1977). *Solvent Extraction Chemistry, Fundamentals and Applications*, p. 386. Marcel Dekker, New York.
SHERRINGTON, L. G. (1983). *Handbook of Solvent Extraction* (ed. T. C. Lo, M. H. I. BAIRD and C. HANSON.) Chapter 25.6, p. 717. Wiley, New York.
SHERRINGTON, L. G. and KEMP, W. P. (1966). British Patent 1026791, amended 1969.
SILAEV, Y. N., KOKURKINA, S. P., REZNIK, A. M. and GLUBOKOV, Y. M. (1980). *Izv. Vyssh. Ucheb. Zaved. Khim. Khim. Technol.* **23**, 534, *CA* **93** 81369x.
SOISSON, D. J., McLAFFERTY, J. J. and PIERRET, J. A. (1961). *Ind. Eng. Chem.* (Nov) p. 99.
SPECKER, H. and WERDING, G. (1964). *Z. anal. Chem.* **200**, 337.
STREAT, M. and NADEN, D. (1987). *Ion Exchange and Sorption Processes, CRAC*, Vol. 19, Soc. Chem. Ind./Wiley.
SUDDERTH, R. B. and KORDOWSY, G. A. (1986) *Chemical Reagents in the Mineral Processing Industry* (ed. D. MALHOTRA and W. F. RIGGS), Chapter 20, p. 181. SME/TMS.
SUETSUNA, A., ONO, N., IIO, T. and YAMADA, K. (1980). *Met. Soc. AIME paper selection A80–2*.
Sumitomo Chemical Co. (1984). Japanese Patent 59/50024.
SWANSON, R. R. (1965). US Patent 3 224 873.
SWANSON, R. R. (1969). US Patent 3 428 449.
SZYMANOWSKI, J., COX, M. and HIRONS, C. G. (1984). *J. Chem. Technol. Biotechnol.* **34A**, 218.
TAKAKASHI, M., OGATA, T., OKINO, H. and ABE, Y. (1983). *Proc. ISEC83, Denver, Colorado*, p. 359. AIChE, New York.
TAMMI, T. (1979). *Tutkimus Tek*, **4**, 44.
TARASOV, V. V. and YAGODIN, G. A. (1988). *Ion Exchange and Solvent Extraction* (ed. J. A. MARINSKY and Y. MARCUS), Vol. 10, p. 141. Marcel Dekker, New York.
TSEREKOVA, A. M., GIGANOV, G. P., KOLOMINA, G. K. and FOKINA, V. I. (1970) *Sb. Nauk. Tv. Vses. Nauch–Issled. Gornomet. Inst. Tsvet. Metal.* no. 19, 119. *CA.* **74**, 33756g.
THORNHILL, P. G., WIGSTOL, E. and VAN WEERT, G. (1971), *J. Metal*, July, p. 13.
TROUVE, G., MALHER, E., COLINART, P. and RENON, H. (1982). *Chem. Eng. Sci.* **37**, 1225.
TUMILTY, J. A., DALTON, R. F. and MASSAM, J. P. (1977a). *Advances in Extraction Metallurgy* (ed. M. J. JONES), p. 123. Inst. Min. Metall., London.
TUMILTY, J. A., SEWARD, G. W. and MASSAM, J. P. (1977b). *Proc. ISEC77, Toronto*, CIM Spec. Pub. 21, vol. 2, p. 542.
TUNLEY, T. H. and NEL, V. W. (1974). *Proc. ISEC74, Lyon*, Vol. II, p. 1519. Soc. Chem. Ind., London.

Unitika Ltd (1984). Japanese Patent 57/186683.

VIRNIG, M. J. (1979). *Proc. ISEC77, Toronto*, CIM Spec. Pub. 21, Vol. 2, p. 535.

WARSHAWSKY, A. (1983). *Separ. Purif. Meth.* **12**, 37.

WHEWELL, R. J. and HANSON, C., (1981). *Ion Exchange and Solvent Extraction* (ed. J. MARINSKY and Y. MARCUS), Vol. 8, p. 1. Marcel Dekker, New York.

WIGSTOL, E. and FROYLAND, K. (1972). *Het. Ingenieursblad.* **41**, 476.

WYNN, N. P. (1985). PCT Int. Appl. WO85/2171.A1.

WYNN, N. P. and ZABELKA, M. (1985). European Patent Application EP 143749.

YAGODIN, G. A. and SINEGRIBOVA, O. A. (1983). *Handbook of Solvent Extraction* (ed. T. C. LO, M. H. I. BAIRD and C. HANSON), Chapter 25.14, p. 805. Wiley, New York.

YAMADA, H. and TANAKA, M. (1985). *Advances in Inorganic and Radiochemistry*, Vol. 29, p. 143. Academic Press, New York.

VAN DER ZEEUW, A. J. and KOK, R. R. (1979). *Proc. ISEC77, Toronto*, CIM Spec. Pub. 21, Vol. 1, p. 17.

ZHOU, T., ZHANG, X., MA, R., HUANG, Z., QU, M., and ZHOU, Z. (1982). *Hydromet.* **8**, 379.

2

NUCLEAR FUEL REPROCESSING USING SOLVENT EXTRACTION

W. BATEY

2.1 Introduction

The traditional sources of energy used in the production of electricity, coal and oil have a finite availability. Consequently, research and development into alternative forms of energy will be needed if the present high living standards in the developed countries are to be maintained in the future and also to raise the living standards in the developing countries.

The renewable energy sources such as wind, solar, tide and geothermal power are unlikely to be able to meet more than a small fraction of the world's energy needs. In Britain, for example, hydroelectric power has been almost fully exploited, whilst the use of other renewable energy sources is limited. In contrast, nuclear power provides an energy source for the production of electricity which is both available and reliable. Nuclear power is derived from uranium, which has no significant commercial use other than as a fuel for nuclear electricity generation. The expansion of the present world's nuclear-generated electricity programmes would release the valuable raw materials, coal and oil, for more important uses such as feed stocks for the production of chemicals and fertilizers and as a source of liquid fuels and gases. One tonne of natural uranium used in a modern nuclear reactor can produce as much electricity as that obtained burning 25 000 tonnes of coal in a conventional power station.

The fuel in a modern nuclear reactor comprises uranium oxide or a mixture of plutonium and uranium oxides. The nuclear fission of the fissile heavy atoms releases heat which can then be used to generate steam and drive a turbine to produce electricity as in any conventional power station. The nuclear fission process also produces gamma radiation, neutrons, and fission products.

Most of the nuclear power stations in the world are thermal reactors. Thermal reactors use uranium-235 as their fuel, an isototpe that makes up some 0.7 per cent of the naturally occurring uranium mined from the earth's crust, the remainder being the isotope uranium-238 which does not fission readily. In order to achieve the most efficient fission of uranium-235 a thermal reactor contains a moderator such as water or graphite which slows down the fast neutrons produced in the fission process by repeated elastic collisions. The proportion of uranium-235 is normally enhanced in a thermal reactor fuel but even so the reactor still only uses some 2 per cent of the available uranium.

A fast reactor differs from a thermal reactor in that it uses plutonium as a fuel which needs fast neutrons to sustain the fission process and has no moderator present which would slow down the neutrons. Fast reactors are fuelled with mixed plutonium and uranium-238 oxides and have a further blanket of uranium-238 (the radial breeder) around the reactor core. The plutonium is consumed in the reactor by the fission process to provide the heat to generate electricity whilst at the same time the radial breeder uranium-238 captures some of the neutrons produced in the fission process and is transmuted into plutonium. The amount of plutonium produced in the blanket by careful design and reactor operation can exceed, be equal to, or be less than the amount consumed in the fission process. The fact that fast reactors burn uranium-238 which is unusable in thermal reactors permits an enormous gain in fuel utilization, producing some sixty times more energy from uranium than can be obtained from thermal reactors.

This high efficiency of uranium utilization can, however, only be achieved in practice if the plutonium and uranium are recovered from the spent fuel, the plutonium from the thermal reactor fuel being used to launch a fast reactor programme which would eventually become self-sufficient in plutonium once the fast reactor fuel cycle becomes established. Nuclear fuel reprocessing using solvent extraction, therefore, is an essential component of the system required to maintain fuel supplies to current and future electricity-producing operating reactors and in the longer term to commercial scale fast reactor electricity generating units.

2.2 Types and selection of contactors

2.2.1 Introduction

An essential part of the nuclear fuel cycle for present and planned nuclear energy programmes is the ability to separate for recycling the fissile and fertile fuel from the fission products and the minor actinides. Solvent extraction is used to perform these separations. The technique of liquid–liquid solvent extraction for the separation of multicomponent mixtures involves the partitioning of two or more solutes between two substantially immiscible solvents and such processes in general require a greater number of theoretical stages than is necessary for the extraction of a single solute. The key piece of equipment for any solvent-extraction process is the contactor in which the two immiscible phases are mixed together. The purpose of the contactor is to mix intimately the two phases to provide sufficient interfacial area across which the transfer of the required solute takes place.

Many diverse types of contactor have been described in the literature and a number of reviews are available. Few of these contactors, however, are suitable for reprocessing nuclear fuels. The major requirements for this purpose are no moving parts, a geometry which from a criticality point of view is ever-safe, and reliability over long periods of time. The types of contactor that have been used to date (Vassalo 1981; Mills 1984) in reprocessing plants comprise box and pump mixer-settlers, pulsed columns, pulsed packed and plate columns, and centrifugal types.

2.2.2 Classification of contactors

[See also Pratt and Stevens, Vol. 1, Chap. 8]

Extractors may be classified by the force applied for interdispersing the phases. The force applied, gravitational or mechanical, acts on the density difference between the phases. All continuous multistage contactors may also be further subdivided according to type of operation: discrete stagewise contact or differential contact (Fig. 2.1).

Stage-wise contactors provide a number of discrete stages in which the two phases are brought to equilibrium, separated, and passed countercurrent to the adjoining stages. Whilst such contactors avoid the problems of axial mixing found in the differential contactors, the necessity for separating the phases between each stage adds significantly to the overall size of the contac-

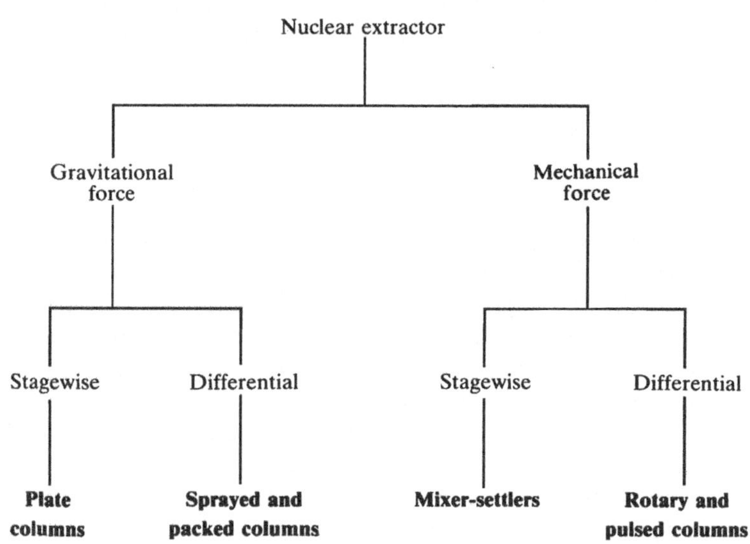

Fig. 2.1. Classification of nuclear extractors.

tor equipment and to the solvent inventory. Phase separation is also required for countercurrent flow but does not contribute to the mass transfer process. The settlers are usually considerably larger than the mixers where the actual solute extraction occurs.

Differential contactors, however, provide conditions for mass transfer throughout their entire length. Equilibrium between the immiscible phases is not established at any point and phase separation occurs only at the ends of the contactor. If a vertical column is employed, as is usually the case, much less ground space is required than is the case for mixer-settlers. In addition, the absence of interstage settlers may lead to low inventories of the process streams.

The size of contactor required for a given separation duty is some function of the rate of transfer of solute between the two immiscible phases. This in turn is proportional to the interfacial area per unit volume and the solute concentration driving force. Mixer-settlers are suitable for a system exhibiting slow mass transfer rates since the residence time may be increased to compensate for the slower kinetic rates. Differential contactors, on the other hand, are only really suitable for use when mass transfer rates are rapid.

Mixer-settlers are the most generally used type of stage-wise contactors. Relatively reliable design procedures are available and they are versatile in operation. More uncertainties arise in the design and performance of columns, particularly in respect of scale-up. The greater degree of confidence, however, in the design of mixer-settlers, certainly in the absence of pilot plant experiments, has to be paid for at the cost of excess capacity. Nevertheless, mixer-settlers have the great attraction that for most processes over-design does not impair their performance.

The centrifugal contactors which offer short contact times and efficient phase separation can be either stage-wise contactors like the Robatel or on the small scale the AKUFVE or differential units such as the Quadronic: the hydrocyclones fall into the stage-wise group and the electrostatic extractor fits into the differential contactor group.

2.2.2.1 *Basic principles of extractors*

The basic requirements of an industrial contactor may be summarized as follows:

(1) efficient extraction;

(2) high throughput;

(3) low volume;

(4) low capital, operating and maintenance costs.

The factors affecting the efficiency of extraction may be detailed as follows:

(1) phase composition;

(2) phase ratio;

(3) phase properties;

(4) temperature;

(5) energy input;

(6) back extraction.

The above factors govern the type of diffusion, eddy or molecular, interfacial area, tension and turbulence, the latter known as Marangoni instabilities. The degree of the mechanical or pneumatic energy input applied is found to have an important bearing on the rate of mass transfer. As the energy input is increased, the increase in interfacial area falls and the fluid circulation within the dispersed phase droplets is suppressed as eddy diffusion is replaced by molecular diffusion. The net effect is a decrease in the rate of mass transfer. Consequently, for efficient extraction, an optimum degree of energy input is required. Back extraction is found to occur in all types of contactors with the differential types the most prone to it, and is a consequence of the axial and/or radial mixing of the dispersed phase within the continuous phase.

2.2.2.2 *Mixer-settlers*

The mixer-settler is a stage-wise device in two parts: the mixer where one phase is dispersed in the other by mechanical or pneumatic energy input to permit solute extraction between two immiscible phases; and the settler where one phase is disengaged from the other. Mixer-settlers may be stacked vertically or horizontally with flows generally countercurrent within as well as between devices. Pumping is effected within or between devices by mechanical, hydraulic, or gravitational means. The design principles for mixer-settlers are largely empirical in nature. The pertinent design parameters are: phase properties, geometry of mixer and settling chambers, internals of mixer and settling chambers, and the energy input into the mixer.

The aim of any design of mixer-settler is to make the total equipment volume as small as possible, commensurate with the flowsheet requirements. In addition, the mixer chamber should be of sufficient size to give an adequate residence time to bring the two phases close to equilibrium and the settler should be big enough to ensure complete phase disengagement.

With most mixer-settler systems residence times in the mixer need not exceed a minute or two although some systems with much slower rates of mass transfer require much longer times. The actual performance of a stage is measured in terms of a stage efficiency, the approach to equilibrium being an exponential function with respect to time.

The stage efficiency will be reduced if the phases are not completely separated in the settler since recycling will occur. This may well be insignificant in the intermediate stages if the only entrainment is due to a secondary haze. In terminal stages, however, any entrainment could lead to impurities in the product. Mixer-settlers invariably have a large hold-up and a relatively slow response to changes in feed conditions.

Ever-safe designs for mixer-settlers are few, safe concentrations or masses of fissile material being the usual method of control. The ever-safe designs which have been proposed are severely limited in one dimension because the maximum allowable thickness for an infinite slab of fissile material is small.

The KAPL Pump-Mix mixer settler (Davidson *et al.* 1957) comprises a mixing chamber in two parts stacked horizontally to the settling chamber (Fig. 2.2). The two phases are fed to the mixing chamber via ports from the settlers of the adjacent KAPL units. The heavy phase is hydraulically lifted from the pre-mixing into the mixing chamber by the impeller, which also controls mixing intensity and interface height in the preceding settler. The phases then flow from the mixing chamber via horizontal baffles into the settler. To prevent flooding, the pumping rate of the impeller is greater than the heavy phase flow. Thus the interface on the final stage is best controlled by a weir box or instrument control valves.

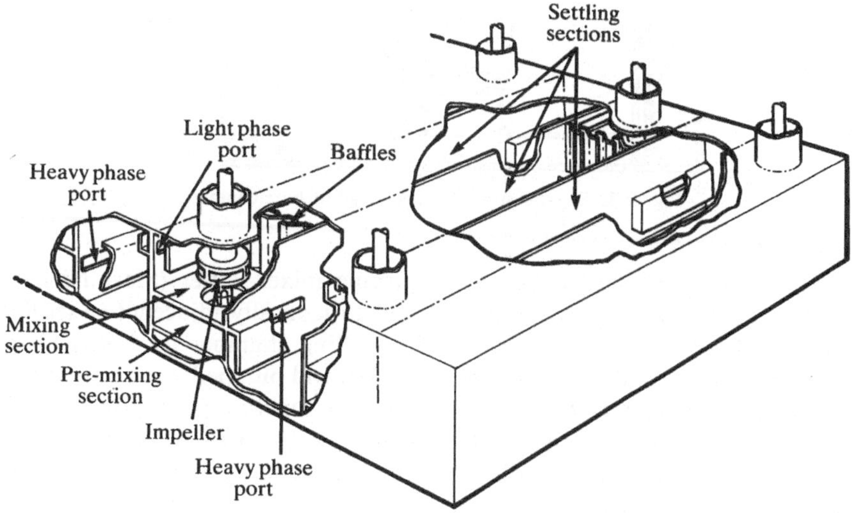

Fig. 2.2. Pump-mix mixer-settler.

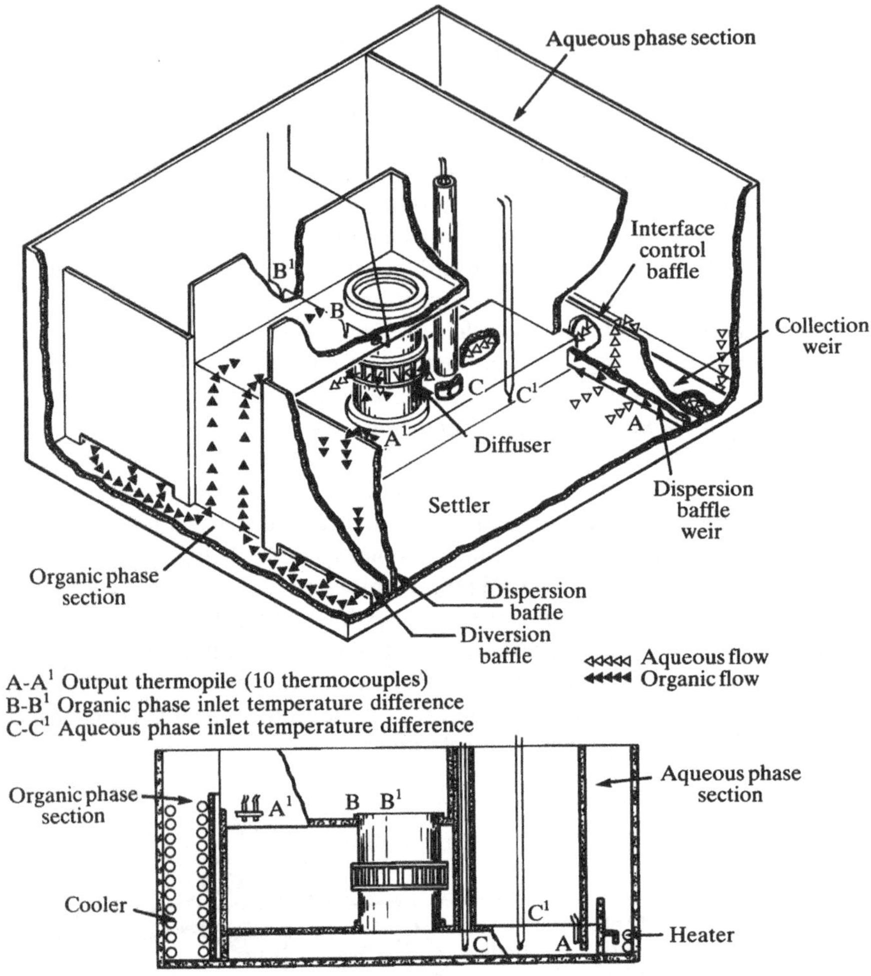

Fig. 2.3. Savannah river single-stage turbine contactor.

The Savannah River Laboratory Turbine mixer-settler (Starr Nicholls 1958) comprises a mixing chamber stacked horizontally adjacent to a settling chamber for light phase removal and a settling chamber for heavy phase removal (Fig. 2.3). The two phases are mixed and pumped into their respective settling chambers by a turbine stirrer. The function of the top baffle in the channel between the two settling chambers is to prevent the light phase from being pumped into the mixer and the bottom baffle to maintain a minimum heavy phase height in the mixer.

The Simple Box Type mixer-settler (Lowes *et al.* 1958) comprises an antechamber, mixing chamber and settling chamber stacked horizontally toge-

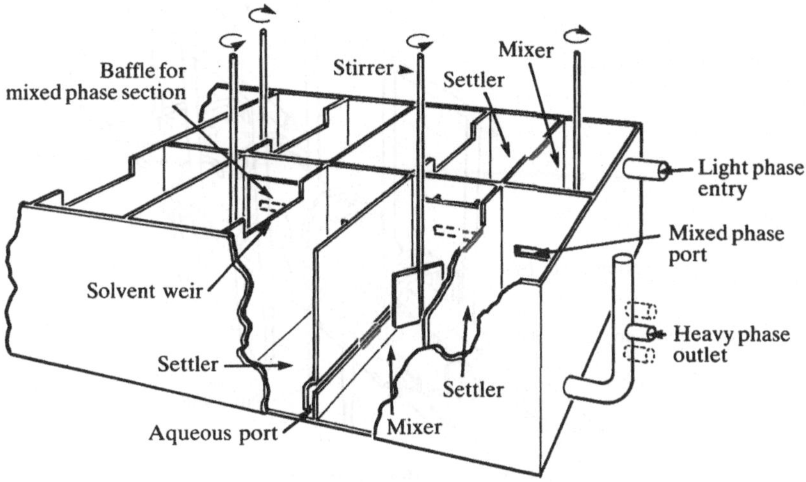

Fig. 2.4. Simple box type mixer-settler.

ther (Fig. 2.4). The two phases are mixed by a simple rotating turbine impeller and the driving force between chambers is by density difference aided by a suitable arrangement of a network of baffles. In the mixing chamber, the baffles also act to nullify hydraulic instabilities generated by the impeller. To overcome frictional resistance and to assist interface control, the device is tilted towards the heavy phase outlet end.

The use of mechanical impellers in an active environment sets limitations to the use of conventional mixer-settlers and, for this reason, units have been developed in which the impeller is replaced by air pulsing devices (Thornton 1955; Logsdail and Thornton 1959).

The UKAEA Air-Pulsed mixer-settler (Buck *et al.* 1959) comprises a mixing chamber with dividing baffle, stacked horizontally to the settling chamber (Fig. 2.5). The two phases are mixed by air surges applied alternately to first one side of the baffle and then to the other. Thus there is no pumping action and the countercurrent flow of the phases through the stages depends solely upon the level differences established by the differing densities between mixed and unmixed phases. The mixer volume is determined by the residence time required and its depth set by the hydraulic requirements of the unit as a whole, but in practice an approximately cubical form is usually the best. A settler area is made sufficiently large to contain the emulsion band of specified thickness below the mixed phase port.

2.2.2.3 *Pulsed columns*

The column is a differential contactor device in which one phase is dispersed in and flows countercurrent to the other by the action of gravitational force

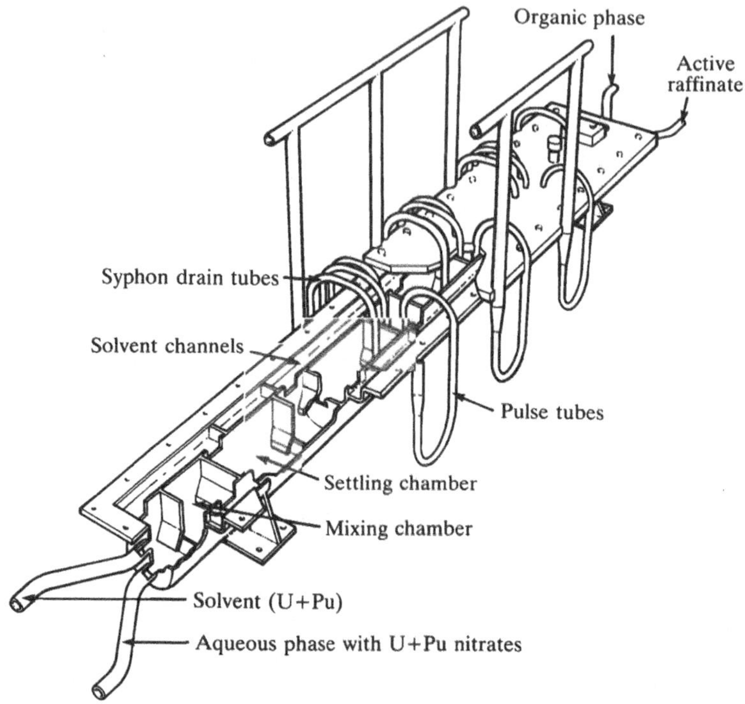

Fig. 2.5. Extractor mixer-settler box.

on the density difference between the phases. With the light phase dispersed, phase disengagement is at the top of the column, whereas with the heavy phase dispersed it is at the bottom. To reduce column height, internals are fitted to improve contact between the two immiscible phases. Column height may be further reduced by pneumatic energy input. Once again the design principles for columns are largely empirical in nature. The pertinent design parameters are: phase properties, geometry of columns, and the energy input.

The packed Pulsed Column (Purex Technical Manual 1955) comprises a vertical column with sections of packing separated by short lengths of column to permit phase introduction, phase removal, or phase redistribution (Fig. 2.6). The light phase is introduced at the bottom of the column via a nozzle type distributor and the heavy phase at the top. Either the light phase is dispersed in the heavy phase or vice versa. Flow is countercurrent and the type of packing selected is either Raschig rings, Lessing rings, Partition rings, Pall rings, Berl saddles, or Intalox saddles (Fig. 2.7). For a sufficiently high non-dispersed or continuous phase flow, the dispersed phase is entrained and prevented from entering the column. The condition of 'flooding' is then said to have occurred.

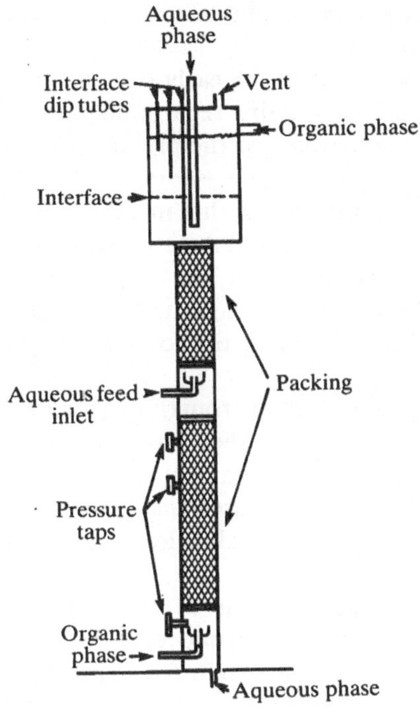

Fig. 2.6. Schematic drawing of a packed column.

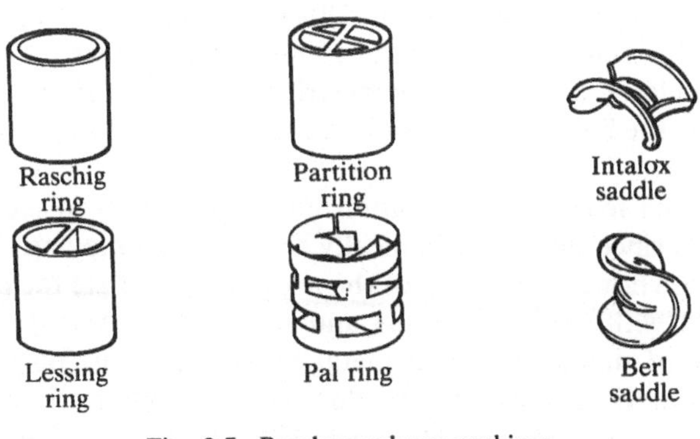

Fig. 2.7. Random column packings.

The column actually functions by the large drops breaking down as they impinge upon the packing, thus maintaining a high interfacial area and promoting mixing of the drop contents. The packing also tends to reduce axial mixing of the continuous phase compared with a spray column. In contradistinction, the packing reduces the mass sectional area available for liquid flow and the volumetric throughputs are as a consequence lower than in a similar sized spray column.

Packed columns were used in the early days of the nuclear industry for reprocessing irradiated fuel. The first separation plant at Windscale used columns packed with Lessing rings for the primary separation process (Burns *et al.* 1949).

The Hanford Redox plant was the first nuclear production solvent extraction plant built in the United States of America and was capable of reprocessing 3–4 tons per day of heavy metal. The original columns were 6 inches in diameter and packed with Raschig rings, and they had an upper disengaging section 12 inches in diameter. Later the column diameter was enlarged to 8 inches.

Packed columns in general suffer from the disadvantage of relatively large stage heights and for this reason they have virtually been superseded by mechanical columns in which mechanical energy is used to produce small drops and correspondingly large interfacial areas for mass transfer. The stage heights in these units are a third or less than the corresponding values for packed columns.

Mechanical columns fall into one of two categories depending upon whether the additional energy is introduced by a rotary or a reciprocating motion. The former are not attractive for many applications because of the necessity to provide mechanical seals around rotating shafts carrying impellers or similar devices. Reciprocating motion on the other hand can be transmitted to the column contents without the need for seals. The first column of this kind was devised by van Dijck (1935) and was termed a pulsed column. Energy was imparted to the column fluids either by reciprocating a set of sieve plates carried on an axial stringer or by reciprocating the fluids by connecting the base of the column to a cylinder closed by a reciprocating piston.

The preferred method of operation has always been to reciprocate the column fluids since this is mechanically simpler than moving the plates. A difficulty arose however because of the necessity to isolate the prime mover from contact with corrosive fluids and much effort was devoted to the development of diaphragms and metal bellows for this purpose. Unfortunately such devices tended to fail due to fatigue after a million or more oscillations and it was not until the early 1950s that the problem was solved by isolating the pulsing unit from the process fluids by a pocket of air or inert gas (Thornton 1954a, b). Further developments of the air-pulsing principle included a method whereby the pulsed column could be installed horizontally instead of vertically, thereby economizing in head room (Thornton 1958; Logsdail and Thornton 1959).

A typical laboratory pulsed column with sampling points is shown in Fig. 2.8. This comprises a vertical column with sieve or nozzle plates

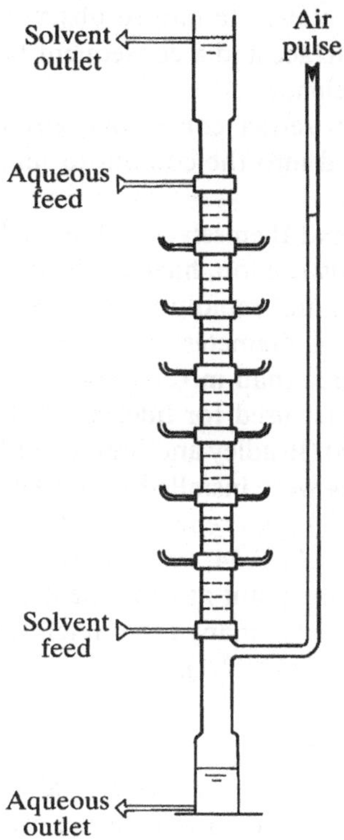

Fig. 2.8. Pulsed plate column (operating in solvent continuous mode).

separated by short lengths of column section. The light and heavy phases are introduced via distributors at the bottom and top of the column respectively and either phase may be dispersed. The overall flow is countercurrent and the continuous phase is pulsed via a limb connected to the bottom of the column; the top of the limb is insulated from the source of fluctuating pressure by means of a pocket of air or inert gas.

The type of flow regime in the column is determined by the rate of energy input and, for a sinusoidal pulse waveform, is a function of the product of the pulse frequency and amplitude. At low values of this product, the column operates in the so-called mixer-settler regime where the dispersed phase coalesces between each pulse cycle; at higher values, the column enters the dispersed flow regime where coalescence is less marked and the column approximates to true dispered phase flow.

Under any set of operating conditions, there is a limiting value of the dispersed phase flowrate for a given value of the continuous phase flowrate. This is termed the flood-point value and is usually determined by experiment. It is important to ensure that, under all conditions, the dispersed phase does not wet the column plates; failure to observe this will result in drop coalescence at the plate surface and a consequent fall off in interfacial area and hence extraction efficiency.

Similarly, as with mixer-settler contactors, electroreduction (Schmieder 1984) may be incorporated into the column to improve solvent extraction performance.

Pulsed plate columns lend themselves well to radioactive fuel reprocessing. They are simple in construction, have no moving parts in them, are flexible in operation, and respond rapidly to process changes. They are much smaller in both height and diameter compared to packed columns, and occupy much less floor area than mixer-settlers.

Pulsed columns were first used for nuclear applications in the Uranium Recovery Plant at Hanford (Bradley and Stevenson 1951) in 1950. Five years later pulsed plate columns were installed in the Hanford Purex Plant, one of the major uranium plutonium separation plants in the United States of America (Sege and Woodfield 1954; Geier 1957). Pulsed columns were also used in the fuel reprocessing plant at Mol (Centeno *et al.* 1972) and are in use or are planned to be used at many other nuclear fuel reprocessing plants throughout the world (see Table 2.1).

2.2.2.4 *Centrifugal devices*

The centrifuge can be a stage-wise or differential device in which one phase is dispersed in and flows countercurrent to the other by the action of centrifugal force on the density difference between two phases. As for mixer-settlers and columns, the design principles for centrifuges are largely empirical in nature. The pertinent design parameters are: phase properties, geometry of the centrifuge, internals of the centrifuge, and the rotational force applied.

Centrifugal contactors can take various forms such as single or multi-stage and be mounted either in the horizontal or vertical modes. Centrifugal contactors are said to have the following advantages over other contactor types:

(1) short residence times, hence less solvent damage;

(2) lower hold-up of material;

(3) smaller space requirements (hence lower shielding and building costs);

(4) easier wash out;

(5) greater flexibility with respect to physical properties of solutions (e.g. density, viscosity).

The Savannah River Laboratory Centrifuge (McKibben and Orth 1969) is a stage-wise device comprising a single stage unit with a separate impeller to

TABLE 2.1 *Extractors in industrial reprocessing plants*

Location		Fuel	Extractor
France	(Marcoule)	Thermal	Mixer-settlers
		Fast (proposed)	Pulsed columns
France	(Cap la Hague)	Thermal	Centrifuges/mixer-settlers
Germany	(Karlsruhe)	Thermal	Mixer-settlers
India	(Trombay)	Thermal	Pulsed columns
Japan	(Tokai-Mura)	Thermal	Mixer-settlers
UK	(Dounreay)	Fast	Mixer-settlers
UK	(Sellafield)	Thermal	Packed columns (plant closed)
UK	(Sellafield)	Thermal	Mixer-settlers
UK	(Sellafield)	Thermal	Pulsed columns
UK	(Dounreay)	Fast (proposed)	Pulsed columns
USA	(Savannah River)	Thermal	Centrifuges/mixer-settlers
USA	(Hanford)	Thermal	Pulsed columns
USA	(Idaho Falls)	Thermal	Pulsed columns

mix the two phases beneath the centrifuge bowl (Fig. 2.9). An inlet diversion baffle, located a short distance up the bowl, prevents any further axial motion of the phases and diverts them radially. The phases are moderated by a number of radial vanes that extend the full length of the bowl. Both phases exit from the top of the bowl: the heavy phase adjacent to the rotor and the light phase adjacent to the periphery of the bowl. The interface position is controlled by pressurizing the aqueous phase weir section hence a reasonable degree of variation in feeds can be processed. The bowl diameter is 25 cm, the speed of rotation is 1745 rpm, the centrifugal acceleration 300–500 g, hold-up 15 l and total flow 37.5 to 250 l per minute.

The Argonne Centrifugal Contactor (Berstein 1973) is a stage-wise device developed from the Savannah River Laboratory contactor. The mixing chamber, impeller, and shaft have been dispensed with (Fig. 2.10) and the two phases are fed directly into the annulus between the rotor fitted with radial vanes and the bowl where they are dispersed by the action of shear forces. The two phases exit from the top of the annular bowl via a weir chamber above.

The Robatel Centrifuge has been developed in France for nuclear applications from a commercially available unit designed for non-nuclear liquid–liquid extraction processes. It comprises a rotary bowl mounted vertically and divided into several stages with a central fixed shaft, fitted with agitating discs (Fig. 2.11). Circular plates are fixed perpendicularly to the shaft and form a series of compartments which act as the stages. Each stage has annular plates fitted to the rotor which divide the stage into mixing and settling sections. The disc in the mixing section and attached to the stationary

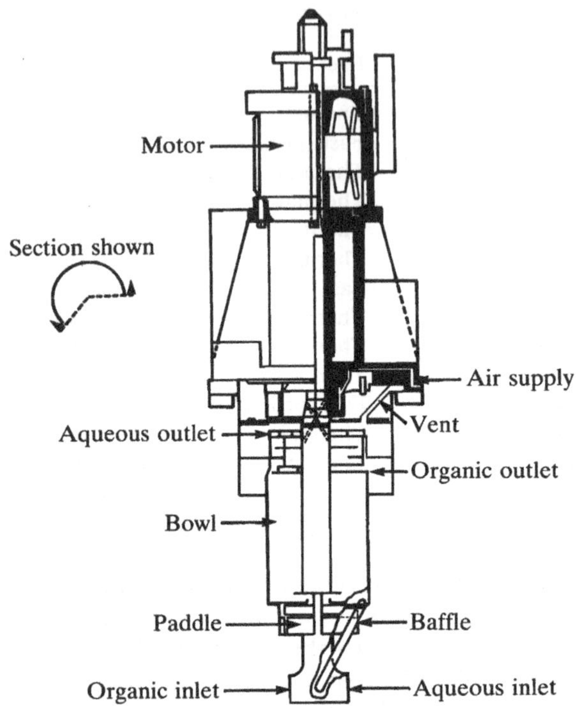

Fig. 2.9. SRL centrifugal contactor.

circular plate is immersed in the liquid and on rotation of the rotor provides the necessary agitation. The unit is suspended from the electric motor driving the rotor with the motor external to the main biological shield. The light phase is introduced at the bottom of the bowl and the heavy phase at the top. The transfer of light phase from each settler to successive mixer is over a circular weir and that for the heavy phase through a series of axial and radial channels drilled in the periphery of the bowl. The transfer of phases between mixer and settler for each stage is by the centrifugal force set-up between the rotating blade and the fixed disc where phase disengagement occurs under centrifugal force.

The Hydroyclone contactor (Breese and Simms 1959) has been developed only to the experimental stage to date at Oak Ridge National Laboratory in the United States of America. The device comprises a series of hydrocyclones cascaded such that the underflow part of each hydrocyclone communicates with the overflow of the one below by an underflow chamber (Fig. 2.12). Each stage is made up of a hydrocyclone, an underflow chamber and a pump. There is one control point at the top interface.

The main attractions of this device are its small size and short residence time; its drawbacks at the present time are the need for one pump per stage;

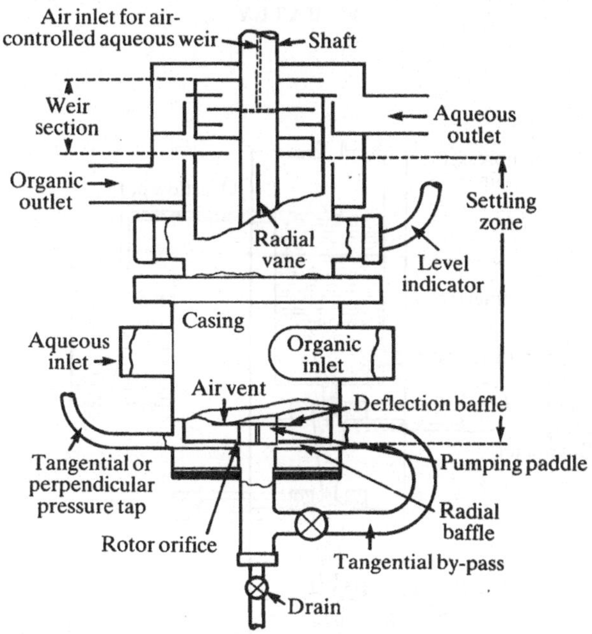

Fig. 2.10. ANL annular contactor.

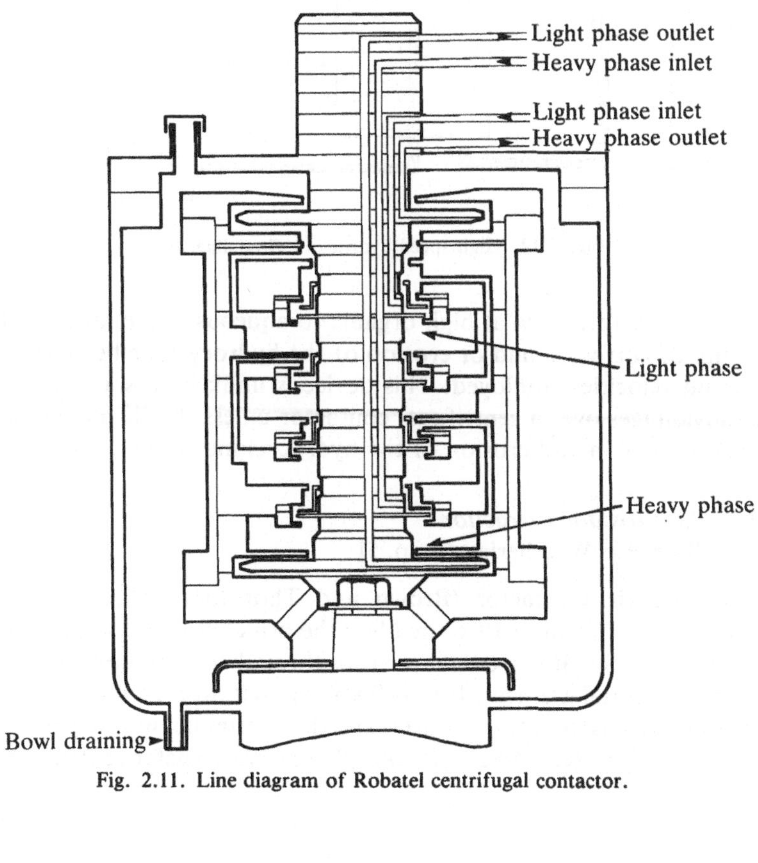

Fig. 2.11. Line diagram of Robatel centrifugal contactor.

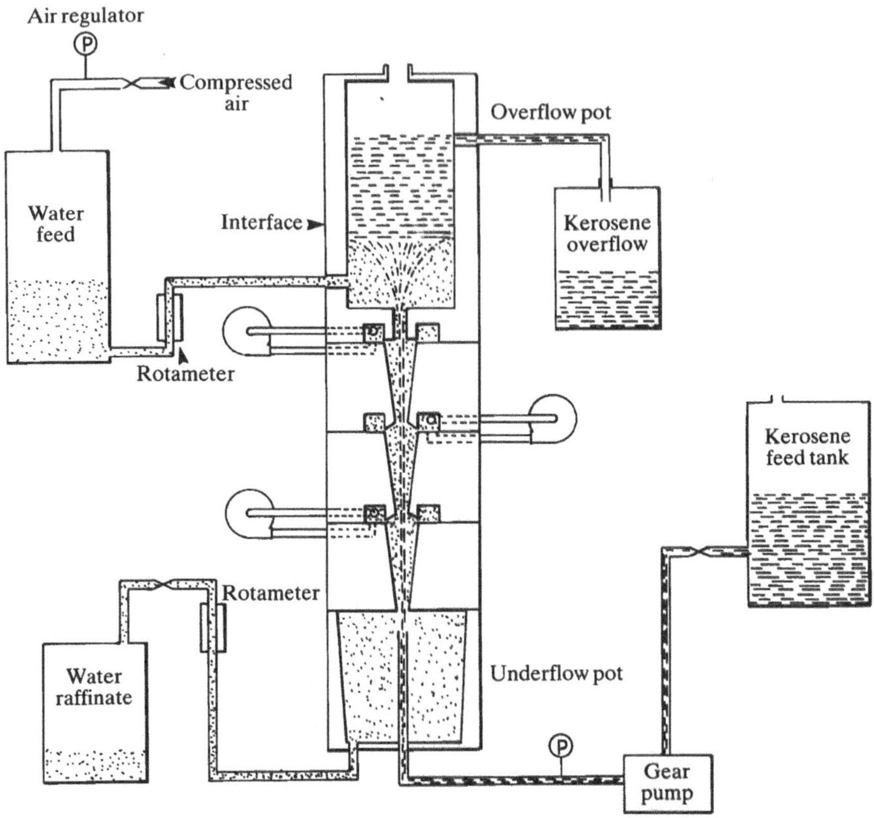

Fig. 2.12. High speed hydroclone flow diagram.

its unknown performance at high organic to aqueous phase ratios and the possibility of corrosion and or erosion of the hydrocyclone because of the high liquid velocities employed. This device is unlikely to show any significant advantages over a centrifugal contactor unless fluidic pumps can be used and corrosion and erosion of the hydrocyclone are negligible.

2.2.2.5 *Electrostatic extraction*
[See also Weatherley Chap. 5]

The Electrostatic Contactor (Brown and Thornton 1966; Stewart and Thornton 1967; Thornton 1976) involves the principle of electrically charging and accelerating dispersed phase drops through a continuous phase subjected to a potential gradient. It is well known that drop oscillations greatly enhance mass transfer between a drop and the continuum in which it is moving. Both the dispersed and continuous phase mass transfer coefficients are increased by oscillations and it has been demonstrated that the dispersed phase coefficients for oscillating drops are dependent on the turbulent flow within the drop.

When a d.c. voltage is applied to a nozzle and an earthed electrode placed below it the drops forming at the nozzle become charged. On detachment from the nozzle the drops are subjected to gravitational and electrical forces. Consequently the drop velocity will be greater in the presence of the electric field than when the drop is subject to the gravitational field alone. In addition, the charge on the drop will generate a mechanical force which is directed normally outwards to the drop surface that is directly opposing inward-acting interfacial forces. Hence the interfacial tension between the drop and the continuum is reduced. The result of these two effects is to increase drop oscillations leading to enhanced mass transfer.

In a liquid–liquid extraction column the rate of mass transfer at constant driving force is determined by the product of the mass transfer coefficient and the interfacial area. Thus the smaller the drops the greater the interfacial area and hence the rate of mass transfer. Small drops, however, have low coefficients and low terminal velocities. The ideal contactor should, therefore, have small oscillating drops moving at high velocities and a conceptual design is shown in Fig. 2.13.

2.2.3 Selection of nuclear contactors

The ideal contactor for fuel reprocessing would be low in cost, require minimum maintenance, the minimum building space and equipment. It

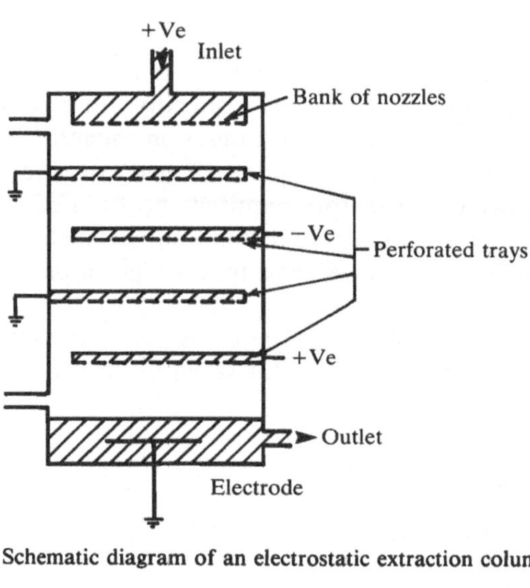

Fig. 2.13. Schematic diagram of an electrostatic extraction column.

would be reliable, maintain a high efficiency over a wide range of operating conditions, and process a high throughput with low hold-up.

The actual cost of the contactor itself is usually only a small part of the total unit cost because the capital investment in the shielding, buildings, and ancillary equipment is much greater. The size and shape of the contactor, however, reflects these costs since they determine the amount of shielding and the size and shape of the buildings needed to contain the contactor. Many criteria need to be considered during the selection of a contactor for radioactive fuel reprocessing and not all of these can be satisfied at the same time.

The types of contactor in use or planned for the Western world's major reprocessing plants are listed in Table 2.1 and comprise mixer-settlers, pulsed columns, and centrifugal contactors.

The criteria governing the selection of nuclear contactors are listed below in order of importance (Mills 1984; Whatley 1982).

(1) criticality;

(2) throughput;

(3) performance;

(4) reagent degradation;

(5) maintenance;

(6) shielding;

(7) control;

(8) cost.

In essence, these criteria require the physical size of the contactor to be as small as possible, within the bounds of criticality, for a given throughput so that the residence time for reagent degradation, quantity of shielding, and cost can be minimized. They also require efficient extraction performance to reduce the number of contactors required, no or minimal moving parts to reduce remote maintenance and handling problems, and good self-regulatory control to minimize sensitivity to perturbations in the phase flows.

Based on this, the particular merits of the various contactors discussed would indicate the following order for selection (Mills 1984).

(1) pulsed columns;

(2) all types of centrifuge;

(3) all types of mixer-settler;

(4) packed columns.

The particular merits of the pulsed columns are a critically safe geometry for all throughputs, high performance together with the ability to cope with particulate matter, low maintenance because of the absence of moving parts, and minimal and low cost as a result of the simplicity of its design. The advantages of all types of centrifuge are their safe criticality as a result of low hold-up, coping with a broad range of throughputs, high performance, minimal reagent degradation as a consequence of short residence times, and minimal shielding because of their small size. The benefits of all types of mixer-settlers are their reliable performance, good control characteristics as a result of their sluggish response to flow perturbations, and their suitability for extraction governed by slow kinetics because of the long residence times involved. The merit of the packed column is its reliable performance.

2.2.4 Applications

In the reprocessing of thermal reactor fuels, characterized by their relatively low fissile content, all three types of nuclear contactor have found application (Table 2.1). The reprocessing of fast reactor fuels–because of their relatively high fissile content–criticality and the need for shorter residence times have tended to preclude the use of the mixer-settler in favour of the pulsed column, the centrifuge, or a combination of both (Table 2.1).

Excluding the original pulsed column, the type of nuclear contactor most widely used at the present time in fuel reprocessing plants is the mixer-settler, because most of the fuel being reprocessed comes from thermal reactors. Its use, however, is precluded when high throughputs are required with short residence times, provided the kinetics of extraction are not too slow and criticality control become the paramount factors such as in the reprocessing of fast reactor fuels. In this instance the preferred contactor would be the pulsed column because it meets all these criteria. An alternative is the centrifuge: it has even shorter residence times, lower disperse phase hold-up, and lower shielding requirements. For fast reactor fuel reprocessing, a centrifuge type contactor for first cycle extraction should be chosen where reagent degradation by the highly radioactive feed can be minimized by the shorter residence time incurred and the selection of pulsed columns for the second and third cycles where such short residence times are not of prime importance.

2.3 Extractants

[See also Cox Chap. 1]

The extractant must meet many requirements, most of which are common to all comparable industrial separation processes but a few of which are specific to the nuclear industry. The process for the extraction of the fissile/fertile elements (plutonium and uranium) by an organic extractant must result in their removal from the aqueous phase and at the same time leave the fission product impurities behind. The selected extractant should be stable against both chemical and radiolytic degradation; any such degradation should not interfere significantly with the whole process. The volatility and solubility of the extractant in the aqueous phase should be low enough to prevent undue losses and also to limit the possibility of secondary reactions in either the aqueous or vapour phase. The flash point of the extractant should be at such a level to obviate the possibility of a fire which, apart from the usual problems, could, in addition, lead to the uncontrolled spread of radioactive contamination. The density, viscosity, and surface tension should allow ready mixing with the aqueous phase and rapid phase disengagement on settling. Finally, to permit ready handling, the extractant should not be highly toxic and must be readily available and not too expensive in the amounts required in the nuclear reprocessing plant.

The most important classes of extractants that have been used for the recovery of plutonium and uranium are ketones, ethers, organophosphorus compounds, and the organonitrogen compounds.

2.3.1 Ketones

Two classes of ketones, the simple aliphatic ($RCOR^1$) and the 1,3 diketones ($RCOCH_2COR^1$), have been employed as extractants for plutonium and uranium.

Methyl-iso-butyl-ketone (MiBK or Hexone) is an aliphatic ketone which extracts by solvating the neutral heavy metal nitrate complex. The tetravalent plutonium and the hexavalent uranium species in aqueous nitric acid solutions readily extract in the presence of salting agents with extraction enhanced by low acid strength. The tetravalent species are less extractable than the hexavalent species. The fission products, together with any trivalent (M (III)) fissile/fertile species, however, remain relatively inextractable. The degradation of hexone by nitric acid produces the di-nitroso-methyl-isobutyl-ketone which in turn hydrolyses to yield methyl-isopropyl-di-ketone and di-nitro-isobutane or it decomposes to produce nitric acid again. These organic degradation products, however, may cause changes in the extraction behaviour of the metal ions present in the aqueous phase.

Other simple aliphatic ketones with similar extraction behaviour to hexone are methyl-n-propyl-ketone, methyl-n-amyl-ketone, methyl-n-hexyl-ketone, di-n-propyl-ketone, and ethyl-n-butyl-ketone. The unsymmetrical ketones are better extractants than their more symmetrical isomers, with extraction efficiency decreasing with increasing alkyl chain length, together with the presence of branched chains.

The diketones extract by forming chelates with the metal cation. Because of steric conditions only 1,3 di-ketones but not the 1,2 diketones can form organically soluble chelates which are chemically stable. Thenoyl-tri-fluoro-acetone (TTA) has been found to extract the trivalent (M(III)), tetravalent (M(IV)), and hexavalent (M(VI)) fissile/fertile species by chelation involving three, four, and six molecules of TTA, respectively. The rate of extraction, however, is slow as is also the case with other 1,3 diketones such as acetyl-acetone, benzyl-acetone and acetyl-tri-fluoro-acetone.

2.3.2 Ethers

The simple aliphatic ethers extract by ionic complex formation but require a salting agent to improve their extraction efficiency. Typical aliphatic ethers are di-ethyl-ether and di-butyl-ether.

The polyethers also extract by ionic complex formations. Their extraction efficiency is observed to decrease with increasing chain length and to increase with increasing spacing between the oxygen atoms. Typical polyethers are di-butyl-di-ethylene-glycol (Butex) and tri-ethylene-glycol-dichloride.

Crown ethers, cyclic polyethers, have found many applications in solvent extraction particularly in the recovery of alkali and alkaline earth metals (Frensdorff 1971; Rais and Selucky 1971; Chiarizia *et al.* 1975). Several approaches were investigated to increase the solubility of the anion associated with the metal cation to be extracted by the crown ether in the organic phase. One of the approaches adopted was the use of the organic-soluble-aqueous-insoluble cation exchanger and this concept represents a case of synergism, if the crown ether is regarded as a neutral donor and the liquid cation exchanger as the acidic extractant. In their study of the use of crown ethers as synergistic ligands in the solvent extraction of plutonium (IV) by thenoyl-tri-fluoro-acetone in a chloroform diluent, Aly *et al.* (1985) used the thenoyl-tri-fluoro-acetone as the organic soluble ligand to be associated with the metal cation and the crown ether as the synergic ligand.

2.3.3 Organophosphorus compounds

Several classes of organophosphorus compounds, alkyl phosphates (RO_3PO) phosphonates (RO_2RPO), phosphinates (ROR_2PO), and phosphine oxides (R_3PO), all have a phosphoryl ($P = 0$) group to which the metal cation can attach via the oxygen atom and are efficient extractants for plutonium and uranium.

2.3.4 Alkyl phosphates

Tri-butyl-phosphate ($(C_4H_9O)_3P = O$) (TBP) is an alkyl phosphate which in the temperature range -78 to 289°C is an odourless, colourless liquid, only slightly miscible in the aqueous phase, chemically stable, even towards concentrated nitric acid, and is hydrolysed only under very vigorous conditions. Its high density (973 kg/m^3 at 25°C) and viscosity (3.41 centipoises at 25°C) just less than the aqueous phase necessitate its use in an inert diluent. Tri-butyl-phosphate is found to extract selectively the tetravalent plutonium (fissile) and hexavalent uranium (fertile) compounds of the dissolved fuel into the diluent phase as the neutral, unhydrated complexes $Pu(NO_3)_4$ 2TBP and $UO_2(NO_3)_2$ 2TBP respectively. The trivalent plutonium component of the fuel in the aqueous phase is relatively inextractable, but differs in that it extracts as the neutral, unhydrated complex $Pu(NO_3)_3$ 3TBP. The fission products extracted above trace amounts by TBP are ruthenium, cerium, zirconium, and niobium. Again they are found to extract as neutral unhydrated complexes in which the fission product is itself complexed with a fissile/fertile species or with another fission product, such as technetium. For both the fissile/fertile and fission product species, the degree of extraction is controlled by the concentration of hydrogen ions (H^+), of nitrate ions ($NO_3^=$), and of other salting-out ions in the aqueous phase together with that of uncomplexed tri-butyl-phosphate in the diluent. The degradation of tri-butyl-phosphate by hydrolysis and radiolysis is found to yield di-butyl-phosphate (DBP), mono-butyl-phosphate (MBP), and finally phosphoric acid (H_3PO_4). The presence of the degradation products di-butyl-phosphate and mono-butyl-phosphate is, however, deleterious to the extraction by tri-butyl-phosphate of the fissile/fertile species, for they both tend to extract the tetravalent and hexavalent fissile/fertile species as neutral insoluble complexes. This makes subsequent recovery of these species difficult. Such insoluble complexes can also adsorb fission product species. This, together with the extraction of fission products by both di-butyl-phosphate and mono-butyl-phosphate and the high alpha/beta/gamma activity transferred to the diluent, necessitates the periodic removal of di-butyl-phosphate and mono-butyl-phosphate from the used solvent by extraction with both alkaline wash and acid wash prior to solvent reuse in the plant.

The other known alkyl phosphates which exhibit similar plutonium (IV) and uranium (VI) extraction behaviour to tri-butyl-phosphate are those with the alkyl groups iso-butyl, n-amyl, iso-amyl, n-hexyl, n-octyl, 2-ethylhexyl, 2-butyl, 3-amyl, 3-methyl-2-butyl, and 4-methyl-2-amyl.

The dialkyl-phosphoric acids are also effective extractants for tetravalent plutonium and hexavalent uranium species. The extractants di-butyl-phosphoric acid (DBP), di-octyl-phosphoric acid (DNPA), di-decyl-phosphoric acid (DDPA), and di-2-ethyl-hexyl-phosphoric acid (DEHPA) were examined by Gureev *et al.* (1964) and it was concluded that increasing alkyl chain length caused only a minor increase in plutonium (IV) and uranium (VI) extractability whilst branching of the chain resulted in an appreciable decrease in U(VI) extractability.

2.3.5 Alkyl phosphonates, phosphinates, and phosphoric oxides

The alkyl phosphonates ($(RO)_2RPO$), phosphinates ($(RO)R_2PO$), and phosphoric oxides (R_3PO) are all somewhat better extractants than the tri-alkyl-phosphates for plutonium and uranium. Baldwin *et al.* (1952) noted that, these compounds could be arranged in order of increasing extracting power, as follows:

phosphoric oxides > phosphinates > phosphonates > tri-alkyl-phosphates

Alkyl phosphonates which exhibit similar tetravalent plutonium and hexavalent uranium species extraction behaviour to TBP are: di-n-butyl-n-butyl-phosphonate (DBBP), di-2-amyl-2-butyl-phosphonate (D2A2BP), di-n-butyl-phenyl-phosphonate (DBPP), and di-sec-butyl-phenyl-phosphate (DsBPP).

Alkyl phosphinates which exhibit similar tetravalent plutonium and hexavalent uranium species extraction behaviour to TPB are: n-butyl-di-n-butyl-phosphinate (BDBP) and ethyl-di-n-hexyl-phosphinate (EDHP).

Phosphine oxides which exhibit similar tetravalent plutonium and hexavalent uranium species extraction behaviour to tri-butyl-phosphate are: tri-n-octyl phosphine oxide (TOPO) and tri-hexyl-phosphine oxide (THPO).

In the TRUEX Process (Horwitz 1986), which is a generic extraction/recovery process for the removal of transuranic elements in nitric acid and hydrochloric acid nuclear waste streams, the key chemical is the new actinide-lanthanide selective, bi-functional extractant octylphenyl-N-N-di-iso-butyl-carbamoyl-methyl-phosphine oxide (CMPO). This extractant contains a unique combination of donor and substituent groups which enables it to extract trivalent actinides which are normally much less extractable than the tetravalent and hexavalent actinides over a wide range of nitric acid concentrations and over moderate to high hydrochloric acid concentrations. In nitrate media CMPO has good selectivity over non-lanthanide fission products, especially in the presence of a small concentration (0.1 M) of oxalic acid. CMPO has good stability towards hydrolytic and radiolytic degradation and, in addition, its behaviour towards hydrolysis and radiolysis is favourably augmented by the fact that the primary degradation product, methyl-octyl-phenyl-phosphine oxide, is capable of replacing a portion of the CMPO molecules co-ordinated to the americium nitrato complex (Chiaizia and Horwitz 1986). The net effect of this replacement is that the distribution ratio of americium does not decrease as rapidly as expected based on the loss of CMPO. The neutral bi-functional organophosphorus extractants, octyl-phenyl-N-N-di-iso-butyl-carbamoyl-methyl-phosphine oxide (CMPO) and di-hexyl-N-N-di-ethyl-carbamoyl-methyl-phosphonate have again been studied in the USA (Navratil 1986) to remove plutonium and americium from 7 M nitric acid nuclear waste. These compounds both extract trivalent (M(III)) actinides from strong nitric acid, a property which distinguishes them from mono-functional organophosphorus reagents. Furthermore, these reagents were also shown to extract hydrolytic tetravalent plutonium polymer which is also present in the nuclear acid waste streams.

2.3.6 Organonitrogen compounds

These are several classes or organonitrogen alkyl amines: primary (RNH_2), secondary (R_2NH), tertiary and quaternary ($R_4(N)X$); N, N-di-substituted amides ($RC(O))NR_2$); phenyl-hydroxylamine derivatives ($RN\text{-}NO\text{-}ONH_4$) and amine oxides ($RO_3N$), all of which have a nitrogen atom and/or oxygen atoms to which the metal cation can attach. They are efficient extractants for the tetravalent plutonium and hexavalent uranium species.

2.3.7 Alkyl amines

The salt formed by the primary, secondary, or tertiary amine and the metal cation is relatively insoluble in the aqueous phase but is soluble in a suitable organic phase such as benzene, xylene, chloroform, or kerosene. The tetravalent plutonium species in the aqueous phase extracts to a greater extent than the hexavalent uranium species with the trivalent plutonium species being relatively inextractable.

In the case of the alkyl amines the extraction efficiency varies with the degree of substitution of the amine, the order being for nitrate solutions above 3M nitrate (Blanco *et al.* 1959) being quaternary > tertiary > secondary and primary; and below this acidity: tertiary > quaternary > secondary and primary. In extraction from sulphate solutions, however, the extractability is primary ≫ secondary > tertiary.

Typical primary amines investigated, particularly in tetravalent plutonium species extraction from sulphate solutions, were those containing alkyl groups containing six to nine carbon atoms.

The extraction of tetravalent plutonium species by secondary amines is consideraly weaker than by tertiary amines, as has already been stated. Amberlite LA-1, a tertiary-alkyl secondary amine, has, however, been investigated as an extractant at relatively high acidities (Winchester 1957). This amine, although it had a much lower extraction for tetravalent plutonium than tertiary or quaternary amines, had a much higher decontamination factor for the more common fission products. Other secondary amines investigated were n-benzyl-1-(3-ethyl-pentyl)-4-ethyl-octylamine, bis-(1-isobutyl-3, 5-dimethyl- hexyl)-amine, and di(tridecyl)-amine.

The most efficient tertiary alkylamine extractants for tetravalent plutonium have been shown (Cleveland 1979) to be those containing eight to twelve carbon atoms in each alkyl group. The propylamines and butylamines form nitrate salts that are too soluble in the aqueous nitric acid phase to allow efficient extraction, whilst the amylamines are unsatisfactory because of the production of two organic phases. The aromatic amines are also poor extractants of tetravalent plutonium from nitrate solutions. Some typical tertiary amine extractants are: tri-n-heptylamine, tri-n-octylamine (TOA), tri-iso-octylamine (TiOA), tri-iso-nonylamine (TiNA), and tri-laurylamine (TLA).

Quaternary compounds extract tetravalent plutonium species much more strongly than the hexavalent uranium species. In fact, di-iso-butyl-phenoxy-ethoxy-ethyl-dimethyl-benzyl-ammonium chloride has been used to separate plutonium from a large excess of uranium (Powell 1958). Quaternary amine extractions are less prone to third phase formation than are tertiary amine extractions which, in turn, have less tendency to form a third phase than do extractions with secondary and primary amines (Bac 1968). Quaternary ammonium compounds are also reportedly more resistant to degradation by nitric acid than are the tertiary amines. Tri-lauryl-methyl-ammonium nitrate (TLMA) was found to be stable in contact with 3M nitric acid at 50°C for some 110 days whereas a tertiary amine, tri-laurylamine (TLA), was degraded under similar conditions (Bac 1968). The higher polarity of the quaternary ammonium compounds does, however, impose some disadvantages including the greater tendency to form three phases and the slow rate of phase separation particularly when the acidity of the aqueous phase is low. These effects are only partially mitigated by the modification of the diluent by the addition of a high molecular weight aliphatic alcohol such as nonyl alcohol.

Some other quaternary amines investigated were di-dodecyl-dimethyl-ammonium nitrate, tri-capryl-methyl-ammonium nitrate (TCMA), and hexadecyl-dimethyl-benzyl-ammonium nitrate (HDDMBA).

2.3.8 N, N-disubstituted amides

A number of N, N-disubstituted amides of the type $RC(O)NR_2$ have been shown to extract plutonium (Sidall, III 1960). Casarci *et al.* (1986) have also studied the use of long chain N, N^1-dialkyl aliphatic amides as extractants in the reprocessing of high burnup fuels, and compounds where R and R^1 are aliphatic chains containing four to eight carbon atoms. They pointed out the low complexing capacity of the degradation products which were easily removed because of their solubility in water. Other advantages were a better coalescence time and the easier stripping of uranium. They did, however, point out that long chain aliphatic amines, especially those with a branched chain, had a tendency to third phase formation.

More recently, Guillaume *et al.* (1987) have advocated their use as actinide extractants because these carboxylic acid amides are completely incinerable and potentially, therefore, they allow the extraction process to produce lower amounts of radioactive waste. The monocarboxylic amides main degradation products, carboxylic acid and a secondary amine, do not affect the separation of plutonium and uranium as does the di-butyl-phosphate in the PUREX process. They also suggest that the distinctive properties of these amides such as the rigidity of the O = C = N bond could be exploited to improve the separation coefficients of plutonium and uranium.

Guillaume *et al.* (1987), in addition to exploring the behaviour of the N, N-dialkylamides, also report on the behaviour as actinide extractants of the N, N^1-tetra-alkyl-malonamides and penta-alkyl-propane-diamides. They concluded that at the present time the penta-alkyl-propane-diamides look to be more promising extractants for actinide recovery from nitric acid solutions than CMPO or other organophosphorus molecules. They concluded, in addition, that they were relatively cheap to prepare and that tetravalent plutonium and hexavalent uranium could be stripped without chemical transformation such as complexation or the change in chemical state necessary when CMPO is employed. Their disadvantage, however, is the large amount of research and development needed to confirm their usefulness in solving the waste problems.

Other typical N, N-di-substituted amides are: N, N-di-hexyl-formamide, N, N-di-butyl-acetamide, N, N-di-cyclo-hexyl-formamide, N, N-di-butyl-benzamide and N, N-di-benzyl-acetamide.

2.3.9 Phenyl-hydroxyl-amines

Phenyl-hydroxyl-amine derivatives extract tetravalent plutonium and trivalent plutonium from aqueous solutions containing either nitric acid or hydrochloric acid. Greater than 99 per cent extraction of tetravalent plutonium from 1M nitric acid solution using cupferron (the ammonium salt of nitroso-phenyl-hydroxylamine) has been reported (Fryxell and Kleinschmidt 1945).

Other typical phenyl-hydroxylamine derivatives are benzyl-phenyl-hydroxylamine, naphthyl-phenyl-hydroxylamine, and n-benzoyl-phenyl-hydroxylamine.

2.3.10 1-phenyl-3-methyl-4-benzoylpyrazolone

The compound 1-phenyl-3-methyl-4-benzoylpyrazolone (PMBP) is an excellent extractant for tetravalent plutonium from nitric acid media. Benzene has most commonly been used as the diluent but other diluents such as chloroform, amyl acetate, and iso-amyl acetate work equally well (Chmutova *et al.* 1968). In addition, Chmutova *et al.* (1968) found that this compound was a poor extractant for trivalent americium and have in fact suggested its use as an extractant for the separation of tetravalent plutonium from trivalent americium, hexavalent uranium, and certain fission products.

2.3.11 Amine oxides

A brief study was made by Kennedy and Perkins (1964) of the extraction of tetravalent plutonium from nitric acid solution by a mixture of long chain (C_8 to C_{10}) tertiary amine oxides ($(RO)_3N$) in solution in di-butyl-carbitol. Their data would suggest that the amine oxides are not as good extractants for tetravalent plutonium as the corresponding tertiary amine.

2.3.12 Organosulphur compounds

Sulphones. The sulphones, a class of organosulphur compounds with the general formula ($R.SO_2.R^1$), were evaluated as extractants for tetravalent plutonium and hexavalent uranium species (Baldwin *et al.* 1951). Their data would indicate that n-propyl-sulphone and isobutylsuphone both have a selectivity for tetravalent plutonium, extracting it some six to ten times more efficiently than hexavalent uranium from 3 M nitric acid solution.

Sulphoxides. The dialkyl sulphoxides, a class of organosulphur compounds with the general formula ($R.SO.R^1$), have been extensively studied as extractants for tetravalent plutonium and hexavalent uranium species in India by Pai, Shukla, and Subramanian (1979). Their study of the solvent extraction of actinides and some fission products with long chain aliphatic sulphoxides indicated that, whereas the actinides were essentially completely extracted from nitric acid solution, the fission products showed negligible extraction. Among the various sulphoxides investigated, di-n-hexyl sulphoxide (DHSO) and di-n-octyl sulphoxide (DOSO) were the most superior extractants. The extractive power of both di-n-hexyl sulphoxide and di-n-octyl-sulphoxide were shown to be superior to that of tri-butyl-phosphate, the most commonly used extractant in the nuclear industry, which is in keeping with their basicity order:

di-n-octyl-sulphoxide > di-n-hexyl-sulphoxide > tri-butyl-phosphate.

Their preliminary studies showed, in addition, that irradiated sulphoxide solutions extracted zirconium to a lesser extent than irradiated tri-butyl-phosphate, suggesting their possible future use as promising extractants in high radiation fields. Among the sulphoxides examined the irradiated di-hexyl-sulphoxide was found to extract zirconium to a lesser extent than di-n-octyl-sulphoxide.

In further work (Shukla and Subramanian 1980), the extraction behaviour of tetravalent plutonium from nitric acid media by di-n-hexyl sulphoxide and di-n-octyl sulphoxide in the presence of several water-miscible organic solvents was investigated to study their possible synergistic effect on metal ion extraction. Methanol, ethanol, n-propanol, isopropanol, dioxane, acetone, and acetonitrile were used as the organic components of the mixed (polar) phase. These additives were found to affect the extraction of the tetravalent plutonium by the sulphoxide extractants to varying degrees. Thus, the extractability of tetravalent plutonium increased two- to three-fold with increasing concentration (up to 20 per cent) of acetonitrile, acetone, methanol, and ethanol, while it decreased with increasing concentration of n-propanol and isopropanol.

2.4 Chemistry of extraction

[See also Cox Chap. 1]

2.4.1 Extraction with ketones

In general, the efficiency of the extraction of a metallic salt by oxygen-containing organic compounds is determined primarily by the electron density on the oxygen atom and with steric effects. The stronger the inductive effect of substituents the more is the electron density displaced towards the oxygen atom. The relationship between the extractive power of ketones towards tetravalent plutonium and the inductive effect of their substituents is quite good : the inductive effect becomes weaker as the length of the alkyl chain is increased, if the alkyl chain is branched, and if a double bond is introduced into the alkyl chain (Kolarik 1975). In the case of steric effects, both the branching of an alkyl chain in a ketone molecule and an increasing molecular symmetry have been reported to reduce the accessibility of the oxygen atom for the formation of a co-ordinative metal-to-oxygen bond (Kuca 1962).

Two classes of ketones, the simple ketones $R R^1CO$ and the 1,3 diketones $RCOCH_2COR$, have been employed as extractants for plutonium from nitric acid solution. The simple ketones extract plutonium by simple solvation of a neutral species whereas the 1,3 diketones extract plutonium by the completely different mechanism of chelate formation.

The solvent extraction of plutonium from nitric acid solution by the simple ketone, methyl-iso-butyl-ketone (hexone), used in the pure form, that is, without the addition of a diluent, is the basis of the Redox process. In this process, simple solvation of the plutonium species takes place without the formation of a complex. The plutonium species extracted from the aqueous phase into the organic phase are the neutral nitrates:

$$Pu^{4+}_{(aq)} + 4NO_3^-(aq) \rightleftharpoons Pu\,(NO_3)_{4(aq)} \longrightarrow Pu(NO_3)_{4(org)}$$

and

$$PuO^{2+}_{2(aq)} + 2NO^-_{3(aq)} \rightleftharpoons PuO_2(NO_3)_{2(aq)} \longrightarrow PuO_2(NO_3)_{2(org)}$$

With increasing nitric acid concentration the higher nitrate complexes of tetravalent plutonium are formed but the requirement of electrical neutrality prevents the extraction of these species in their anionic form. Kolarik (1975), in fact, has suggested that at higher nitric acid concentrations the tetravalent and hexavalent plutonium species are extracted by methyl-iso-butyl-ketone as the $H_2Pu(NO_3)_6$ and $H\ PuO_2(NO_3)_3$ complex acids.

2-Thenoyl-tri-fluoro-acetone (TTA) is a 1,3 diketone that has been used in a process for the large scale separation of plutonium from uranium using benzene as the diluent (Culler 1956).

The chelation reaction involved in the extraction of tetravalent plutonium from the nitric acid solution by thenoyl-tri-fluoro-acetone proceeds through the tautomerization of its keto form to the enol form, the acidic enol hydrogen then being replaced by one equivalent of tetravalent plutonium to form the chelate,

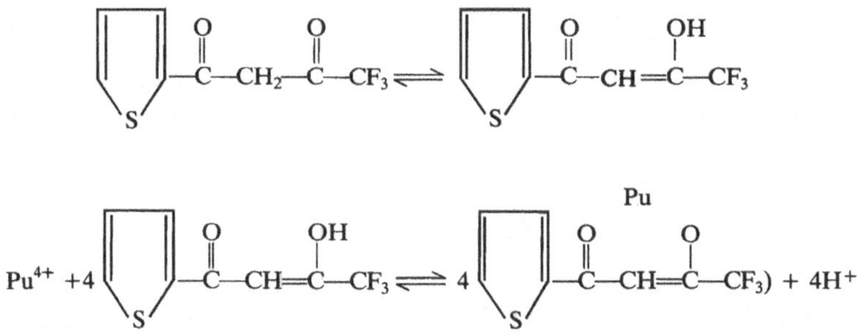

The chelation of trivalent, quadrivalent, and hexavalent plutonium, therefore, involves three, four, and two molecules of thenoyl-tri-fluoro-acetone respectively.

2.4.2 Extraction with Polyethers

Di-butyl-di-ethylene glycol or di-butyl carbitol (Butex) and tri-ethylene glycol di-chloride (Trigly) were used as the extractants for plutonium from nitric acid solutions in the early British and Canadian separation processes.

2.4.2.1 *Di-butyl-di-ethylene glycol*

The extraction of plutonium and uranium from nitric acid solution by di-butyl carbitol is based on the formation of organic soluble complexes. In an aqueous solution of uranyl nitrate, tetravalent plutonium, and hexavalent plutonium, there are present not only the hydrated ions UO_2^{2+}, Pu^{4+} and PuO_2^{2+} but also a series of hydrated nitrate complexes, $(UO_2NO_3)^+$, $(UO_2(NO_3)_2)$, $(UO_2(NO_3)_3)^-$, $(Pu(NO_3)_4)$, $(Pu(NO_3)_5)^-$, $(Pu(NO_3)_6)^{2-}$, $(PuO_2(NO_3)_2)$, $(PuO_2(NO_3)_3)^-$: at room temperature there is rapid equilibrium between the nitrate species. The formation of the higher nitrato species is promoted by the presence of nitrate ions which in the feed solution for a solvent extraction process may be present as nitric acid or as nitrates added specifically for this purpose, the higher complexes having the greater tendency to solvate with the di-butyl carbitol phase to form compounds of the type $UO_2(NO_3)_2.xH_2O.y$(di-butyl carbitol). The extraction of both uranyl nitrate and plutonium nitrates, therefore, may be presented as step-wise reactions. The anion replaces water in the co-ordination sphere of the cation and the di-butyl carbitol and water molecules compete for the remaining positions.

No direct analysis of any ether solution has been made which would establish the metal-to-nitrate ratio in extracted metal species, but the existence of $H_2Pu(NO_3)_6$ in ethers has been demonstrated by a comparison of the spectra of tetravalent plutonium in methyl ethyl ketone, in which the presence of $H_2Pu(NO_3)_6$ has been proven, and the spectra of tetravalent plutonium in di-butyl carbitol. A correlation of partition and spectrophotometric data would suggest that hexavalent plutonium is extracted preferentially as the hexavalent plutonium trinitrato species (Kolarik 1975).

2.4.2.2 *Tri-ethylene glycol dichloride*

The extraction of plutonium by tri-ethylene glycol dichloride (Trigly) has several features in common with its extraction by di-butyl carbitol. Tetravalent plutonium is more readily extracted than hexavalent plutonium and the extraction of both these plutonium valency states from nitrate solutions may be enhanced with the use of a salting-out agent such as ammonium nitrate. The ultra-violet absorption spectrum of tetravalent plutonium is very similar to that observed for tetravalent plutonium in di-butyl carbitol (Aiken *et al.* 1951). Comparison with the spectrum of nitric acid solutions indicated that the absorption spectrum of tetravalent plutonium in tri-ethylene glycol dichloride was the same as that of tetravalent plutonium in 11 M nitric acid, thus suggesting that the extracted tetravalent plutonium species was $H_2Pu(NO_3)_6$.

The spectrum of hexavalent plutonium in tri-ethylene glycol dichloride is, likewise, very similar to that of hexavalent plutonium in dibutyl carbitol.

The absorption spectrum was, however, unlike that of any hexavalent plutonium spectrum in nitric acid. Transference measurements of hexavalent plutonium solutions demonstrated in nitric acid solutions above 9 M the existence of a plutonium anion. If the behaviour of the plutonyl ion is analogous to that of the uranyl ion it would suggest that the anion was plutonyl trinitrate and it was therefore concluded that the extracted hexavalent plutonium species was probably $HPuO_2\ (NO_3)_3$.

2.4.3 Extraction with organophosphorus compounds

2.4.3.1 *Tri-butyl-phosphate*

The extractive power of tri-butyl-phosphate derives mainly from its phosphoryl group, the oxygen atom forming the co-ordinate link with the cation:

$$(C_4H_9O)_3\,P = O \longrightarrow M$$

That is, tri-butyl-phosphate forms solvates or adducts with the metal cation and may, therefore, be classified in the category of neutral extractants. Thus, the solvent extraction of the metals by tri-butyl-phosphate is in many respects typical of a large number of organic compounds including ketones, ethers, and esters. The extraction of uranyl nitrate can again, therefore, be presented as a stepwise reaction. The anion replaces water in the co-ordination sphere of the cation and the tri-butyl-phosphate and the water molecules compete for the remaining positions. In the case of uranyl nitrate, for example, the complex formed is $UO_2(NO_3)_2$ 2TBP based on the evidence that (1) reasonably uniform equilibrium constants may be written for the extraction process involving $(TBP)^2$; (2) the limiting solubility of uranyl nitrate in tri-butyl-phosphate approaches the ratio 2TBP/1 $UO_2(NO_3)_2$; and (3) solid complexes of $UO_2/NO_3)_2$ and esters conforming to the above structure have been isolated (Burger 1958). In contrast to the complexes formed with ketones and ethers, water is completely eliminated from the extracted complex which, in view of the extreme difficulty in removing the last two molecules of water from uranyl nitrate molecule, their replacement by tri-butyl-phosphate molecules is indicative of a rather stable complex.

Tri-butyl-phosphate has, indeed, established itself as one of the most versatile solvent extractants for many inorganic salts and has been made the basis for several processes involving various actinide elements. In the form of the Purex process, tri-butyl-phosphate has been used world-wide for the separation of plutonium from uranium and fission products in nitric acid solution.

The actinides most readily extracted from nitric acid solution by tri-butyl-phosphate are those that form triply and quadruply charged cations. The hexavalent actinide cations of the type MO_2^{2+} are also extracted from nitric acid solution by tri-butyl-phosphate, presumably as a result of the electrons being displaced towards the oxygen atoms leaving the metal atom with an effective charge greater than + 2.

The nature of the actinide species extracted by tri-butyl-phosphate has been well established (Healy and McKay 1956; Hahn 1957; Alcock *et al.* 1958; Hesford and McKay 1958; Hesford *et al.* 1959; Naito 1960; McKay and Streeton 1965; Baumgartner and Finsterwalder 1970). A neutral un-ionized cation free of water of hydration but solvated by a fixed number of tri-butyl-phosphate molecules is the species extracted from nitric acid solution. The reactions of uranyl, tetravalent, and hexavalent plutonium nitrates with tri-butyl-phosphate have been shown to proceed as follows:

$$UO^{2+}_{2(aq)} + 2\,NO^{-}_{3(aq)} + 2TBP_{(org)} \longrightarrow UO_2(NO_3)_2 2TBP_{(org)}$$

$$Pu^{4+}_{(aq)} + 4NO^{-}_{3(aq)} + 2TBP_{(org)} \longrightarrow Pu(NO_3)_4\,2TBP_{(org)}$$

$$PuO^{2+}_{2(aq)} + 2NO^{-}_{3(aq)} + 2TBP_{(org)} \longrightarrow PuO_2(NO_3)_2 2TBP_{(org)}$$

From an examination of extraction data obtained using aqueous solutions containing the same concentration of trivalent plutonium into tri-butyl-phosphate solutions at two different concentrations it was inferred (Shevchenko *et al.* 1960) that trivalent plutonium in the organic phase was present as the complex $Pu(NO_3)_3.3TBP$.

The effect of nitric acid concentration on the extraction of the actinides by tri-butyl-phosphate is quite complex. At first with increasing nitric acid concentration the extraction coefficient increases because of the increase in the concentration of the extractable actinide species. Tri-butyl-phosphate, however, forms with nitric acid the solvent soluble species HNO_3TBP and thus at even higher nitric acid concentrations competition between the actinide ions and the nitric acid for the tri-butyl-phosphate molecules takes place with a resultant decrease in actinide extraction.

2.4.3.2 *Phosphonates, phosphinates, and phosphine oxides*

It has been well demonstrated that the extraction of the tetravalent and hexavalent actinides increases in the series phosphates, phosphonates, phosphinates, and phosphine oxides (Baldwin *et al.* 1952; Burger 1958). This effect of enhanced extraction has been attributed to the increase in electronegativity of the phosphoryl group as the ester radicals are replaced by alkyl radicals.

The mechanism of complex formation of hexavalent uranium from nitric acid solution (0.2–1.1 M) with organophosphorus compounds (Nikitina *et al.* 1959, 1960) does not change in the series phosphate–phosphinate and phosphine oxide. Thus, on the basis of the data obtained from an examination of the extraction mechanism of metal nitrates by tri-butyl-phosphate, it was assumed that in the above organophosphorus extractant systems the mechanism of the process is analogous. Therefore, the extraction of hexavalent uranium and plutonium may be shown as follows:

$$MeO_2^{2+} + 2NO_3^- + 2T \longrightarrow MeO_2(NO_3)_2.2T$$

and for tetravalent plutonium,

$$Pu^{4+} + 4NO_3^- + 2T \longrightarrow Pu(NO_3)_4.2T$$

where T represents phosphate, phosphinate or phosphine oxide.

The compound that has received the most study in this class of compounds for the extraction of plutonium and uranium is tri-n-octylphosphine oxide (TOPO). Foreman *et al.* (1961) studied the extraction of trivalent, tetravalent, and hexavalent plutonium from nitric and hydrochloric acid solutions of various strengths. Their work on the chemical nature of the complexes extracted into cyclohexane by tri-n-octyl-phosphine oxide was restricted to those derived from tetravalent and hexavalent plutonium nitrates.

From the variation of the tetravalent and hexavalent plutonium extraction coefficients with tri-n-octyl-phosphine oxide concentration and by an examination of the absorption spectra of the tetravalent and hexavalent plutonium–tri-n-octyl-phosphine oxide complexes extracted into cyclohexane it was concluded that the complex species formed in the organic phase had plutonium : extractant ratios of 1 : 2. The extraction equations may be written as:

$$Pu(NO_3)_4 + 2TOPO \rightleftharpoons Pu(NO_3)_4.2TOPO$$

and

$$PuO_2(NO_3)_2 + 2TOPO \rightleftharpoons PuO_2(NO_3)_2.2TOPO$$

The formula of the tetravalent plutonium tri-n-octylphosphine oxide complex was confirmed by employing an application of Job's continuous variation method to a series of spectrophotometric measurements carried out on the plutonium extractant solutions.

2.4.4 Extraction with organonitrogen compounds

High molecular weight amines will react with plutonium anionic complexes to form salts that are relatively insoluble in the aqueous phase but which are soluble in the organic diluents benzene, chloroform, and kerosene (Coleman 1963). The extraction reactions may be written as:

$$nR_{3-m}NH_{(o)} + nH^+_{(a)} + A^{n-}_{(a)} \rightleftharpoons (R_{3-m}NH^+_{m+1})_nA^{n-}_{(o)}$$

where R_{3-m} NH represents either primary, secondary or tertiary amine and A^- represents a plutonium anionic complex, and

$$nR_4N^+_{(a)} + A^{n-}_{(a)} \rightleftharpoons (R_4N^+)_nA^{n-}_{(o)}$$

for the quanternary amines.

Dimethyl-didecyl-ammonium nitrate dissolved in benzene was evaluated by Goldschmidt *et al.* (1956) for the separation of tetravalent plutonium from hexavalent uranium in nitric acid solution. The composition of the tetravalent plutonium complex was quoted as $(R_4N)_2\ Pu(NO_3)_6$. For the same separation the extractant proposed by Wilson (1959) was tri-lauryl-amine on the basis of data obtained by Sheppard (1957) during the extraction of tetravalent plutonium and neptunium by tri-n-octyl-amine in xylene as diluent from nitrate solutions. Tri-lauryl-amine has also been used as an extractant in France (Bathellier *et al.* 1963) to separate tetravalent plutonium from uranium. This was because of its higher selective extractability for tetravalent plutonium at low acidities and because of the greater radiation stability of this extractant in comparison to that of the organo-phosphorus extractants.

A comparison of the properties of the tertiary and the quaternary ammonium salts which show as extractants a high selectivity for tetravalent plutonium was carried out by Lopez-Menchero and Gehem (1963). The extraction coefficients for tetravalent plutonium increased for both types of extractant with nitric acid concentration to show a maximum at 4 M nitric acid.

The sequence for the extraction of tetravalent actinides by the tertiary ammonium salts is: plutonium > neptunium > uranium > thorium. The extraction coefficients of these tetravalent actinides vary essentially with the second power of the extractant concentration, suggesting that the ratio of extractant to actinide in the extracted complex is two to one. By a comparison of the ultra-violet absorption spectra of the organic phase from the extraction of the tetravalent actinide by tri-n-octyl-amine-xylene solutions with the spectra of concentrated nitric acid solutions, or with the spectra of organic solutions of the corresponding tetra-ethyl-ammonium-hexanitrato-actinide (IV) complexes of known compositon, the complex was shown (Keder *et al.* 1961) to contain the hexanitratometallate ion, $M(NO_3)_6^{2-}$.

A drawback to the use of quaternary ammonium compounds for tetravalent plutonium extractions is their limited solubility in organic diluents. In general, ammonium compounds show an order of solubility of tertiary > secondary > primary amine salt.

When the concentration of the metal salt in the organic quaternary ammonium-containing phase reaches a certain limit separation into heavy and light organic phases may occur and this phenonemon of third phase formation is usually attributed to the low polarity of the aliphatic diluents used.

The stability of the organic phase may be affected by virtually every variable parameter present in the extraction system. Organic phase stability is generally assisted by increase in molecular weight of the amine, by increasing the branching in the alkyl groups of the amine, by changing the type of amine from primary to secondary to tertiary, by using a more polar diluent and, in some instances, by increasing the acidity of the aqueous phase. Since the selection of the amine used and the acidity are usually determined by flowsheet considerations, adjustment of the diluent is the principal technique for controlling phase stability.

Koch (1965) successfully used adiponitrile as a modifier to minimize the third phase formation in the use of amine-based extractants for actinide extraction. Bac (1968), to prevent third phase formation during the extraction of tetravalent plutonium from the nitrate solution, used mixed quaternary ammonium nitrates with a tertiary amine and noted that this technique appreciably enhanced the solubility of both the quaternary ammonium salt and the plutonium nitrate–quaternary ammonium complexes in the organic phase.

Hexavalent actinides are extracted from nitric acid solutions by tertiary or quaternary ammonium salts. Keder *et al.* (1960) extracted from the nitric acid solutions the three hexavalent actinides, plutonium, neptunium, and uranium, using tri-octyl-ammonium-nitrate in a xylene diluent. The hexavalent plutonium (PuO_2^{2+}) and hexavalent neptunium (NpO_2^{2+}) are extracted to the same extent and considerably more so than hexavalent uranium (UO_2^{2+}).

The complex ion species of the hexavalent actinides extracted by tri-octyl-ammonium nitrate from nitric acid solutions were identified by comparing their ultra-violet absorption spectra with those of solutions of known complex composition (Keder *et al.* 1961). The spectrum of hexavalent uranium extracted into tri-octyl-amine-xylene solution from 4 M nitric acid was found to be essentially the same as that of the trinitratodioxouranate (VI) spectrum measured previously. The spectrum of this ion had been identified by Job's method of continuous variations in which acetone solutions of uranyl nitrate and tetra-butyl-ammonium nitrate gave a maximum absorbance at a 1 : 1 ratio, indicating the presence of the $UO_2(NO_3)_3^-$ ion (Adar *et al.* 1956). The spectra of the tetra-ethyl-ammonium trinitrato salts of hexavalent plutonium, neptunium, and uranium in acetone and/or nitromethane solutions, when compared with the spectra of the same metal ions in tri-octyl-amine-xylene solutions, were found to be almost identical, thus confirming the presence of the trinitrato complex of each of these three hexavalent actinides in tri-octyl-amine solutions.

2.4.5 Extraction with sulphur compounds

The solvent extraction of tetravalent plutonium and hexavalent uranium from nitric acid solution using di-n-hexyl sulphoxide and di-n-octyl sulphoxide has shown that the metal ion species are extracted as disolvates (Pai *et al.* 1979).

The extractive power of the sulphoxides for plutonium and uranium was shown to decrease with increased temperature, which would indicate that the reactions are exothermic in nature. They also found that the extraction of plutonium and uranium decreased with increase in radiation dose in tests performed using a Co^{60} irradiation source.

2.5 Choice of extractant in nuclear fuel reprocessing

Many organic compounds were screened for use as possible extractants for plutonium and uranium extraction from irradiated fuel by the Americans, the British, and the Canadians in the mid 1940s. Their initial choices were:

US: Methyl-iso-butyl-ketone-Hexone (MIBK).

UK: $2,2^1$-di-butoxy-diethyl-ether-Butex or Di-butyl-carbitol.

Canada: Tri-ethylene-glycol-dichloride-Trigly.

The first extractant to be used for the large-scale reprocessing of irradiated fuel was Hexone, methyl-iso-butyl-ketone, at Hanford in the United States (Lawroski and Levenson 1957). This extractant was unsatisfactory because it was soluble in the aqueous phase to the extent of 20 g/l with the production of unwanted degradation products: it also reacted vigorously with nitric acid concentrations in excess of 3 M and the process required the addition of 1–2 M aluminium nitrate as a salting-out agent for efficient plutonium and uranium extraction. The addition of aluminium nitrate meant that the highly active raffinate could not be reduced by evaporation to small bulk.

The triple ether Butex, $2,2^1$-di-butoxy-diethyl-ether was used at Sellafield, UK from 1952 to 1964 in the Magnox fuel reprocessing plant (Howells *et al.* 1959). This solvent had the advantages of a high flash point (low fire hazard) and a low solubility in the aqueous phase. Its disadvantages were found to be (in addition to its high cost) high viscosity, its density (near to that of water), the formation with uranyl nitrate under certain conditions of crystalline complexes and that it was reactive with nitric acid.

Tri-ethylene-glycol-dichloride (Trigly), $Cl(CH_2)_2O(CH_2)_2O(CH_2)_2Cl$, was used in a small batch plant at Chalk River, Canada from 1948 until 1954 (Navratil *et al.* 1984). Unfortunately this extractant recovered plutonium only, leaving the uranium together with the fission products in the highly active raffinate–clearly an undesirable process. Furthermore, the chlorine in the extractant molecule was released during radiolytic degradation, such effect having the potential to cause corrosion problems in all stainless steel reprocessing plants.

Tri-n-butyl-phosphate (TBP) is cheaper and more stable and permits better plutonium and uranium extraction than do either Hexone or Butex,

TABLE 2.2 *Influence of diluent on formation of HDBP in irradiated two phase Systems (20 vol. % TBP in diluent)*

Diluent	Plant	G-value (*n*-dodecane = 1)
n-Dodecane (S)	-	1
Odourless kerosene (S,B,C)	Magnox Plant (UK)	0.32
n-Alkane (S)	WAK (Germany)	0.51
Adakane 12 (USA) (S,B)	Savannah River	0.61
HPT (B)		
Hydrogenated Propylene Tetramer	Marcoule (France) Eurochemic (Belgium)	2.37

Letters in brackets describe hydrocarbon type.
S = straight-chain, B = branched-chain, C = cyclic.

at least in the first cycle of recovery, and is now universally used in a flowsheet generally known as the PUREX process (Cooper and Walling Jr 1959). All major reprocessing plants for which information is available use a solvent consisting of tri-butyl-phosphate as the extractant and a hydrocarbon mixture as the diluent (Table 2.2). Sellafield and Dounreay in the UK continue to find 20 per cent TBP in odourless kerosene (OK) a satisfactory solvent but all other major plants use 30 per cent TPB in a mixture of alkanes as the diluent. In the past other types of diluents were widely used and Marcoule used 40 per cent TBP until 1961 (Jonannaud 1964), but these alternatives have now been superseded. Too low a tri-butyl-phosphate concentration could give rise to inadequate extraction of plutonium and uranium, whereas too high a percentage raises the viscosity and density of the solvent to levels that are too high for efficient contactor operation. Twenty or 30 per cent tri-butyl-phosphate may be regarded as a suitable compromise, for the choice between these concentrations of tri-butyl-phosphate generally appears to be a marginal one which can only be decided on the basis of detailed flowsheet compilation. Furthermore, so much experience has now been built up over the past 30 years or so with these tri-butyl-phosphate concentrations, particularly in respect of contactor design, that there is a strong vested interest in retaining these concentration values.

2.6 Diluent selection for nuclear fuel reprocessing

The Purex process, which employs tri-butyl-phosphate–diluents as extractants in the various stages of the nuclear fuel reprocessing plants used to recover plutonium and uranium, necessarily involves the exposure of both tri-butyl-phosphate and the diluent to nitric acid and ionizing radiation. These conditions cause both the chemical and radiolytic degradation of the tri-butyl-phosphate/diluent solvent. Such degradation manifests itself in poor process performance as exemplified by poor decontamination of plutonium and uranium from the fission products and increasing loss of plutonium and uranium to various waste streams.

As a consequence, then, since the first large-scale nuclear fuel reprocessing plant usage of tri-butyl-phosphate/diluent mixtures some 30 years ago, considerable research and development effort has been expended in an attempt to understand the nature of extractant degradation and also of the role of the diluents in this process. The result of these studies has been to find diluents which exhibit maximum stability to chemical and radiolytic attack while meeting all the following criteria:

(1) miscible with the selected extractant;

(2) immiscible with the aqueous phase;

(3) non-polar;

(4) high boiling point;

(5) low viscosity;

(6) high density difference with respect to the aqueous phase;

(7) non-flammable or high flash point;

(8) low cost.

It should be noted, further, that in nuclear fuel reprocessing the diluent modifies the physical properties of the extractant (for example viscosity and density to promote phase disengagement), the extraction power and selectivity of the extractant, and avoids criticality problems by reducing the concentration of the fissile material extracted into the organic phase.

The result of some 30 years' testing has been to narrow the list of candidates as diluents for the tri-butyl-phosphate extractant in nuclear fuel reprocessing plants to four principal types:

(1) normal paraffinic hydrocarbons;

(2) saturated aliphatic hydrocarbons;

(3) aromatic hydrocarbons;

(4) halocarbons.

2.6.1 Normal paraffinic hydrocarbons

Normal paraffinic hydrocarbons (NPH) as used in the Hanford, USA, nuclear fuel reprocessing plant are mixtures of 10 to 16 carbon atoms containing straight chain compounds (Navratil *et al.* 1984). The effect of the carbon chain length in the normal paraffinic hydrocarbon diluent on its chemical and radioloytic stability is not completely understood. The free radical yield of *n*-paraffin hydrocarbons is known (Egorov *et al.* 1979) to increase with increasing chain length and it would be expected, therefore, that their yields of oxidation and nitration products, which are used as the principal criteria of radiochemical stability, would be similar.

2.6.2 Saturated hydrocarbon diluents

Branched chain aliphatic hydrocarbon compounds are more susceptible to chemical and radiolytic degradation than are *n*-paraffin compounds and the cyclic naphthene hydrocarbons are even more susceptible to such degradation. Despite the fact that cyclic naphthenes and branched chain alkanes are not as stable to nitric acid as are straight chain alkane compounds, both the Sellafield and Dounreay reprocessing plants in the UK both find odourless kerosene a very satisfactory diluent.

This odourless kerosene diluent is not of completely fixed composition and is, in fact, a complex mixture of normal hydrocarbons and branched chain hydrocarbons, monocyclic and bicyclic paraffinic hydrocarbons. The carbon number of the constituents ranges from 12 to 16 with the most abundant (8–10 per cent) species being the n-C_{13} and n-C_{14} hydrocarbons. In fact, the odourless kerosene used at Sellafield and Dounreay contains some 60 per cent straight chain alkanes, 10 per cent branched chain alkanes, and 30 per cent naphthenes (McKay 1979) and some 180 different compounds have been identified in this mixture.

2.6.3 Aromatic hydrocarbons

A considerable number of aromatic compounds have been examined with a view to their use as a diluent in the reprocessing of nuclear fuels. These aromatic compounds have several properties which make them potentially more attractive for use as diluents than paraffinic type compounds. Third phase formation does not occur in tri-butyl-phosphate-aromatic hydrocarbon diluent solutions and, in addition, plutonium and uranium are more extractable. Furthermore, because of the resonance structure of benzene, which acts as an energy sink, the aromatic diluents in combination with the tri-butyl-phosphate reduce its susceptibility to radiolytic degradation more so than do the paraffinic diluents. In contrast to these desirable properties, the aromatic hydrocarbon diluents suffer from the disadvantage that they are generally very much more susceptible to degradation by nitric acid than the aliphatic paraffin hydrocarbons. In general, the aromatic hydrocarbon compounds also have higher densities compared with those of the candidate paraffinic hydrocarbon diluents.

2.6.4 Halogenated hydrocarbons

Carbon tetrachloride has been used for many years as a diluent for tri-butyl-phosphate in both Russia (Nikipelov *et al.* 1974) and in the United States (McKay 1979).

The advantages claimed for the use of carbon tetrachloride as a diluent for tri-butyl-phosphate were:

(1) non-flammability;

(2) high density;

(3) low viscosity;

(4) high fission product decontamination factors;

(5) easy purification by distillation;

(6) excellent phase disengagement;

(7) low cost.

But it has major disadvantages in that it sensitizes the tri-butyl-phosphate extractant to radiolytically enhanced degradation and, in addition, that the chloride ion produced by the radiolysis of carbon tetrachloride is an extremely corrosive element in the presence of nitric acid to the stainless steel equipment and pipework used in nuclear fuel reprocessing plants.

For use in the Purex type processes the only diluents that are useful are those in class 1, the normal paraffinic hydrocarbons, and class 2, saturated aliphatic hydrocarbons. Currently, the diluents used in all countries operating the Purex process are of the following three types (Table 2.2).

1. Normal paraffin hydrocarbon–either pure compounds (n-dodecane) or mixtures of 9 to 14 long carbon chains (NPH).
2. Saturated petroleum fractions (odourless kerosene used at Sellafield and Dounreay).
3. Alkylated and polymerization products (hydrogenated propylene tetramer).

2.6.5 Third phase formation

Plutonium extracts into tri-butyl-phosphate as the complex $Pu(NO_3)_4$.2TBP which does not mix in all proportions with hydrocarbons. At high plutonium levels there is, consequently, a risk of the organic phase separating into a tri-butyl-phosphate-rich phase containing essentially $Pu(NO_3)_4$.2TBP and a hydrocarbon-rich phase. Even uranium under sufficiently extreme conditions can produce two organic phases. The formation of a third phase in this way cannot be tolerated in a solvent-extraction contactor.

The phenomenon of third phase formation with tri-butyl-phosphate is normally avoided by restricting the plutonium concentration in the organic phase, although an increase in temperature has a similar effect. It may be possible to reduce third phase formation by increasing the lengths of the carbon chain in the extractant or by shortening the length of the carbon chains in the diluent (Kertes 1965) but none of these suggestions for changing the composition of the extractant or diluent molecules has been tested in an actual nuclear fuel reprocessing plant.

2.7 Decontamination from fission products

The main defects of the solvent systems in current use are their failure to decontaminate the actinides effectively of fission products, principally ruthenium, cerium, zirconium, and niobium, and the susceptibility of the solvent systems to both chemical and radiolytic degradation. There are two main consequences of these defects in the operation of nuclear fuel reprocessing plants. The first is that several solvent extraction cycles, each involving extraction, scrub, and stripping stages, are necessary to obtain the actinides sufficiently free of the fission products to enable their re-use in the complete fuel cycle. The second is that, in order to recycle the solvent continuously in the reprocessing plant, it must be cleaned of the degradation products after it has passed through the plant by a washing process prior to its re-use. In consequence of these operational problems, large amounts of contaminated raffinates are produced, requiring treatment before they can be disposed of in a suitable manner.

In the first extraction cycle of the reprocessing plant some 99.9 per cent of the actinides is extracted by the tri-butyl-phosphate-diluent system, leaving behind at least 99.9 per cent of the fission products which, in the form of the highly active raffinate, are kept for indefinite storage. The small proportion of the fission products extracted by the solvent consist essentially of ruthenium, cerium, zirconium, niobium, iodine, antimony, and technetium together with some transuranic elements, in particular neptunium. These extracted fission products are among those with sufficiently high activity and relatively long half-lives to require their subsequent removal from the plutonium and uranium product streams before the latter are recycled. The iodine I^{131} (half-life of 8.05 days) is not of great significance provided that the fuel has been cooled (that is, stored) for several months prior to reprocessing. Despite the fact that the Zr^{95} and Nb^{95} isotopes are relatively short-lived (half-lives of 65 and 35 days respectively), they are present in sufficient quantities again to affect plutonium and uranium product qualities unless the fuel is long cooled. The other fission products Ru^{106} (half-life I year), I^{129} (half-life 1.6×10^7 years), Sb^{125} (half-life 2.8 years), Np^{237} (half-life 2.1×10^6 years), and Tc^{99} (half-life 2.3×10^5years) are the long-lived isotopes with sufficient impact on the reprocessing cycles, product qualities, product recycle, and environmental discharge limitations to warrant specific attention during the solvent-extraction process. The problem with I^{129} is confined mainly to the first extraction cycle and the waste streams arising from it, although a good proportion of the iodine is removed during or immediately after the fuel dissolution stage so that the actual amounts of iodine that come into contact with the solvent are relatively low.

2.8 Effects of radiolysis and degradation

2.8.1 Diluent effects on solvent radiolysis

The radiolytic and chemical degradation of tri-n-butyl phosphate have been well studied (Burger 1958; Lopez-Menchero and Salomon 1967) and the yield values or G values for the two principal degradation products di-butyl-phosphoric acid (DBP) and mono-butyl-phosphoric acid (MBP) are well known, with considerable variations in the G values quoted, however, in the literature.

Although both the tri-butyl-phosphate and the diluent used in reprocessing have been selected for their high chemical and radiation stability, they are slowly affected by a combination of acid hydrolysis, thermal degradation, and radiolysis in contact with nitric acid and the irradiated fuel species in solution. This effect is of particular concern in the first extraction cycle of a nuclear fuel reprocessing plant where the fission product activity is highest and where radiolysis accounts for most of the tri-butyl-phosphate and diluent degradation. During the second and third extraction cycles the alpha radiation from the plutonium is more significant and the amount of degradation encountered is proportional to the plutonium concentration and its residence time in contact with the organic phase. With the current aim for higher burn-up in both thermal and fast reactor fuels, both the fission product and plutonium concentrations will increase and lead to greater solvent degradation. The decomposition of tri-butyl-phosphate and the diluent in the solvent extraction cycles of nuclear fuel reprocessing plants has created a number of problems because the degradation products can form strong complexes with plutonium and uranium which cannot be back extracted from the organic phase and this can lead to the loss of these elements. In addition, these solvent degradation products can form extractable complexes with certain of the fission products such as zirconium, ruthenium, and technetium and reduce their decontamination factors. Other problems may be encountered due to the formation of surfactants which can cause foaming and the subsequent loss of interface control as well as the possible formation of precipitates which may cause blocking of the contactors.

In fact, operational perturbations have been reported from several nuclear fuel reprocessing plants especially at the start-up of reprocessing. Problems that arose during the recycling of the solvent have been reported at the Savannah River plant, USA (Olcott and Orth 1963; Orth 1965) where the ruthenium activity in the washed solvent during a run of several months' duration increased irreversibly, leading in turn to an increase in the fission product content of the plutonium and uranium product streams. Centrifugal contactors were installed in the first extraction cycle of the plant and a long chained alkane was used as the diluent to reduce this effect. Problems were encountered in the reprocessing plant at Marcoule, France, where the appearance of a zirconium-containing phase at the interface was noted (Faugeras and Talmont 1967; Breschet and Miquel 1971). Difficulties in plant operations periodically arose at the Eurochemic plant, Belgium, due to the appearance of precipitants and heavy organic phases (Centeno *et al.* 1971). Perturbations were further noted in the first extraction cycle of the WAK reprocessing plant at Karlsruhe, Germany during a run with high burn-up fuel (Huppert *et al.* 1974).

2.8.2 Solvent/diluent interactions during radiolysis

A considerable amount of data exists in the literature on diluent effects on radiolysed tri-butyl-phosphate-diluent systems but the results are somewhat conflicting. Becker and Stieglitz (1981) have examined the G values for di-butyl-phosphoric acid formation in the tri-butyl-phosphate diluent mixtures used in a number of reprocessing plants around the world (Table 2.2). These results would suggest that the use of *n*-paraffins as the diluent results in between two to seven times more di-butyl-phosphoric acid formation than if odourless kerosene is used as the diluent. The situation is, however, complicated by the different process environments in each of the reprocessing plants where the solvent systems were irradiated and the different analytical techniques used to determine the amount of di-butyl-phosphoric acid present. Data taken from work reported by Navratil (1984) (Table 2.3) would indicate a somewhat different relationship between diluent and di-butyl-phosphoric acid formation on radiolysis. In this case the *n*-paraffins would

TABLE 2.3 *Influence of diluent on HDBP formation*

Diluent	TBP %	G-value	Type of radiation
n-Paraffin mixture (96%)	30	0.59	Discharged fuel elements
Iso-octane	30	1.4	Co-60
Benzene	20	0.46	Co-60
Amsco 125-82*	25	0.71	1 MeV electrons
Amsco 125-82†	25	0.68	1 MeV electrons
Odourless kerosene	20.7	0.8‡	1 MeV electrons

*Amsco 125-82 is described as a mixture of 5- and 6-carbon naphthas that have various short normal and branched chain substituents, branched chain hydrocarbons, a few per cent *n*-paraffins, and *c* 1 per cent olefines. Dry solution irradiated.
†Water saturated solution irradiated.
‡Total acid determined (i.e. HDBP + H_2MBP).

appear to be superior to the branched chain paraffins and the cyclic paraffins, although the qualifications raised above about radiation environment and analytical techniques used are again pertinent.

The situation is further complicated by the fact that odourless kerosene, used as the diluent in the Sellafield and Dounreay nuclear fuel reprocessing plants in the UK, is not of completely fixed composition. The claimed protective effect of odourless kerosene on tri-butyl-phosphate degradation during irradiation has been ascribed to the presence of the cyclic hydrocarbons which comprise some 30–40 per cent of the composition of this diluent.

Michalik *et al.* (1979), Nowark and Nowark (1973), and Nowark *et al.* (1979) produced a series of papers on the degradation of irradiated tri-butyl-phosphate-nitric acid-dodecane systems. The first paper (1972) reported the detection of stable radicals during an electron spin resonance study of the system, with the highest concentration for the system irradiated in the presence of 1 M nitric acid. These radicals were not detected when the system was irradiated in the absence of nitric acid. The second publication (1973) included the effects of ultra-violet light and thermal degradation. Furthermore, at low nitric acid concentrations (<2 M nitric acid) using gamma irradition more than half of the potential zirconium complexing compounds remained in the solvent after an alkaline wash, whereas in the case of ultra-violet or thermally degraded solvent systems most of the zirconium complexing compounds were removed by the alkaline wash over the whole range of initial nitric acid concentrations employed. In all three degradation techniques employed, gamma irradiation, thermal degradation or ultra violet treatment, the major degradation species were assumed to be di-butyl-phosphoric acid and mono-butyl-phosphoric acid.

The effect of the presence of the different diluents dodecane, mesitylene, and carbon tetrachloride with nitric acid present on the degradation of tri-butyl-phosphate was also investigated by Nowak *et al.* (1974). Using thin layer chromatography, they detected the presence of nitration and oxidation products as well as di-butyl-phosphoric acid, mono-butyl-phosphoric acid, and phosphoric acid. They demonstrated that the highest yield of degradation products occurred in the TBP-CCl_4-HNO_3 system and the lowest yield with the TBP-mesitylene-HNO_3 system. They suggested that the considerable amounts of hydrochloric acid produced in the former system accelerated the process of degradation, whereas in the tri-butyl-phosphate-mesitylene system the aromatic nature of the diluent decreased the yield of degradation products. In addition, the highest yield of phosphorus-based radiolytic degradation products occurred at low (0.5 M) nitric acid concentrations although, once again, it was observed that a considerable portion of the radiolysis products produced were not removed from the solvent by washing with 1 M sodium hydroxide solution.

An even more detailed paper by Nowak (1977) on the presence of degradation products produced by the radiolysis of tri-butyl-phosphate-diluent-nitric acid systems indicated that the amount of nitro compounds produced increased with nitric acid concentration, whereas the yield of carbonyl compounds peaked at a 3 M nitric acid concentration: only above a 7 M nitric acid concentration did the yield of nitro compounds produced exceed that of carbonyl compounds produced. At low (< 1 M) nitric acid concentrations the formation of hydroxamic acids was observed but their presence was not detected at higher nitric acidities, presumably because they were oxidized by the nitrous acid present. The effect of diluent on the yields of nitro and carbonyl compounds was examined, in addition, and it was reported that low yields of carbonyl compounds were obtained using mesitylene as a diluent whilst mesitylene and dodecane diluents both gave low yields of nitro compounds. Carbon tetrachloride used as diluent, on the other hand, resulted in a high yield of carbonyl compounds but a very low yield of nitro compounds.

Work by Healy and Pilbeam (1974) resulted in the suggestion that metal retention in irradiated tri-butyl-phosphate solutions could be caused by the presence of hydroxamic acids, which was in disagreement with earlier work by Stieglitz (1971), who had proposed that carbonyl compounds were responsible for this effect. There is now good evidence that the principal metal retention ligands in irradiated tri-butyl-phosphate solutions are long-chain alkyl-phosphoric acids and possibly to a lesser extent polycarbonyl compounds: the suggestion that they might be hydroxamic acids would appear to be incorrect (McKay 1979).

All the above data refer to the radiolysis and hydrolysis of tri-butyl-phosphate in the organic phase: no consideration has been given to the presence of metal ions and little attention has been paid to the aqueous phase radiolysis of tri-butyl-phosphate.

Clay and Witort (1974) examined the radiolysis of tri-butyl-phosphate in the aqueous phase and again found that the major degradation product was di-butyl-phosphoric acid. Although the presence of a number of carbonyl compounds were qualitatively detected only the compound butyraldehyde was positively identified. The presence of carbonyl compounds was no longer detected when the solutions were made up to 4 M nitric acid, indicating their susceptibility to post-radiolytic attack in the presence of this concentration of nitric acid. Furthermore the G values for di-butyl-phosphoric acid decreased as the nitric acid concentration increased. They proposed that the chemical attack of the tri-butyl-phosphate was caused by the production of hydroxyl radicals by radiolysis in the aqueous solution:

$$\cdot OH + CH_3(CH_2)_3{-}O{-}\underset{\underset{O}{\|}}{P}{-}(OBu)_2 \longrightarrow CH_3(CH_2)_2\dot{C}HO\underset{\underset{O}{\|}}{P}(OBu)_2$$

Their postulated reactions are purely speculative, that in oxygenated solution these reactions could proceed by the addition of oxygen to the radical site which results ultimately in the decomposition of the tri-butyl-phosphate radical to produce di-butyl-phosphoric acid and other products arising from the butyl chain such as butyraldehyde. In the absence of oxygen, the radicals must react by alternative routes to produce a lower yield of di-butyl-phosphoric acid. In the absence of nitric acid, however, no such effect of oxygen was noted, the yields of di-butyl-phosphoric acid being similar in both oxygenated and de-oxygenated solutions. This observation could be associated with the reactions of NO_2 produced by the radiolysis of nitric acid, for example:

$$CH_3CH_2CH_2CH{-}O{-}\overset{\overset{O}{\|}}{P}{-}(OBu)_2 + NO_2 + H_2O \longrightarrow CH_3CH_2CH_2CHO$$
$$+ HO - \underset{\underset{O}{\|}}{P} - (OBu)_2 + H^+ + NO_2^-.$$

The decrease in the yield of di-butyl-phosphoric acid observed at their higher nitric acid concentrations was attributed to the reaction of the OH radicals with undissociated nitric acid in competition with the OH radical attack of the tri-butyl-phosphate:

$$\cdot OH + HNO_3 \longrightarrow H_2O + \cdot NO_3.$$

Experiments by Baumgartner *et al.* (1976) and Bautz *et al.* (1978) were made in an attempt to isolate the complexing agents responsible for the reduced efficiency of the solvent extraction system. After irradiation of a 20 per cent tri-butyl-phosphate–dodecane mixture in contact with nitric acid, the acid degradation products were removed by alkali washing and the lighter volatile residues isolated by distillation. Any complexing species were retained in the residue and separated by liquid chromatography. It proved impossible, however, to achieve isolation of the individual complexing species by this chromatographic technique although separation into three individual groups was attained. These three groups were then subjected to gas chromatographic and mass spectrometric analysis. The first two groups comprised non-polar species (45 per cent, such as alkanes, alkenes and ketones) and neutral phosphate esters (35 per cent, such as monomeric and dimeric phosphate esters and pyrophosphate esters, and, to a lesser extent, polycarbonyl compounds).

Nowak *et al.* (1979) have reported the influence of dose-rate, temperature, nitric acid concentration, and metallic ions (uranium, molybdenum, palladium, and zirconium) on the radiolytic degradation of tri-butyl-phosphate-dodecane-nitric acid systems. They observed that the yields of di-butyl-phosphoric acid and mono-butyl-phosphoric acid were little influenced by dose-rate, temperature, or nitric acid concentrations but that the presence of uranyl ions markedly increased the yields of these two acidic phosphate degradation products. In contrast, with an increase from 0.5 M to 3 M in nitric acid concentration the yields of nitro and carbonyl compounds showed an increase.

2.8.3 Thermal degradation of tri-butyl-phosphate

The choice of tri-butyl-phosphate because of its chemical inertness as an extractant in nuclear fuel reprocessing has already been noted. Its stability towards oxidation—in particular its resistance to nitric acid—is excellent (Burger 1958), although at temperatures above 100°C it may react with nitric acid to form nitro compounds and at even higher temperatures react violently. Acid hydrolysis may occur in either the organic phase or the aqueous phase, and although the rate constants in the organic phase were found to be several orders of magnitude lower than in the aqueous phase it may be shown that in a two-phase mixture most of the hydrolysis will occur in the organic phase.

The conclusions of early studies by Wagner (1951), who investigated the reactions of tri-butyl-phosphate when mixed with equal volumes of 3 M and 6 M nitric acid at elevated (up to 105°C) temperatures, were queried by Burger (1955) on the basis of certain assumptions made and the kinetic analysis calculational technique employed. Davis and Kibbey (1970) subsequently studied the rate of decomposition of tri-butyl-phosphate in the aqueous phase with 1–5 M nitric acid at temperatures in the range 35–75°C, in which the initial organic to aqueous phase ratio was about 10^{-4}, but the rate of di-butyl-phosphoric acid formation was found to be the same as for phase ratios of 10^{-2}. The data appeared to indicate that the rate constant for tri-butyl-phosphate decomposition in the aqueous phase was about an order of magnitude greater than that in the organic phase. During nuclear fuel reprocessing, however, the overall tri-butyl-phosphate decomposition rate in the organic phase will be some two orders of magnitude higher than in the aqueous phase.

Nowak and Nowak (1979) have examined, in addition, the thermal degradation of tri-butyl-phosphate systems in which the systems tri-butyl-phosphate, tri-butyl-phosphate-n-dodecane, and tri-butyl-phosphate-mesitylene were heated in contact with an aqueous nitric acid phase for 40 hours using an organic-to-aqueous phase ratio of 2 : 1. Apart from the di-butyl-phosphoric acid and the organic nitrates which were produced at temperatures up to 80°C, nitrates were also detected in the tri-butyl-phosphate-nitric acid and tri-butyl-phosphate-n-dodecane-nitric acid systems but were not found to be present in the tri-butyl-phosphate-nitric acid-mesitylene system. Carbonyl and nitro compounds were not detected in the tri-butyl-phosphate-n-dodecane-nitric acid system when it was heated to 80°C, in contrast to the result obtained during the radiolysis of this mixture (Nowak 1977) when, in systems heated to 100°C, both carbonyl and nitro compounds were detected. They concluded that in general the yields of most degradation products increased significantly only above 90°C, which is in agreement with the much earlier conclusions of Dukes (1957).

2.8.4 Diluent radiolysis

It is possible that the reduced performance of the solvent with time in a nuclear fuel reprocessing plant may not be entirely caused by the formation of tri-butyl-phosphate degradation products and their possible complexing with the actinides and fission products. The second component of the solvent used during the extraction process, the diluent, may also form compounds by reacting with nitric acid or nitrous acid under the influence of radiation or at elevated temperature. Unlike the tri-butyl-phosphate degradation products, the diluent degradation product species may not be removed by an alkaline solvent wash but could slowly accumulate in the organic phase and impair the performance of the solvent. The species mainly involved in the thermal degradation of the diluent was thought to be nitrous acid (Ellerhorst *et al.* 1958), given an induction period sufficient for its formation; its effectiveness can then be reduced by the addition of nitrous acid scavengers (urea), by sparging, or by operating at lower temperatures. Lane (1963) and Huggard and Warner (1963) suggested that the main complexing species from diluent degradation in the presence of nitric acid were hydroxamic acids, RCONHOH; the precursors to these complexing species were postulated as primary nitro-paraffins formed by either nitration of the $-CH_3$ groups in hydrocarbon chains, nitration of alkyl side chains in aromatic and naphthenic compounds, nitration of terminal ethylenic linkages in olefines, or the ring opening of naphthenic compounds. The primary nitro-paraffins could be converted to hydroxamic acids by acidification reactions as follows:

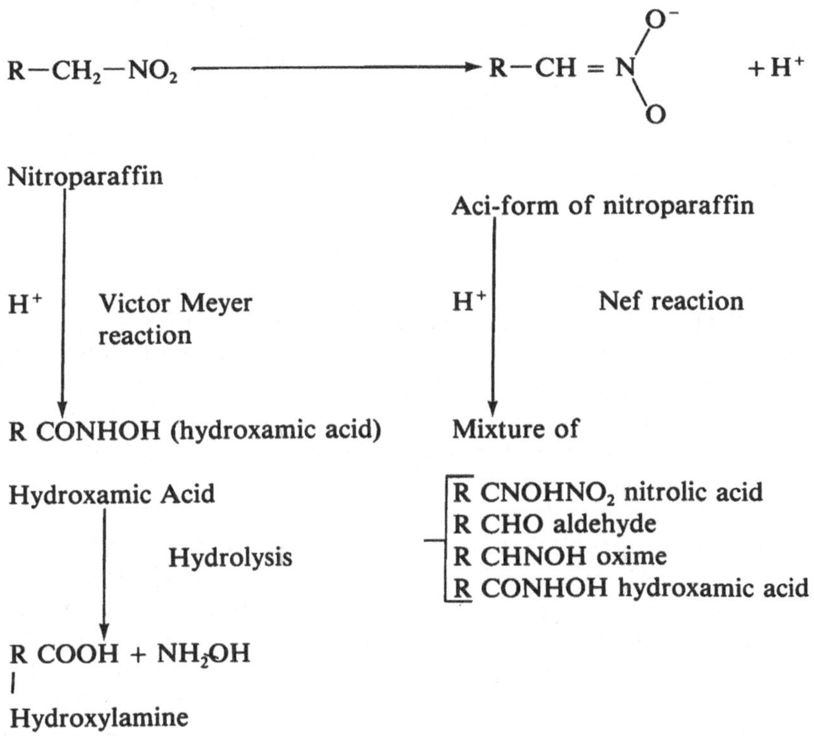

They reported that a dodecyl hydroxamic acid–ferric iron complex had been observed in a degraded solvent containing added iron but that otherwise no hydroxamic acids had been isolated. Lane (1963) concluded that the addition of primary nitro compounds did not produce metal retention until a substantial concentration of hydroxamic acid built up by the mechanism specified above. Hydroxamic acids do accumulate but are lost from the system by hydrolysis.

In contrast, Blake *et al.* (1963) concluded that the presence of nitro paraffins in the enol form were the main metal complexing species, possibly as a result of a synergistic extraction in the presence of tri-butyl-phosphate. The following equilibria were postulated:

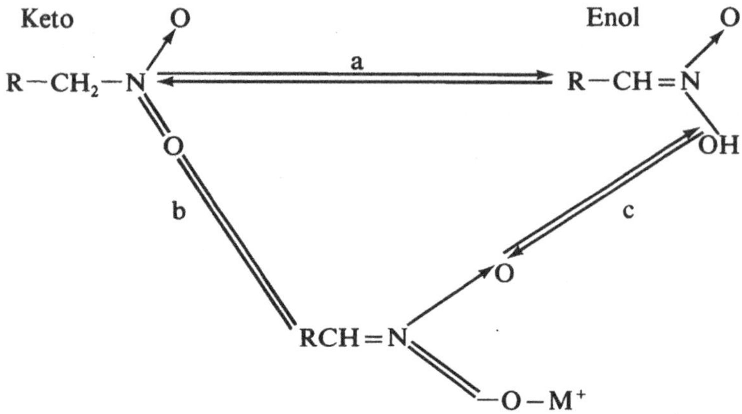

In acid conditions the equilibrium of reaction a favours the keto form and equilibration is slow, whereas the equilibrium of reaction b in basic media favours the enol form and equilibration is moderately fast.

Change to an acid medium, however, produces the free enol form in a much larger amount than via reaction a and the hydroxamic acid may then in turn form metal complexes. Despite the different reaction mechanisms postulated to account for the metal retention by diluent degradation product production, all these groups of workers indicated that the elimination of nitro-paraffins during solvent washing was desirable.

The radiolysis of decalin used as a diluent with tri-butyl-phosphate (Giefer and Merkelin 1975) produced hydrogen and an olefin ($C_{10}H_{16}$) with G values of 4.4 and 0.18 respectively. The yield of hydrogen was found to be dose dependent, although the yield of olefin showed a small inverse dose dependence. The radiolysis products of dodecane were examined by Merkelin and Razvi (1980) using a gas chromatographic technique. Paraffins up to $C_{24}H_{50}$ were detected and their yields appeared to increase with absorbed dose. In the presence of nitrous oxide the yield of intermediate products, C_{17}–C_{20}, and low molecular weight species, C_5–C_{11}, showed significant reductions, but the dimer yield of $C_{24}H_{50}$ was unaffected.

Becker and Stieglitz (1981) have investigated the radiolytic and chemical behaviour of the solvent in the Purex process and have suggested the following reaction scheme:

In the Purex process the diluent undergoes a series of chemical and radiolytic reactions shown schematically above to form dimers class I. The alkyl radicals from the diluent react with the inorganic radicals (OH, NO, O_2) to form substitution and degradation products with variable carbon chain lengths and this class II consists of ketones, carboxylic acids, nitro alkanes, nitroso compounds, and nitrate esters. The presence of these compounds may result in a gradual change in the physico-chemical properties of

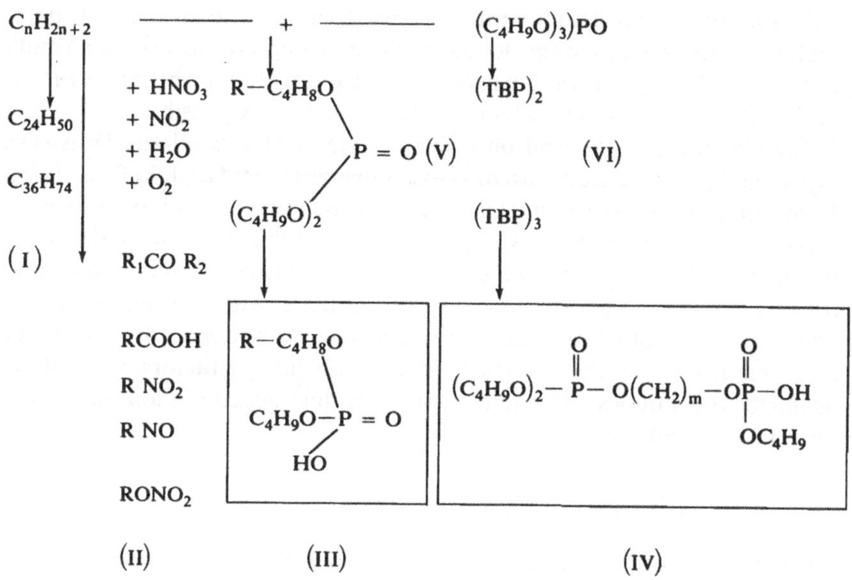

the solvent, especially in the case of a nuclear fuel reprocessing plant where the solvent is recycled. They concluded that these compounds had no deleterious effect on process performance through metal complex formation and fission product retention. Furthermore, they discounted the suggestion of hydroxamic acid formation by rearrangement reactions involving nitro compounds on the grounds that the presence of these acids had never been really proven by chemical analysis. Detailed analysis of irradiated solvent by special enrichment, separation and analytical techniques, however, resulted in the discovery of the two compound classes III and IV which showed high extraction of tetravalent metals. Class III consists of a number of long chain acid phosphates which are formed by a combination of alkyl radicals with tri-butyl-phosphate, followed by the elimination of a butyl group, or directly from di-butyl-phosphoric acid. The aliphatic radical R may vary from C_2H_5 to $C_{12}H_{25}$. In total, more than eighteen different compounds of class III have been identified. Furthermore, eleven isomers of an acid dimeric tri-butyl-phosphate, class IV, were isolated and identified. Again these compounds show a high affinity for tetravalent metals because of the presence of the acid phosphate groups.

In the neutral fractions of the irradiated solvent a series of neutral phosphate esters, class V, was detected which contained alkyl groups with chain lengths of between five and sixteen carbon atoms. Using gas chromatographic and gel chromatographic techniques, the presence of class VI compounds, the oligomeric butyl-phosphates with molecular weights of 530,

800, and 1005, was demonstrated. Thus it can be shown that during the radiolytic process not only do the molecules degrade to produce smaller entities but that by a recombination of alkyl and tri-butyl-phosphate radicals the construction of much larger molecular species is possible.

The chemistry of the acid phosphate esters of classes III and IV is easily explained by comparison with di-butyl-phosphoric acid and, in fact, they do form stable complexes with tetravalent metals. Furthermore, because of the presence of much longer alkyl groups, the solubility of the sodium salts (formed during solvent washing) of these two classes of compounds in the organic phase is high so that they accumulate with solvent recycle. In addition, these compounds are interfacially active and have the potential to form emulsions at the interface in the solvent contactors with all the problems this implies—frothing, fission product accumulation, and loss of interface control.

2.9 Alternative extractants

Despite extensive research no neutral organophosphorus compound clearly superior to tri-butyl-phosphate as an alternative extractant in nuclear fuel reprocessing has been discovered. One of the problems in finding a suitable alternative extractant is that its extractant power should be essentially similar to that of tri-butyl-phosphate so that both the extraction and stripping of plutonium and uranium can be readily achieved. If the replacement is a week extractant it would require a high extractant concentration, at which point the physical properties of the extractant are likely to be unfavourable, whereas if it is a powerful extractant it would require to be highly diluted and, as a consequence, the solvent loadings could be too low for efficient plant operation. The extractant tri-octyl-phosphine-oxide (TOPO) is much too powerful an extractant for easy stripping of the extracted plutonium.

Another possible route to a suitable extractant would be to modify chemically the basic tri-butyl-phosphate molecule either to increase or decrease the extracting power. The replacement of the butoxy groups by butyl groups, giving in turn a phosphonate, a phosphinate, and a phosphine oxide, results in an increase in extraction power by approximately an order of magnitude at each of these steps (McKay 1979). The replacement of the butoxy groups by aromatic groups, on the other hand, causes a decrease in extracting power but has the benefit of enhancing the radiation stability of the parent molecule because such groups act as energy sinks. Further reductions in extracting power may be achieved by replacing the butoxy groups by longer straight chain or branched chain carbon groups than the butyl groups mentioned previously.

One such modified compound, di-sec-butyl-phenyl-phosphonate (DSBPP) was selected for trial at Oak Ridge, USA (Blake and Schmidt 1965). The laboratory tests indicated better fission product decontamination factors than if tri-butyl-phosphate had been used but were not conducted in such a manner to prove the point conclusively. In any case this extractant is unsatisfactory for plant use because it required an aromatic diluent to keep the extracted uranium in solution.

The amines and quaternary ammonium nitrates would appear to be potentially attractive extractants (Muller 1967; Ali and Eberle 1969; Kertes *et al.* 1974) in nuclear fuel reprocessing plants. In nitrate systems, for example, there is an increase in metal extracting power from primary to secondary to tertiary amines and finally to quaternary ammonium nitrates. Tetravalent metals are the most strongly extracted, considerably more so than those in the hexavalent state, and in this way these nitrogen-based extractants differ considerably from tri-butyl-phosphate which extracts both the tetravalent and hexavalent states equally well.

The tertiary amines extract the tetravalent plutonium so powerfully that the plutonium is difficult to strip. Furthermore, the extraction of hexavalent plutonium is not complete and the retention of the hexavalent uranium in the organic phase cannot be guaranteed during plutonium stripping. Employing these extractants is further limited by the relatively low heavy metal loading available, practical values lying in the range 5–20 g/l; even at these values aliphatic diluent is precluded if third phase formation is to be avoided. In addition, before considering tertiary amines in a plant, compensating advantages in the form of better fission product decontamination factors and higher radiation stability would need to be clearly demonstrated.

Quaternary ammonium nitrates, on the other hand, extract plutonium and uranium more readily than the tertiary amines although they have a greater tendency to third phase formation than the tertiary amines which, again, could necessitate using an aromatic diluent. Nitrogen-containing compounds in fuel reprocessing may be feasible but any benefits they currently possess over the use of tri-butyl-phosphate in a nuclear fuel reprocessing plant has not been unequivocally demonstrated.

The possibility of using sulphur-based extractants in nuclear fuel reprocessing for the recovery of plutonium and uranium is at an early stage and clearly a substantial development programme would be required to confirm any claim to any of them being an alternative extractant to tri-butyl-phosphate.

2.10 Reprocessing plants

Historically, the recovery of plutonium was first achieved by precipitation processes. Despite the fact that the concentration of plutonium may be too small to exceed the solubility product of even extremely insoluble compounds, plutonium can be removed from solution by coprecipitation. Coprecipitation of plutonium will, in general, occur if the anion of the bulk precipitate forms an insoluble plutonium salt with plutonium in a particular oxidation state. The decontamination factors of the plutonium and uranium from the fission products achieved in these precipitation processes are, however, too low to allow them to be selected for a primary separation process.

At the present time all the plant scale processes used for separating plutonium from irradiated uranium are based on solvent extraction. These processes yield highly purified plutonium and uranium with respect to the fission products and are, therefore, vastly superior to the original precipitation processes which they replaced.

The solvent extraction reprocessing of plutonium and uranium is now dominated by the Purex process, using a tri-butyl-phosphate–hydrocarbon solution as the organic extractant. Earlier methods such as the 'Trigly', Redox, and Butex processes are no longer employed, but the flowsheet chemistry is of interest. Extraction using amines is currently studied in a primary separation process for high burn-up fast reactor fuels.

2.10.1 The Trigly process

This process (Campbell 1956), using three different extractants, was employed on the plant scale in Chalk River, Canada, and was operated on a batch basis. The irradiated uranium fuel after dissolution in nitric acid in the dissolver was conditioned with ammonium nitrate and then extracted seven times with trigly (tri-ethylene glycol dichloride), the volume of organic extractant each time being one-quarter of the volume of the aqueous phase. Each trigly extract was scrubbed four times with an ammonium nitrate/nitric acid solution to enhance fission product decontamination. The combined trigly extracts contained some 97 per cent of the plutonium and 5 per cent of the uranium with a fission product decontamination factor of 10^4, the bulk of the uranium and the fission products remaining in the highly active raffinate.

The next step in the process was an oxidation-reduction stage using hexone (methyl iso-butyl ketone) as the extractant. This was carried out by first stripping the plutonium from the trigly extractant by a ferrous sulphamate/ammonium nitrate solution which reduced the plutonium (IV) to the organic-insoluble plutonium (III) form. The aqueous solution containing plutonium (III) and residual uranium (VI) was extracted six times with hexone to remove the uranium. The aqueous solution was then treated with sodium dichromate to oxidize the trivalent plutonium to the hexavalent plutonium state and extracted four times with hexone. The solvent extract was then stripped of plutonium using water.

The final plutonium purification consisted of taking the aqueous plutonium solution from the first two stages and precipitating the plutonium as the fluoride to separate it from any organic degradation products. The plutonium fluoride was dissolved in an aluminium nitrate/nitric acid solution from which the plutonium was extracted using 0.5 M thenoyl-tri-acetone in benzene. The plutonium was then stripped from the organic extractant using 10 M nitric acid and precipitated using sodium hydroxide. The precipitated plutonium hydroxide was finally redissolved in nitric acid to yield the plutonium nitrate product solution.

The trigly process involving three different extractants and two separate precipitation–dissolution stages has been abandoned because of the complexity of the process and, in addition, from the fact that it was not designed to recover uranium which remained in the highly active raffinate.

2.10.2 The Halex process

Solvent extraction of plutonium and uranium using tri-butyl-phosphate in carbon tetrachloride as the extractant has been studied in Russia (Pushlenkov *et al.* 1965). Carbon tetrachloride was selected as the diluent because of its ready availability, non-flammability, low viscosity, and easy regeneration.

The flowsheet for reprocessing some 8–12 kg of uranium per day is shown in Fig. 2.14. The feed solution contained uranium at a concentration of 1.1 M in 2 M free nitric acid. The plutonium was in the tetravalent and hexavalent oxidation states.

Extraction operations were performed in banks of mixer-settlers. The first extraction cycle contained eight stages extracting 99.9 per cent of the plutonium (even if it was completely in the hexavalent state), 85 per cent of the uranium, and some 85–90 per cent of the neptunium.

The solvent product was scrubbed with 3 M nitric acid in a four-stage mixer-settler bank giving decontamination factors of 250 for zirconium and niobium, 500–600 for ruthenium, and 1500 for the rare earths.

Splitting of the plutonium from the uranium was accomplished by the use of a reducing agent, ferrous nitrate stabilized by hydrazine, to convert the plutonium into the inextractable trivalent form. Backwashing of the uranium and the neptunium was carried out in an eight stage mixer-settler bank.

The plutonium separated in the first cycle from the uranium and the fission products was further purified from the fission products in two subsequent cycles using the same extractant as in the first cycle. After completing these two further purification cycles, the fission product decontamination factor obtained was some 10^4.

Further purifications of the uranium and the separation of the uranium and neptunium as well as the separation of the trace amounts of plutonium is performed in two subsequent cycles. The uranium is extracted from the second cycle solution containing 0.5 M nitric acid and 0.1 M hydrazine. The

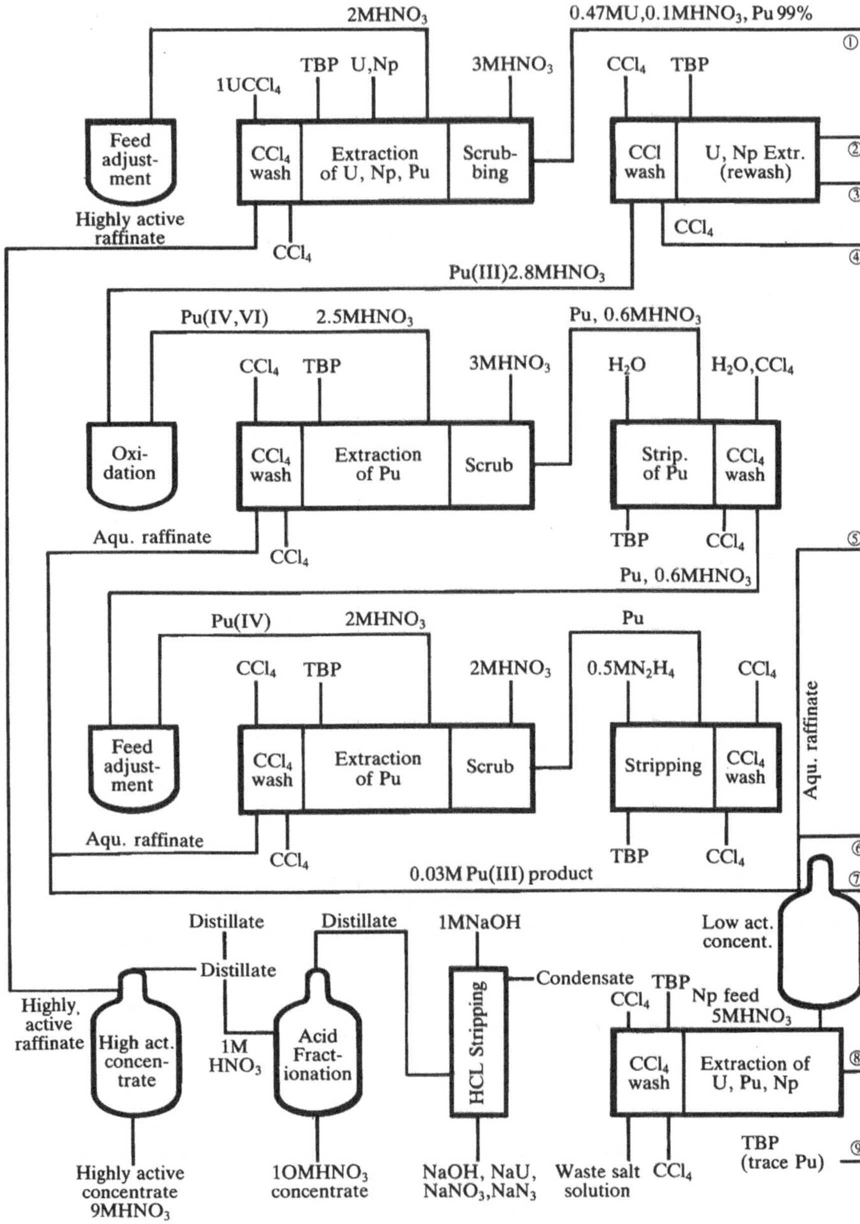

Fig. 2.14. The Halex solvent extraction flowsheet.

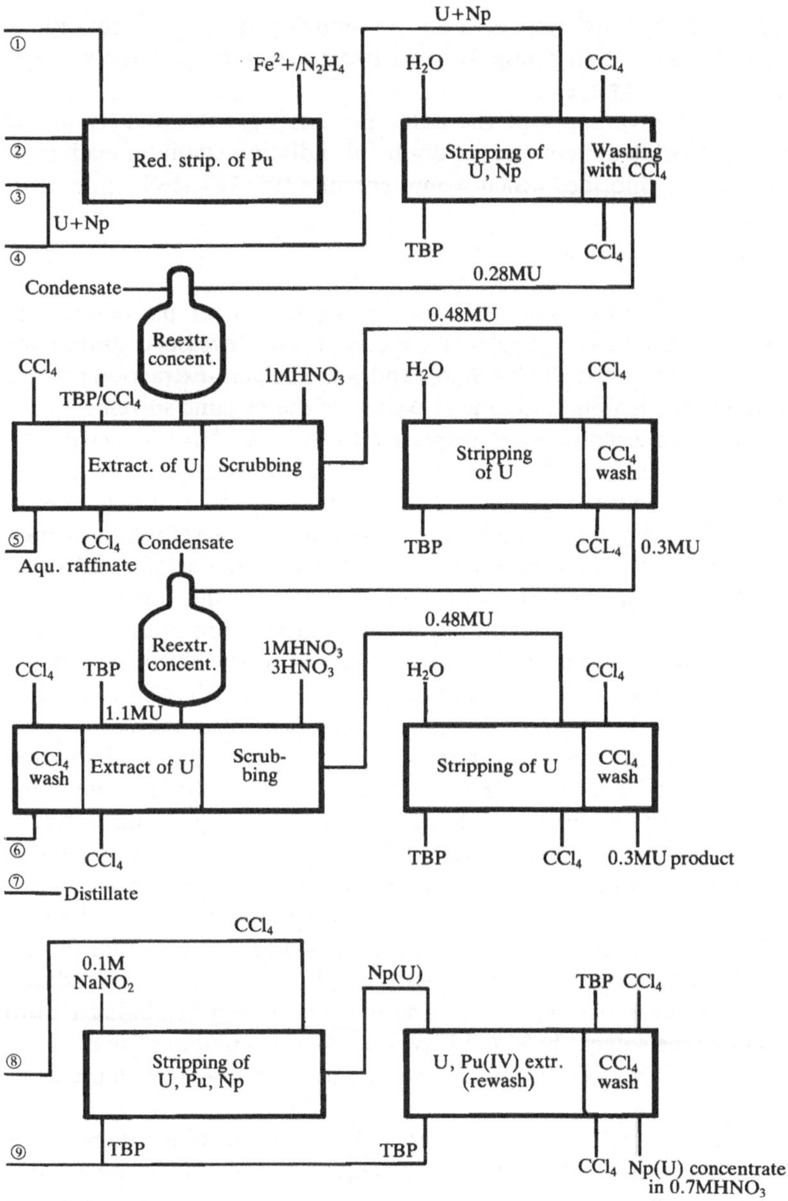
U+Np
①
Fe^{2}+/N_2H_4
H_2O
CCl_4
②
Red. strip. of Pu
Stripping of U, Np
Washing with CCl_4
③
U+Np
④
TBP
CCl_4
0.28MU
Condensate
0.48MU
Reextr. concent.
CCl_4
1MHNO$_3$
H_2O
CCl_4
TBP/CCl_4
Extract. of U
Scrubbing
Stripping of U
CCl_4 wash
⑤
CCl_4
Condensate
TBP
CCL$_4$
0.3MU
Aqu. raffinate
0.48MU
Reextr. concent.
1MHNO$_3$
3HNO$_3$
CCl_4
TBP
H_2O
CCl_4
1.1MU
CCl_4 wash
Extract of U
Scrub-bing
Stripping of U
CCl_4 wash
⑥
CCl_4
TBP
CCl_4
0.3MU product
⑦
Distillate
CCl_4
0.1M
NaNO$_2$
Np(U)
TBP CCl_4
⑧
Stripping of U, Pu, Np
U, Pu(IV) extr. (rewash)
CCl_4 wash
⑨
TBP
TBP
CCl_4
Np(U) concentrate in 0.7MHNO$_3$

feed solution to the third cycle contains no reducing agents. The conditions for the extraction and stripping in these two cycles are again the same as those used in the first cycle.

The major disadvantages of the halex process were the toxicity of the carbon tetrachloride diluent and its doubtful radiation stability, leading to the formation of chlorides which would corrode stainless steel equipment.

2.10.3 The Redox process

The redox process (Ransom 1951) for the separation of plutonium and uranium from fission products was conceived and developed during the Second World War in the United States and was a solvent-extraction process employing methyl iso-butyl-ketone (hexone) as the organic solvent.

Plutonium and uranium are extracted by hexone from concentrated aqueous nitrate solutions while the fission products are only slightly extractable. High nitric acid concentrations, however, cause solvent degradation and increase fission product extraction. It is desirable, therefore, to operate the process at as low an acidity as possible but with the acidity still sufficiently high to prevent hydrolysis of plutonium (IV) to the non-extractable polymeric form. These considerations led to the oxidation of plutonium to the hexavalent state by the sodium dichromate prior to solvent extraction.

The redox process (Culler 1956), illustrated in Fig. 2.15, is a three-column cycle in which plutonium and uranium are separated from each other and, in addition, from the majority of the fission products. The feed solution from the dissolver is an acid-deficient uranyl nitrate solution containing sodium dichromate to oxidize the plutonium present. The plutonium and the uranium are both extracted using acidified hexone and the organic extract is scrubbed in the same column with acid-deficient aluminium nitrate. In a second column the plutonium is stripped from the organic hexone phase by reducing the plutonium to the inextractable trivalent form without affecting the uranium extractability. Uranium stripping is prevented by the addition of aluminium nitrate to the reducing solution and then scrubbing it with additional hexone solvent before it leaves the second column. The uranium is finally stripped from the organic phase in a third column by a dilute nitric acid strip.

The plutonium product from the first cycle containing aluminium nitrate from the reducing strip is oxidized using sodium dichromate to the hexavalent state and is then further purified by additional solvent extraction cycles. The plutonium is extracted with acidified hexone and scrubbed with acid-deficient aluminium nitrate in one column and stripped with dilute nitric acid in a second column. The scrub stream does not now require the presence of a reducing agent as in the first cycle partitioning column where separation from uranium takes place. The plutonium solution from the second cycle

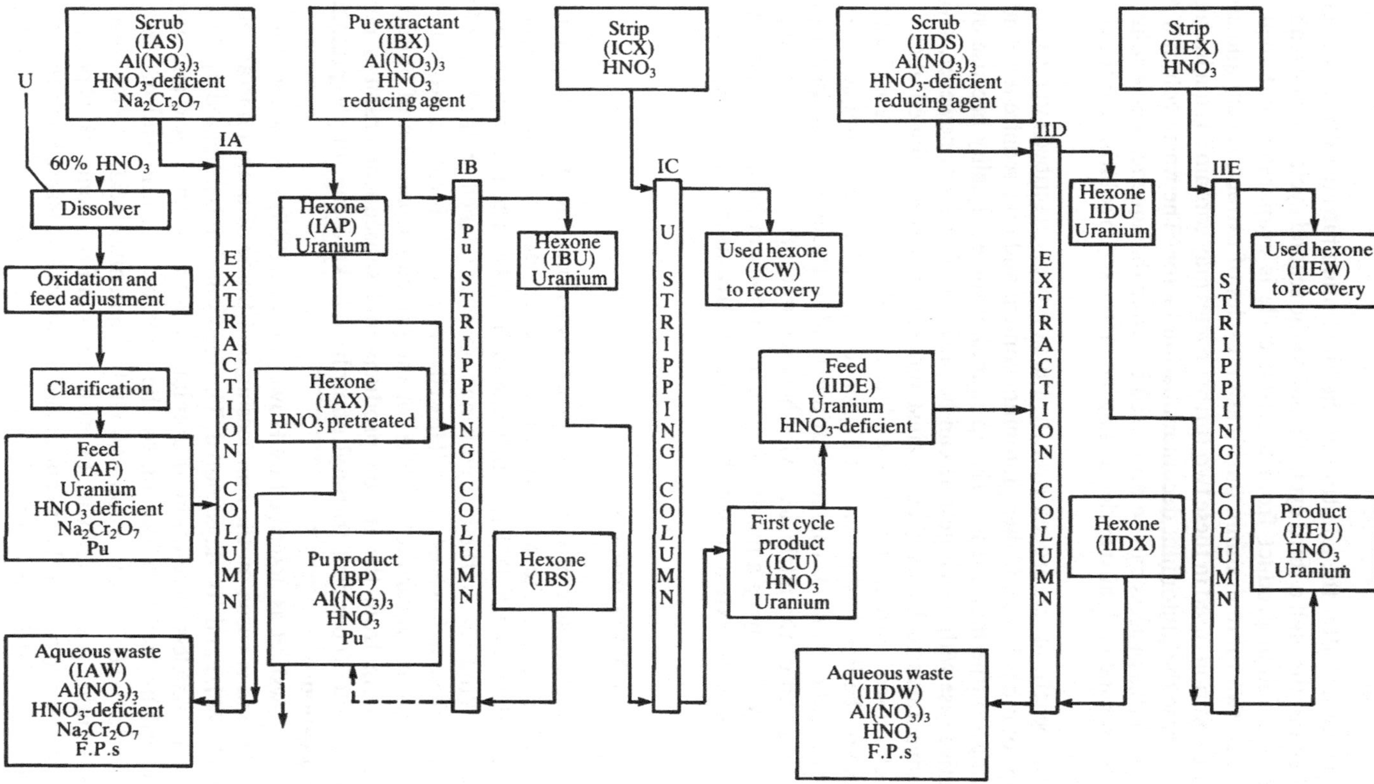

Fig. 2.15. Schematic flowsheet for first redox cycle and second uranium cycle.

strip column, after the addition of the aluminium nitrate salting-out agent, is then put through a third cycle similar to the second cycle described above.

The uranium product from the first cycle is evaporated and then put through two more uranium purification cycles of extract, scrub, and strip. Scrub solutions used in these two further cycles may contain a reducing agent to increase the plutonium decontamination factor. The waste volumes produced in this process can be decreased by using the aqueous waste solutions for the second and third cycles in making up the scrub solution for the earlier cycles.

The benefit of this process over the Purex process described later is that the strip solutions have a higher plutonium content and thus require less concentration. Furthermore, it is easier to maintain solvent quality because of the relative ease with which hexone can be purified. There are, however, several disadvantages to be weighed against the advantages mentioned above and which subsequently caused the abandonment of the process. These disadvantages are the low flashpoint and higher toxicity of hexone compared to the tri-butyl-phosphate/kerosene solutions used in Purex process. In addition, the reaction of the nitric acid with hexone prevents its use as a salting-out agent and employing a metal salt such as aluminium nitrate as a salting-out agent limits the extent to which any waste streams may be concentrated.

2.10.4 The Butex process

The basic chemistry of the butex process was first investigated by a group of workers under the direction of Spence (Spence 1957 and Nichols and Spence 1957) at Chalk River in Canada in 1947. The primary separation plant at Windscale in the United Kingdom was designed on the basis of this work. The actual process used (Howells *et al.* 1958) involved two solvents: undiluted dibutyl carbitol (butex) for the primary separation of plutonium and uranium from the fission products and subsequent uranium purification, and 20 per cent tri-butyl phosphate in kerosene for the plutonium purification.

The flowsheet as described by Howells *et al.* (1959) is shown in Fig. 2.16 to 2.19 and is based on the reprocesing of 1 tonne of uranium irradiated to 1000 M Wd/t and 100 days cooled. The feed solution is prepared by the continuous fumeless dissolution of uranium rods in 6 M nitric acid and oxygen followed by feed conditioning by the addition of nitric acid or water to maintain the feedstock at approximately 3 M in nitric acid. The highly active feed solution is extracted with butex in column PS1. Some 99.98 per cent of the plutonium and 99.9 per cent of the uranium is extracted into the solvent phase and 99.5 per cent of the fission product activity remains in the highly active raffinate. The fission product activity associated with the

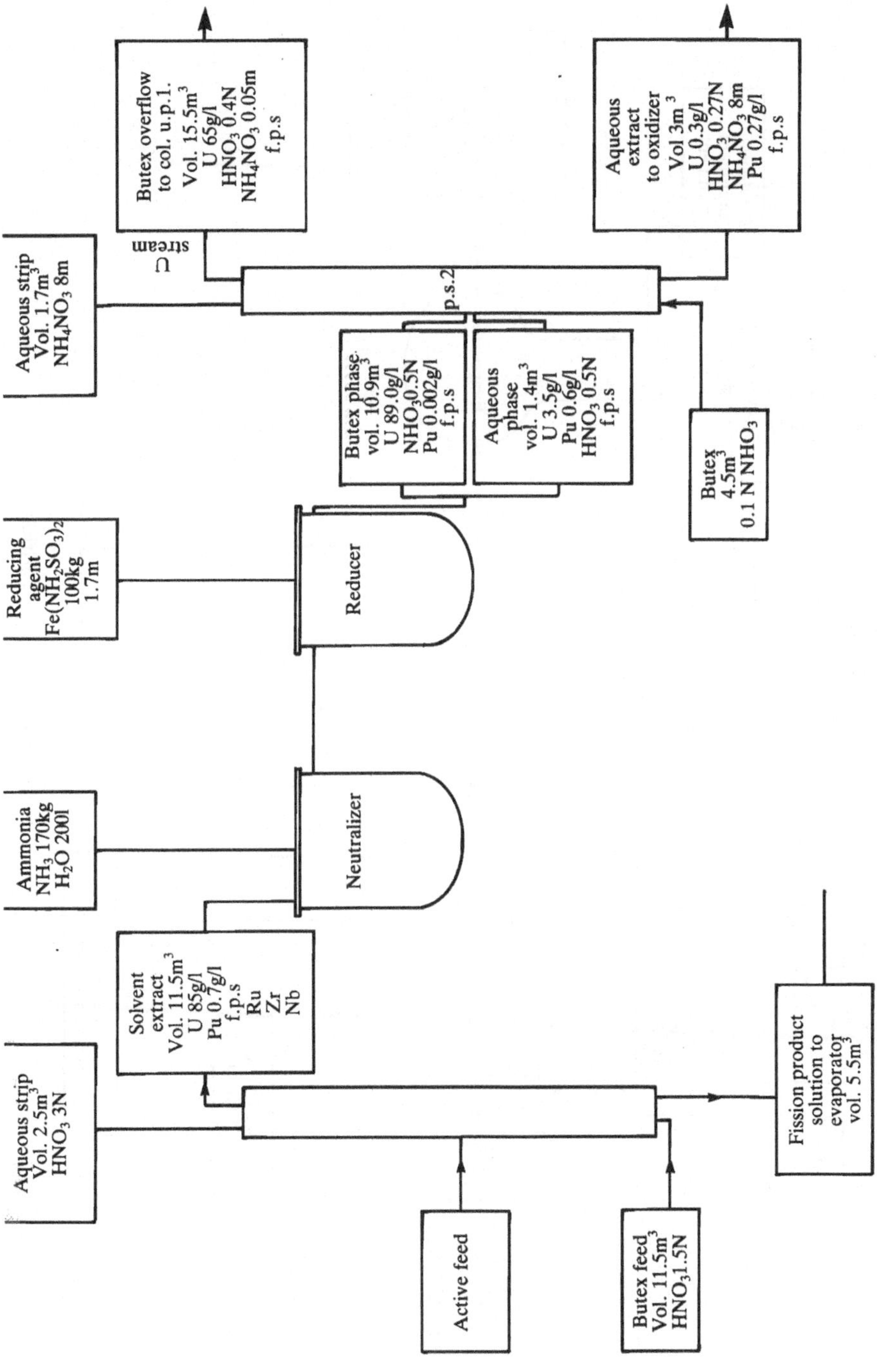

Fig. 2.16. Primary separation of uranium, plutonium and fission products.

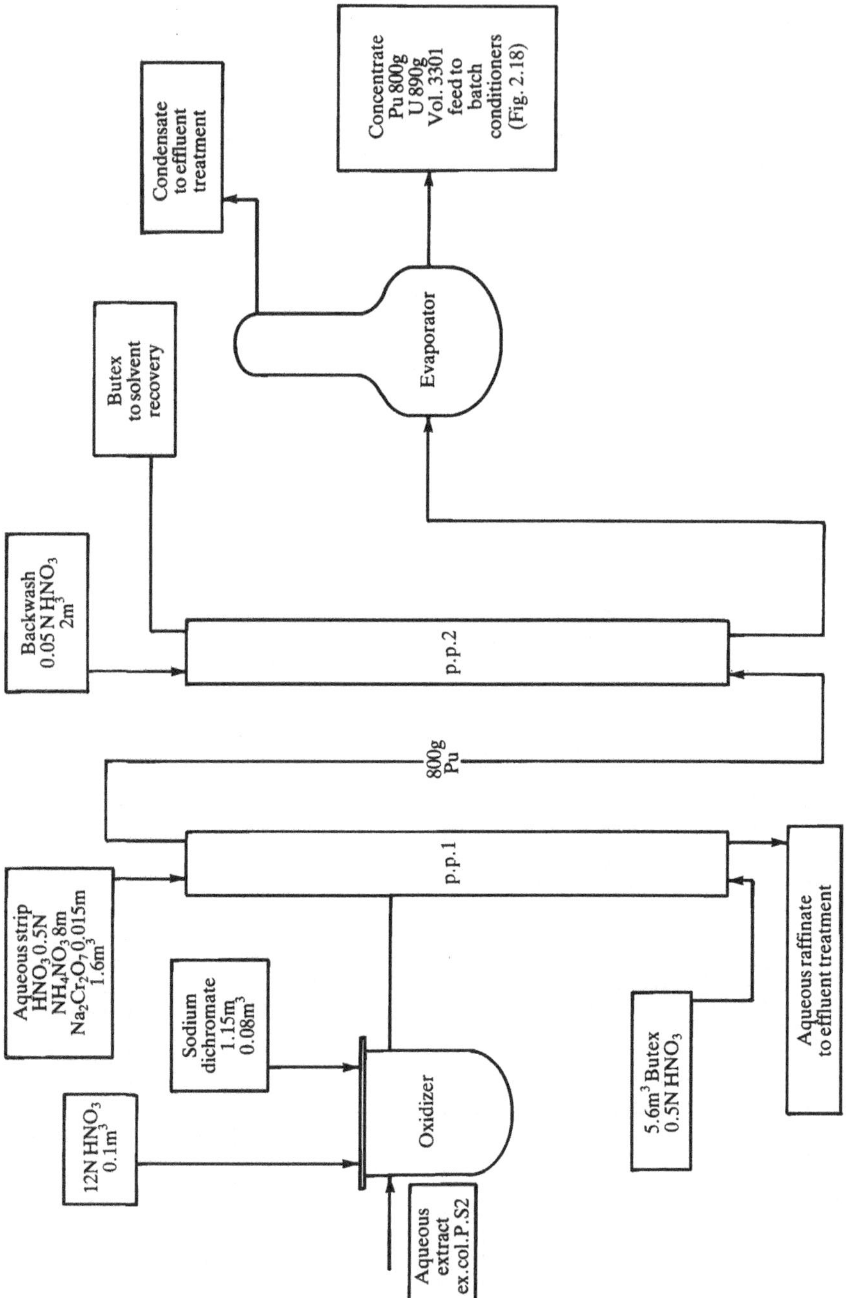

Fig. 2.17. First cycle plutonium purification.

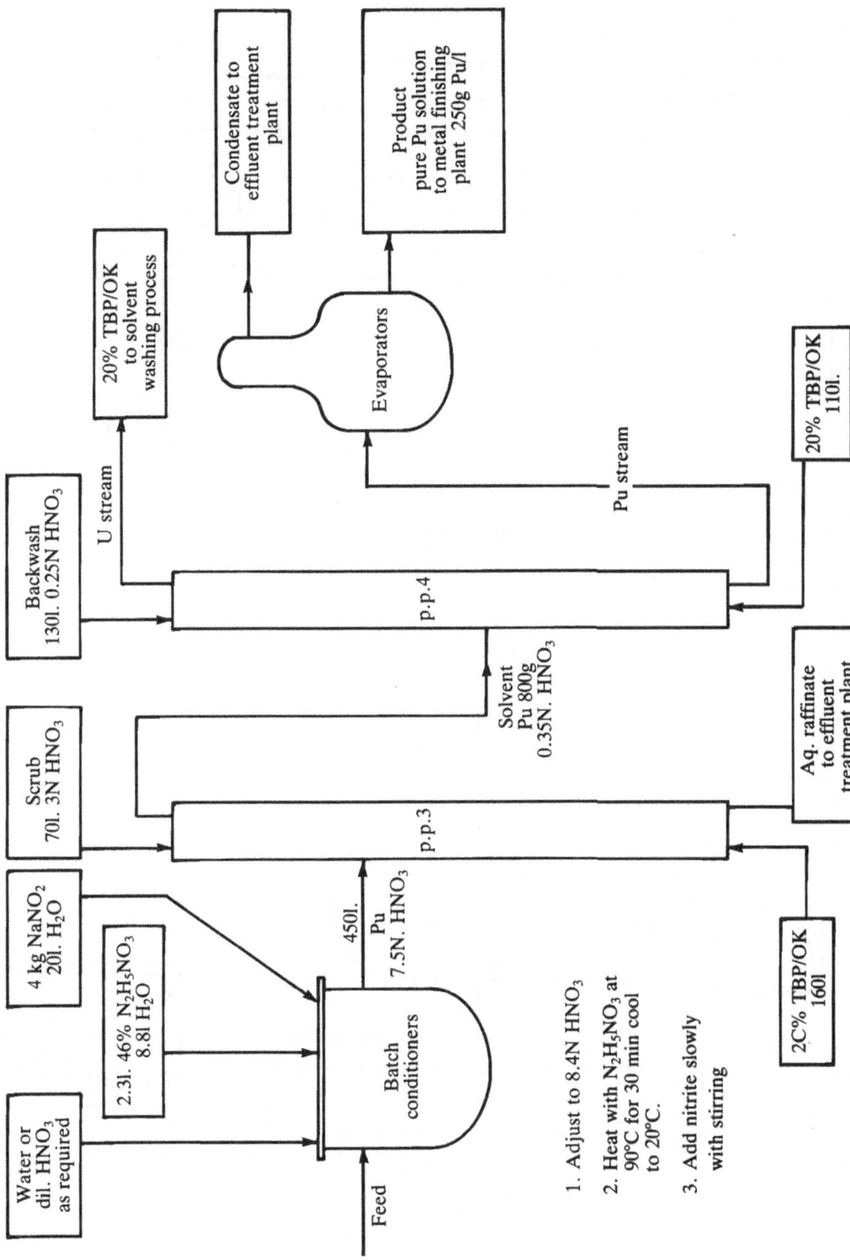

Fig. 2.18. Second cycle plutonium purification.

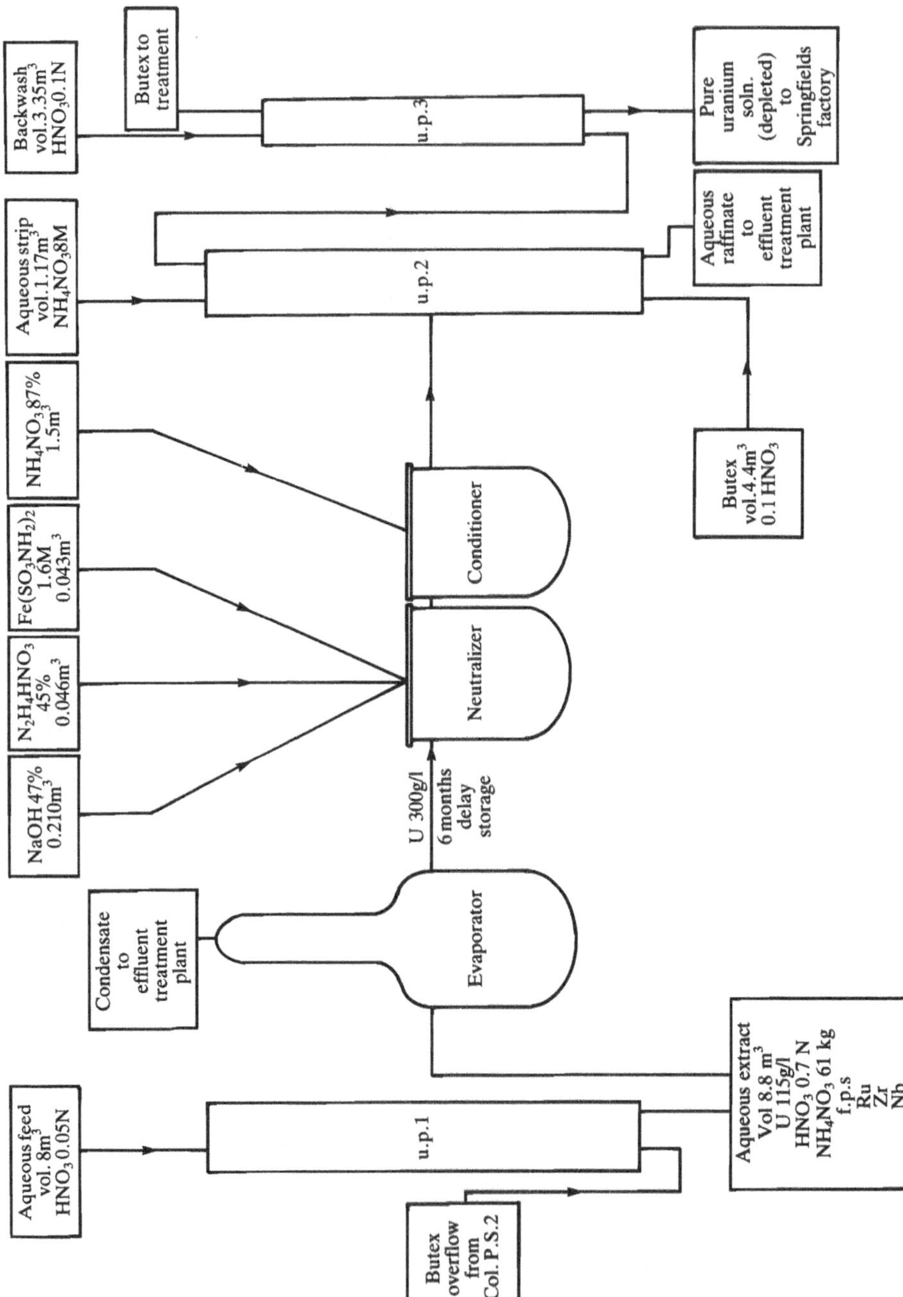

Fig. 2.19. Uranium pu

plutonium and uranium stream is mainly due to ruthenium together with small amounts of zirconium, cerium, and niobium.

The highly active raffinate from PS1 is stream stripped to remove any organic compounds and is then concentrated by evaporation under vacuum and finally stored underground in stainless steel tanks fitted with water cooling coils.

The acidity of the butex phase from PS1 is adjusted by additional ammonia; ferrous sulphamate is then added to reduced the plutonium to the trivalent state which permits it to be stripped into the aqueous phase. The organic (now only uranium containing) and the aqueous (plutonium containing) phases are separated in column PS2. The aqueous phase contains virtually all the plutonium and some uranium and, after treatment with sodium dichromate to oxidize the plutonium to the hexavalent state, the plutonium is extracted into butex in PP1 and stripped with dilute nitric acid in PP2 (Fig. 2.17). The volume reduction achieved in column PP2 depends on the acidity of the solvent from PP1: the lower the acidity the greater the volume reduction achievable; but this is constrained by the need to prevent the formation of colloidal plutonium. The aqueous plutonium product from PP2 is concentrated by evaporation to 2–3 g/l prior to feeding into the second plutonium purification cycle.

The second plutonium purification cycle is designed to purify the plutonium stream further of uranium and fission products. The plutonium is converted to the tetravalent state by treatment with 8.4 M nitric acid followed by the addition of hydrazine nitrate at 90°. Finally, to complete the oxidation, the solution is cooled to 20°C and sodium nitrite solution, which also destroys residual hydrazine, is added. In column PP3 the plutonium and uranium are extracted into 20 per cent tri-butyl phosphate-kerosene; the aqueous raffinate is sent to an effluent treatment plant; the organic stream after scrubbing with 3 M nitric acid is stripped with 0.25 M nitric acid in PP4 (Fig. 2.18) to remove the plutonium; the aqueous plutonium product stream is then concentrated to 250 g/l by evaporation; the solvent after washing is recycled.

The butex solvent containing the uranium from PS2 is stripped with dilute nitric acid in UP1 (Fig. 2.19). The uranium aqueous stream is then evaporated to reduce the volume handled in later stages of the process and also to provide a more concentrated uranium feed to assist in ruthenium decontamination in the subsequent extraction stages. Any trivalent plutonium present is oxidized to tetravalent plutonium and hexavalent plutonium during the evaporation process. The concentrated uranium solution is then stored for some six months to obtain additional ruthenium decontamination by decay before further reprocessing.

The aged solution acidity is adjusted by adding sodium hydroxide followed by further additions of hydrazine and ferrous sulphamate and then heated for 2 hours at 80°C. This treatment reduces the plutonium to the inextractable trivalent state and modifies the ruthenium species present to relatively inextractable forms. Ammonium nitrate is then added to ensure that the conditioned liquor is 6 M in nitrate: this salt concentration, together with that of the sodium nitrate present, ensures efficient uranium extraction. In column UP2 the uranium is extracted into butex, which is scrubbed with ammonium nitrate and fed into column UP3 while the aqueous raffinate, after treatment in the effluent plant, is discarded. In column UP3 the organic solution is stripped with dilute nitric acid to yield the uranium product solution, while the butex after treatment is recycled.

The butex process has the same advantage as the purex process in that it extracts plutonium and uranium from nitric acid solutions without adding a salting-out agent, thus making it possible to concentrate the highly active raffinates by evaporation. Butex is not, however, stable in the presence of concentrated nitric acid and the addition of ammonium nitrate, which is needed later in the process, is an additional disadvantage. Butex itself has the further disadvantages of high viscosity, high density, and a tendency to form crystalline uranium complexes. In addition, because two different solvents, butex and tri-butyl phosphate, are used similar fission product decontamination factors are achieved by employing tri-butyl phosphate alone. Consequently the butex process has been superseded by the purex process.

2.10.5 The Purex process

A number of processes for the separation of plutonium and uranium from fission products are based on the use of the extractant tri-butyl phosphate. With the Purex process, tri-butyl phosphate has provided a very efficient method for the separation of plutonium and uranium from the fission products. Tri-butyl phosphate is normally diluted with a hydrocarbon that reduces the density, viscosity, and extractive power to suitable levels: the proportion of tri-butyl phosphate in the mixture is usually in the range 5–30 per cent by volume.

The first large-scale reprocessing plants that used tri-butyl phosphate were those at Savannah River in 1954 and Hanford in 1956 (Joyce *et al.* 1960; Irish 1959) in the United States. In the United Kingdom tri-butyl phosphate was first used in 1956 for the recovery of plutonium and enriched uranium and then in 1960 for the uranium purification cycles (Clelland and Hughes 1963). In the second separation plant at Windscale, in the United Kingdom, which started operating in 1964, the extractant was 20 per cent tri-butyl phosphate in odourless kerosene (Marshall *et al.* 1965). The highly active cycle at the reprocessing plant at Marcoule in France originally used 40 per cent tri-butyl phosphate (Jonannaud 1965), but more recently they have used 30 per cent tri-butyl phosphate throughout the process (Miquel *et al.* 1974). In both instances the diluent was hydrogenated propylene tetramer. At the reprocessing plant at La Hague in France the solvent is again 30 per cent tri-butyl phosphate–hydrogenated propylene tetramer (Duboz 1965). At Dounreay in the United Kingdom, 6–30 per cent tri-butyl phosphate–odourless kerosene has been used in the reprocessing of fast reactor fuels since 1962 (Lillyman and Mills 1974).

A typical flowsheet for a three-cycle purex process (Irish and Reas 1957) is illustrated in Fig. 2.20.

The earlier three-cycle purex process has been modified to a two-cycle process by employing improved solvent extraction technology and solvent treatment techniques and by the addition of an ion exchange step for the preparation of plutonium.

In the Purex process, the chemistry of the plutonium valency adjustment is important to permit the quantitative recovery of plutonium and to maintain a high degree of separation of plutonium and uranium. To ensure that the plutonium is in the most extractable tetravalent form, nitrite can be added to the solution either as a salt, such as sodium nitrite, or as a gas NO_2, the reaction being rapid and completed in a few minutes at 50°C.

$$PuO_2^{2+} + NO_2^- + 2H^+ = Pu^{4+} + NO_3^- + H_2O$$

The classic reagent for the separation of plutonium from uranium was ferrous sulphamate, the ferrous ion reducing the tetravalent plutonium to the inextractable trivalent state with the sulphamate ion acting as a nitrite suppressor. If the sulphamate ion were not present the low concentrations of nitrite ion would initiate the auto-catalytic oxidation of the ferrous ion to the ferric ion, thus preventing plutonium reduction.

$$Pu^{4+} + Fe^{2+} + (NH_2SO_3^-) = Pu^{3+} + Fe^{3+} + (NH_2SO_3^-)$$

$$NH_2SO_3^- + NO_2^- = N_2 + SO_4^{2-} + H_2O$$

A several-fold excess of ferrous sulphamate is normally used to ensure that reduction of the plutonium does take place. This reagent after oxidation produces, as shown in the equation above, ferric ions and sulphate ions which limit the extent to which the raffinates may be concentrated and which also may enhance the corrosion rate of the waste storage tanks. For this reason, the use of ferrous sulphamate as a reductant for tetravalent plutonium and hexavalent plutonium to trivalent plutonium has been replaced by other reductants such as tetravalent uranium.

The nitrite ion is an ideal oxidizing agent for converting trivalent plutonium into the tetravalent extractable state for further reprocessing.

$$6Pu^{3+} + 2NO_2^- + 8H^+ = 6\ Pu^{4+} + N_2 + 4H_2O$$

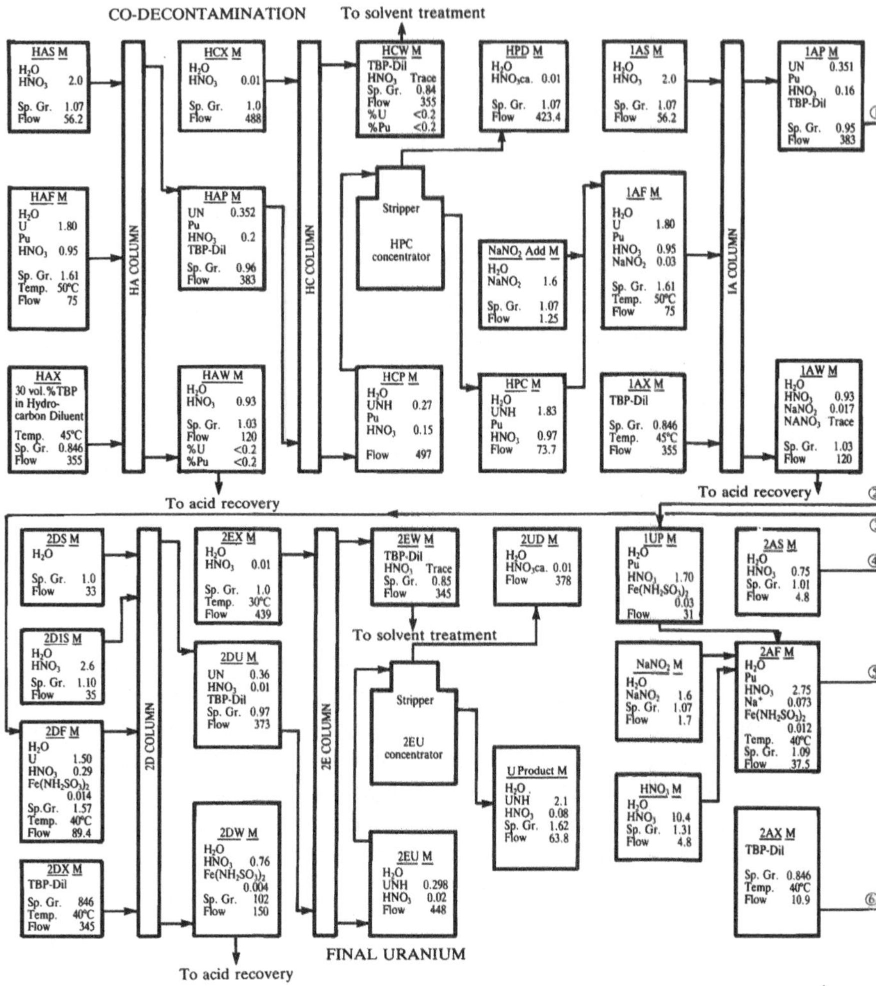

Fig. 2.20. Purex solvent extraction flow sheet.

In practice, however, the use of tetravalent uranium for plutonium and uranium separation is limited in the case of mixer-settlers because of autocatalytic oxidation reaction of trivalent plutonium with nitrous acid and the oxidation of tetravalent uranium in the organic phase. To ensure, therefore, a high separation efficiency a large excess of tetravalent uranium is required (some four- to six-fold stoichiometry) together with at least two tetravalent uranium feed points. This reductant has, however, been successfully used in plants using pulsed column contactors.

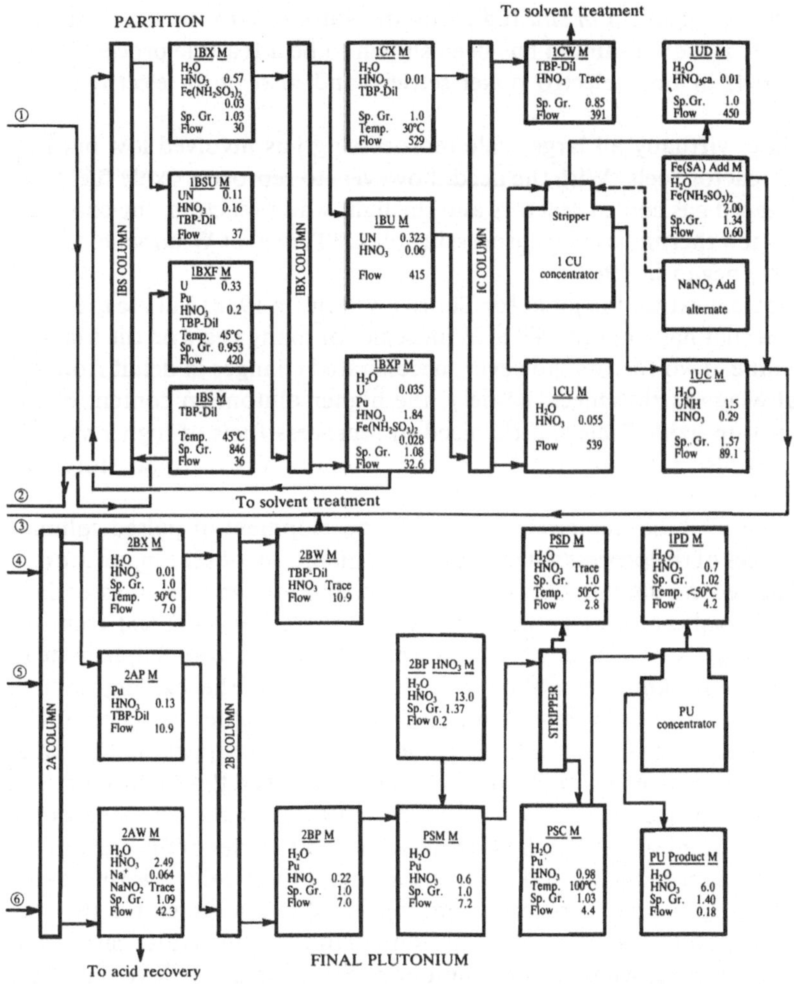

Hydroxylamine nitrate has been used successfully to separate plutonium from uranium (Baumgartner *et al.* 1977), the principal advantages of this reagent being the possibility of its complete decomposition to inert gases and the large reduction rate. The rate of reduction is, however, reduced considerably with the square of the trivalent plutonium concentration and the fourth power of the nitric acid concentration: these are both reaction products. In a commercial process, therefore, acceptable reduction rates could only be achieved using of a large excess of reagent and by employing elevated temperatures.

The electrochemical *in situ* reduction of plutonium to the trivalent state to separate it from uranium has been shown to be a feasible process (Galla *et al.* 1984) in both electro mixer-settlers and also in an electro pulsed column.

To date, virtually all large-scale reprocessing has involved low burn-up thermal reactor fuels. With the need, however, to reprocess oxide fuel from the advanced gas cooled reactors and the light water reactors, the construction of a new thermal oxide reprocessing plant (Thorp) at Windscale, United Kingdom, began in 1983.

The solvent extraction plant for Thorp (Wilkinson 1988) will use the purex flowsheet that has worked well at Windscale for many years on the Magnox programme. Two factors, however, necessitated changes in detail from the original Magnox flowsheet; namely, the higher plutonium content of the high burn-up oxide fuels and the need to reprocess fuels of higher specific activity whilst reducing the quantity of liquid waste to be stored or discharged.

The first of these changes has led to the development of pulsed columns for all stages of the process where significant quantities of plutonium are present. The second change has resulted in the development of an essentially salt-free flowsheet. This Thorp flowsheet is, therefore, different from the original Magnox flowsheet which uses the reductant ferrous sulphamate to effect the plutonium separation from uranium and the oxidant sodium nitrite to convert the trivalent plutonium to the extractable tetravalent form. The oxidation/reduction products of these added salts appear in the liquid raffinate waste streams and either reduce or prevent their concentration and subsequent treatment. In the Thorp flowsheet these two reagents are replaced by uranium nitrate/hydrazine nitrate and oxides of nitrogen as the main plutonium-reducing and oxidizing agents respectively. It is thus based on an early split purex process operating in pulsed columns for the first highly active cycle. The subsequent plutonium purification is also carried out in pulsed columns, whereas the uranium purification is employed in mixer settlers.

The result of operating this salt-free flowsheet will be to produce salt-free medium active raffinates which will be concentrated by evaporation and then combined with the highly active raffinate destined ultimately for vitrification. The present Thorp flowsheet will still use salt-bearing reagents for solvent washing, as does the Magnox fuel reprocessing plant, and these will be concentrated in a salt evaporator plant.

The low level liquid wastes, after treatment in facilities designed to minimize radioactivity releases to the environment, will be discharged.

Fast reactor fuels have a higher burn-up and also a much higher plutonium and fission product content than thermal reactor fuels. Small scale reprocessing plants in both the United Kingdom and France have demonstrated the feasibility of reprocessing these fuel types.

The prototype fast reactor reprocessing plant at Dounreay in the United Kingdom was designed to reprocess irradiated mixed (U, Pu) oxide fuel pins. In this plant the fuel is dissolved using a chop-leach process, and the clarified, conditioned liquor is then fed to the solvent extraction plant. The high plutonium content of the feed liquor (30–40 g/l) requires that criticality restrictions must be applied not only to the mixer settler extractors themselves but to the whole of the plant wherever fuel or liquor is likely to be present. The criticality control in all the sections of the plant (fuel-handling, dissolver, centrifuge, tanks, etc.) is achieved by geometrical means as is the case for the solvent extraction equipment.

The flowsheet used (Fig. 2.21) (Mills 1979) requires that in the first cycle in the strip stages all the plutonium, but only part of the uranium, is stripped from the organic phase for further treatment. The balance of the uranium is stripped from the organic phase after the partial strip and is not a product stream. Complete separation of the plutonium from the uranium is achieved in the second cycle: again the uranium is not a product stream. The plutonium is finally purified in the third cycle and the plutonium nitrate product stream concentrated by evaporation.

The high plutonium content of the prototype fast reactor fuel precludes the use of ferrous sulphamate reductant to achieve the plutonium separation from uranium in the Dounreay plant. The quantity of ferrous sulphamate required would be too great to make the process operable and for similar reasons uranous nitrate cannot be used as a reductant. The reductant hydroxylamine nitrate could be used but the kinetics of the reduction reaction between it and plutonium are too slow to achieve the throughput required in the small Dounreay plant.

In the prototype fast reactor fuel reprocessing plant sulphuric acid is used to carry out both the partial and complete separation of the plutonium from the uranium. The plutonium sulphate formed is not soluble in the organic phase. Some of the uranium present is also converted to uranyl sulphate which, again, is insoluble in the organic phase. The partial or complete separation of plutonium may be controlled by altering the sulphuric acid/nitric acid ratio as shown in the following equations:

$$Pu(NO_3)_4 \cdot 2TBP + 2H_2SO_4 \rightleftharpoons Pu(SO_4)_2 + 4HNO_3 + 2TBP$$

$$UO_2(NO_3)_2 \cdot 2TBP + H_2SO_4 \rightleftharpoons UO_2SO_4 + 2HNO_3 + 2TBP$$

The highly active raffinate from the plant is stored in underground stainless steel tanks fitted with cooling units. The medium and low active raffinates are collected and, after the precipitation of a uranyl floc by the addition of ammonia, which also coprecipitates most of the actinides and fission

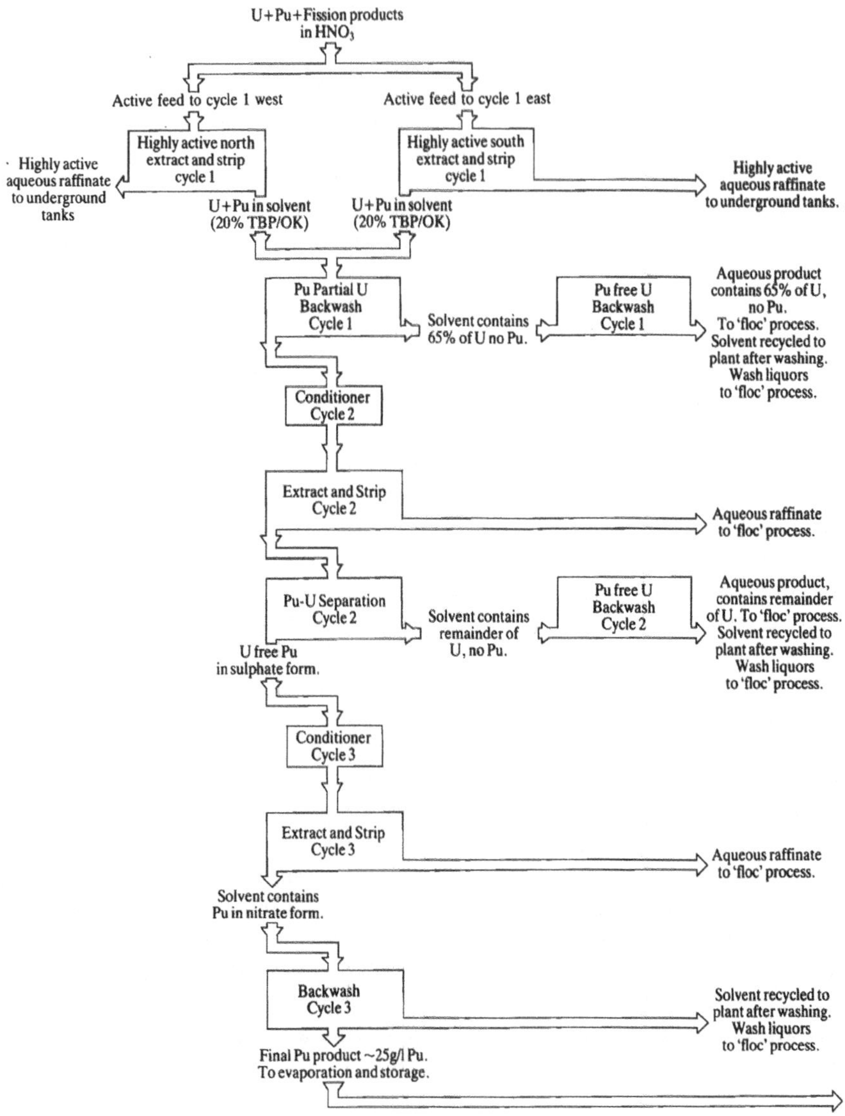

Fig. 2.21. Fast reactor fuel solvent extraction flowsheet.

products present, the supernatant liquid is discharged to sea after monitoring. The solid uranyl floc is retained on the site for future treatment.

2.10.6 Future processes

Amides with high uranium loading factors have been suggested (Moccia 1986) as substitutes for the classic purex extractant, tri-n-butyl phosphate because, although their radiation stability was similar to that of tri-n-butyl phosphate, the degradation products were more easily removable in the solvent recovery process. The amides show, in addition, better decontamination factors in the extraction process—particularly for ruthenium and the rare earths—than those obtained using tri-n-butyl phosphate. Furthermore, because of the very low distribution coefficients for plutonium and uranium at low acidities, costripping is possible without the necessity of reducing the plutonium to the trivalent state and with an aqueous-to-organic ratio of less than one, thus enabling a product stream concentration of 100–150 g/l heavy metal to be achieved. This latter possibility could be of major importance to a reprocessing plant because it would avoid using reductants, high acidity sections, and intercycle evaporators.

Thus, by the careful choice of extractant, Moccia (1986) has shown that if, for example, an N–N alkyl-substituted amide were used it would be possible to separate plutonium and uranium either partially or completely without the need of reductant or plutonium complexing agents, so long as the plutonium was conditioned to the tetravalent state prior to the solvent-extraction process.

2.11 Plant control

Process control in a nuclear fuel reprocessing plant is somewhat different from that in a conventional plant because, in the former, for safety reasons the operators are remote from the actual process. The need, therefore, for the operating staff to be constantly informed of the process operations and to evaluate plant performance is a crucial factor in the control and instrumentation provisions in nuclear fuel reprocessing plant design. Such provisions ensure that the operating staff can analyse the data supplied and control and take the appropriate action to either correct or modify aspects of the processes and equipment used.

Nuclear fuel reprocessing plants have traditionally relied upon conventional measurement and control solutions; for example, instrumentation backed by a comprehensive system of sampling and analysis, rather than adopting the advanced control strategies and techniques increasingly found in the oil industry and in fine chemical plants. Such conservatism has been engendered, amongst other reasons, by the particular difficulties encountered when attempting to obtain on-line measurements (which are a prerequisite for modern control systems) in a radiological environment. On-line measurements, whether invasive or non-invasive when available, have been costly to engineer within the shielding and containment structures found within a nuclear fuel reprocessing plant. Reliance has, therefore, been mainly placed on tolerant flowsheets and a considerable overdesign of the process equipment. The effect of this approach in the past on plant operating costs were not seen as sufficient justification for the use of more modern control strategies.

The more recent modern nuclear fuel reprocessing plants have had an increased priority attached to process quality control, to reductions in waste arisings, and in operational costs because of the requirements of the national nuclear energy programmes of the countries with such plant. To achieve these aims there will be a need for increased process information which will be met by the use of a new generation of largely non-invasive on-line instrumentation. Thus in a modern nuclear fuel reprocessing plant the design of control systems and the measurement of solutions is based on a consideration of the following items:

(1) the nature of the processes and the operating conditions required;

(2) the manning level, the distribution of operating positions, the skills of the operators, and the degree of supervision provided;

(3) the requirement for information records to aid analysis of past performance and to assist the planning of future operations;

(4) the degree to which control and measurement technology exists to meet the operational and environmental requirements.

In a modern nuclear fuel reprocessing plant the design objective is to provide sufficient control and measurement to ensure its safe, reliable, and efficient operation. 'Control' is used in its widest sense and covers operator interactions and managerial requirements as well as automation features. Control is discussed under four main headings:

(1) process considerations;

(2) instrumentation;

(3) nuclear materials accountancy and safeguards;

(4) radiological monitoring.

2.11.1 Process considerations

2.11.1.1 *Fuel receipt and storage*

Fuel receipt and storage operations for a nuclear reprocessing plant are normally batch organized and dictated by fuel deliveries from the reactor, the return of empty flasks, and the movement of sub-assemblies to the dismantling route. As a flask enters the building, preparatory operations and flask manipulations are carried out by the operators using locally positioned equipment such as cranes. When the flask is mated to the loading port in the shielding the flask is opened by remote control, as indeed is the door of the loading port, to permit the removal of the sub-assembly. All these operations are remotely operated and monitored by locally sited operators.

Once the sub-assembly has entered the shielded plant, the operators may remotely control the loading of the sub-assembly handling machine by using a combination of closed circuit television monitoring, visual observation through shielded lead glass windows, and automatic positioning equipment. The identification code of each sub-assembly, once received in cell, is then checked using a closed circuit television monitor; the sub-assembly is then subjected to non-destructive assay to determine its plutonium content. These two parameters are then recorded on a computer data logging system which, in addition, determines the appropriate storage position for each particular sub-assembly. The allocation of a storage space and the movement of each sub-assembly is managed from the main control room. The locally based operators control the movements of the sub-assembly handling machine with the computer system acting as a monitor which, in addition, acts as an alarm if the actual operations and movements are at variance with the instructions detailed from the main control room.

2.11.1.2 *Fuel dismantling and dissolution*

The dismantling of each individual sub-assembly is performed by a series of semi-automatic operations using machines which are able to carry out a series of pre-programmed operations according to the type of sub-assembly. The locally based operators initiate and perform each series of machine operations from the appropriate cell face console using visual observations through shielded lead glass windows.

The fuel pins, as they are removed from the sub-assembly, are counted, batched together, and cropped into suitable lengths for the dissolution process. A known weight of heavy metal from a number of cropped pins is introduced into the dissolver basket. The filled basket is then transferred to the dissolver using an in-cell crane or hoist, again under local visual control. Various interlocks and management control systems protect the dissolver from operator error such as the inadvertent overcharging of the dissolver.

The dissolver is fitted with an automatic heating and cooling cycle which allows the dissolver to be operated to a predetermined temperature profile which is capable of adjustment to suit different operating parameters. The control of the dissolver and the liquor transport systems is again controlled locally.

After the dissolution cycle has been completed the dissolver liquor containing the plutonium, uranium, and fission products in solution is transferred to a hold-up tank prior to the clarification stage. This stage is, again, locally controlled and once started the process is batch continuous in operation by utilizing constant volume feeders to the actual liquor clarifying centrifuges. The clarification stage equipment would be duplicated to enable throughput to be maintained in the event of a centrifuge coming off line.

After the dissolution cycle has been completed the fuel pin clad hulls are washed prior to their removal from the dissolver and, once removed, they are monitored by a non-destructive assay technique to determine the residual plutonium content for accountancy, criticality control purposes, and, further, as a measure of the efficiency of the actual dissolution process. The monitored hulls, provided the plutonium content is below a certain predetermined level, are then consigned as part of the solid waste.

2.11.1.3 *Chemical separation*

The chemical separation process is performed using an immiscible organic solvent in a diluent to extract and separate the plutonium and uranium from the fission products which remain in the aqueous phase. The actual process of liquid–liquid extraction may be carried out in contactors of various types such as mixer-settlers or pulsed columns or centrifugal contactors. This process of chemical separation is a continuous operation remotely controlled from the main control room. After start-up the process should ideally require little operator attention other than to maintain the predetermined operating conditions and to respond to process and equipment malfunctions. In a typical nuclear fuel reprocessing plant three cycles of extraction, scrub, and strip are normally required to produce plutonium of the required purity; these would be designed to operate concurrently and would normally be controlled by very accurate flowrates of the inactive chemical and active feeds.

The quality of the inactive chemical feedstocks is ensured by a comprehensive blending, dilution, and chemical analysis system. The actual quantity of any specific feedstock is accurately measured using, for example, constant volume feed equipment. Furthermore, the rate of delivery of each feedstock is capable of being automatically adjusted so as to respond to any changes in the active feed throughput.

In a modern plant the heavy metal concentration of the highly active clarified feed is monitored by such on-line analytical techniques as density meters, K-X-ray fluorescence analysers, or—perhaps in the longer term—fibre optic spectrophotometry. The feed rate of the highly active feed is again controlled by remotely operated constant volume feed equipment. Further measurements to ensure that the predicted flowsheet performance was being obtained throughout the liquid–liquid extraction part of the plant is carried out using equipment such as ultrasonics, neutron counting, gamma spectrometry, K-absorption-edge densitometry, and density meters. The data obtained would permit the plutonium and uranium contents of the products and raffinate streams and the recycled solvent streams to be continuously assessed.

For the liquid–liquid extraction process to proceed satisfactorily the hydraulic performance of the solvent extraction contactors, whatever their type, would be required to achieve predetermined flowsheet values. This necessitates control and monitoring of all the various streams—organic and aqueous—both entering and leaving the solvent extraction contactors. The flowrates of the various organic and aqueous streams to the contactors is controlled by airlifts, constant volume feed equipment, or fluidic pumps as appropriate.

The plutonium nitrate product, prior to being sent to the oxide conversion plant, is concentrated using an automatically controlled continuous evaporator which would also involve an on-line gamma absorption meter to ensure a plutonium nitrate product of a given concentration. To ensure the safe operation of this equipment and to permit, in the event of some equipment malfunction, a controlled shutdown, a number of interlocks are fitted which would activate in the event of high pressure in the evaporator.

2.11.1.4 *Product conversion process*

The plutonium nitrate and uranyl nitrate products from the separation process would not normally be retained in the liquid form but converted to the solid oxides prior to storage or transport to the fuel fabrication plant.

The conversion processes to manufacture plutonium oxide and uranium oxide from the respective nitrates are similar in their control requirements in that both processes commence with liquid feeds and produce oxide powder as the finished products.

In both conversion processes, which involve precipitation, filtration, drying, and calcination steps to convert the oxalate or carbonate to oxide, all the processes are controlled using conventional measurements such as pH, temperature, flowrate, time, moisture content, and hydrogen concentration.

The modern plutonium oxide production plant would contain a significant amount of remote control and automatic equipment up to and including the product store to reduce the radiation dose to the operators. The uranium oxide production plant, too, would involve the use of semi-automatic equipment so designed that the process is sequential in nature, again with the minimum of operator involvement.

The recycle and aqueous raffinate streams from both these oxide conversion processes would need to be monitored by on-line plutonium and uranium monitors as appropriate. The final handling, packaging, and storage operations would be carried out using dedicated machines. Instrumentation for weighing and, in the case of the plutonium oxide, a non-destructive assay technique for the determination of plutonium would be provided for material control and accountancy purposes.

2.11.1.5 *Liquid waste management*
[See also Weatherley Chap. 5]

Aqueous raffinate streams from the solvent extraction process are stored either for subsequent treatment in the case of the cycle 1 highly active raffinate; or, in the case of cycle 2 medium active and cycle 3 low active raffinates, collected, treated to remove trace residual actinides and fission products in a floc process plant or concentrated by evaporation and, again, stored for subsequent treatment. The supernatant liquor from a floc process or the condensate from an evaporation process is then monitored, sentenced, and disposed of. The highly active liquor storage tanks require comprehensive monitoring facilities to determine liquor levels and temperature and the corrosion rate of the storage tank material itself. In addition, monitoring devices need to be installed inside the double walls of the storage tanks themselves as well as in the tank sumps to alarm in the event of a liquor leak.

The medium and low active waste aqueous raffinate streams going either to a floc plant or to an evaporator would require semi-automatic control to ensure that the batch processing of the liquor through a series of sequential operations proceeds smoothly.

Once again interlocks would need to be provided on both the highly active and medium and low active waste aqueous raffinate streams to prevent the wrong routing of these streams in the event of a maloperation.

On-line gamma spectrometry could be used to provide information on the degree of decontamination achieved in both the floc and evaporation processes. The final sentencing and discharge would, however, be authorized from the main control room after a representative sample followed by chemical analysis had confirmed that the liquor had achieved its required decontamination level. As a further back-up, on-line detectors for fission products and actinides with the ability to activate shutdown valves could be fitted to the discharge lines which would arrest the flow of liquor in the event of a predetermined activity level being achieved.

2.11.1.6 *Solid waste management*

The solid waste management process would be sequential in nature and involve the use of dedicated machines that require, for their operation, the use of locally based operators using the shielded lead glass windows at specified positions.

The principal solid wastes are produced from the sub-assembly dismantling and fuel dissolution stages and comprise sub-assembly wrappers and fuel pin clad hulls from which the fuel has been acid leached.

The main measurements made in this area are for the classification and accurate categorization of the waste arisings to determine their correct waste treatment and storage route. Furthermore, the plutonium levels of the fuel pin clad hulls in particular need to be measured for the purpose of nuclear materials accountancy.

The measurement of the plutonium levels in the solid waste are performed using a non-destructive assay technique based on one of the following measurement techniques:

1. Neutron interrogation, which involves the use of a neutron source to produce neutron emissions from the plutonium. This technique may be used to measure the plutonium content of solid waste in the presence of very high radiation fields.
2. Passive neutron detection, which is based on the measurement of spontaneous fission neutrons emitted from Pu^{240}.
3. Passive gamma detection which is based on the measurement of gamma emission from plutonium.

The solid waste, after its plutonium content has been monitored, is then packed into drums using remotely controlled automatic equipment; the drums are then transported using a shielded waste transfer flask to a shielded storage depository.

2.11.1.7 *Other plant facilities*

The supporting plant areas vary widely in their individual control requirements; the ventilation system, for example, being a plant-wide system controlled from the main control room. The availability of the plant ventilation system and its effect on the whole plant operation in the event of its malfunction leads to a requirement for local fall-back control. Furthermore, the reliability of the interlock and protection systems for the extract system in particular must be high.

A number of other plant areas such as those providing the electrical, water, and steam supplies, and gas storage and industrial waste is in the form of unit-plant complete with their own local control consoles, the appropriate warning signals and alarms from these unit-plants being transmitted to the main control room.

The inactive chemical feed preparation would normally be carried out during the day and be supervised locally.

2.11.1.8 *Main control room*

The main control room is the principal control location for the nuclear reprocessing plant and, therefore, all the various control and information systems are focused at this point. The layout of the control room are such to allow clear unobtrusive operating conditions for the plant operators by the positioning of auxiliary functions and facilities into partitioned-off areas. For example, the main area within the control room caters for the following main plant areas:

(1) fuel receipt through to liquor clarification together with solid waste handling operations;

(2) solvent extraction together with liquid waste treatment;

(3) oxide conversion plants;

(4) service facilities; for example, chemical reagent preparation and ventilation.

These areas are provided with visual display units with keyboards for command input and closed circuit television monitors; they would have facilities for analysis, accountancy, and radiological monitoring data access and display. These data would provide the essential information to the operators for plant control, alarm handling, and the supervision of maintenance operations. In addition, each facility is provided with an independent set of controls and indicators for the sole purpose of ensuring safe shutdown operation.

Ideally the main control room and the local work area control consoles should be connected to each other and to the plant via a distributed control system. This technique would permit the real-time control of the equipment and the processes by allocating functionally and geographically the control and data acquisition equipment to certain plant areas.

In parallel to, but independent of, the plant control and data acquisition systems needed to run the plant, a safety system is installed which is operator initiated and would permit the safe and efficient shutdown of the plant.

2.11.2 Instrumentation

In a modern nuclear fuel reprocessing plant measurement techniques that are non-invasive with respect to the active plant processes are to be preferred because they reduce the amount of waste arisings. In addition, these techniques reduce the exposure of the operators to radiation as well as assisting in meeting the nuclear safeguards criteria. Where invasive techniques have to be adopted care should be taken to ensure that the current practices of containment and purge flow to contain the activity are carefully maintained.

Typical instrumental techniques:

1. *Ultrasonic detectors*. These are used as point level detectors and in the analogue mode for the continuous measurement of such parameters as liquid levels and heavy metal concentration. Using the ultrasonic technique for heavy metal concentration relies on the fact that the velocity of the pulse (related to the delay between transmit and receive) is proportional to the heavy metal concentration. The use of a stainless steel waveguide not only forms the 'glass' of the window but also seals the tube from liquor ingress.

2. *K-edge densitometry*. In this technique an X-ray beam is pulsed through a liquid sample containing plutonium and uranium in nitric acid solution. The spectrum obtained will display various attenuation phenomena, of particular interest being the abrupt changes in the absorption displayed by the photo-electric effect as the incident radiation ejects electrons from the next most tightly bound electronic shell. The abrupt change displayed by the K-edge is of most use for the measurement of plutonium and uranium in active plant solutions; because it occurs at the highest energy it may, therefore, be detected through the walls of a stainless steel vessel. The concentration of heavy metal contained in the sample volume is computed from the ratio of the counts obtained at the onset of the edge and at the finish of the edge and is unique for each individual heavy metal.

3. *Fibre optic spectrophotometry*. The flexibility and specificity of this technique makes it very valuable for actinide valency and concentration measurements in both aqueous and solvent process streams. Radiation damage to the fibre optics would, however, need to be taken into account in sensor designs for highly active plant applications. The particular range and accuracy of the heavy metal measurements depend on the sample cell geometry through which light at the visible/ultraviolet end of the spectrum is passed via the fibre optic cable. The wavelengths absorbed are characteristic of the elements present in the solution.

4. *Neutron counting*. This technique is suitable for the estimation of plutonium concentration in both liquid and solid samples. Neutrons are

generated using a pulsed neutron interrogation tube and the spontaneous fissions from the plutonium present are measured using boron trifluoride or helium-3 filled detector tubes. The measured count rate may then be computed after taking into account the geometry of the system to yield the plutonium inventory.

5. *Gamma spectrometry*. This technique is based on the detection of the spontaneous gamma ray emissions for a sample of gamma-emitting elements. Each element may be estimated by determining the count rate at specific points in the energy spectrum. This technique, therefore, permits the detection and measurement of all the gamma-emitting isotopes present.

In previous reprocessing plant designs, plant control has invariably depended upon the laboratory analysis of samples taken from the plant at frequent intervals, necessitating the transfer of many samples from the plant to the analytical facilities. These exceedingly expensive and labour-intrusive operations eventually led to the development of automatic sampling and transfer systems.

The provision of both on-plant analysis techniques and on-line measurements minimizes the necessity for routine laboratory chemical analysis for the operational control of the reprocessing plant. These techniques would provide significant benefits in plant operation by, for example, the provision of rapid analytical data, the reduction in material transfers from the plant to the laboratories with the consistent saving in time for chemical analysis results, the reduction in waste generation, and a substantial saving in analytical chemical operating cost because of the need for less labour and facilities.

2.11.3 Nuclear materials accountancy and safeguards

Fissile nuclear materials such as plutonium have a high commercial and strategic value. Furthermore, they are regarded as sensitive both in the security sense and also for health and safety reasons. The Treaty on the Non-Proliferation of Nuclear Weapons (United Nations 1969) imposed obligations upon the signatories to place all non-military nuclear materials under the International Atomic Energy Agency safeguards procedures. The safeguards imposed by the International Atomic Energy Agency have as their objective the prevention of nuclear materials being diverted from civilian nuclear energy usage to military usage without declaration. The technique of nuclear materials control, Nuclear Materials Accountancy, involves the maintenance and audit of written or computer recorded records for all the types and quantities of nuclear material at all stages of the process and in storage.

In the United Kingdom, for example, statutory regulations are laid down in the form of legal safeguards documentation and these provide the guidelines for the negotiation of the detailed procedures that are implemented at the various nuclear establishments. These negotiations are held between the two safeguards Inspectorates, that is Euratom and the International Atomic Energy Agency, the Department of Energy, and the Establishment Operators. In all these establishments subject to nuclear materials accountancy and safeguard, the system is designed to permit the International Safeguards Inspectorate to verify the location and quantity of nuclear materials in the plant.

2.11.3.1 Accountancy techniques

The basic accountancy principles are the same for all plants and are designed to provide the information needed for plant management purposes as well as for international safeguards. Mass balances are prepared at monthly intervals to permit a comparison of the physical inventory of nuclear material with the inventory based on the recorded flows of nuclear material to and from the plant. To permit the structuring of the nuclear material mass balance in an orderly manner, it is often beneficial to divide the nuclear plant into several separate areas known as mass balance areas. The boundaries of a specific mass balance area will normally be determined by the physical and process boundaries within the plant, that is, where the nuclear material is subjected to changes in physical or chemical form (the dissolution process) or where changes in process treatment occur (the oxide conversion process). The choice of these mass balance area boundaries will obviously be determined to some extent by the location and relative ease at which sensible measurements on the flow of nuclear material can be made. Thus one plant may contain more than one mass balance area but, on the other hand, a whole plant or several plants may comprise a single mass balance area.

The mass balance of nuclear material is given by:

$$MUF = PB + R - S - PI$$

where

MUF = nuclear material unaccounted for,

PB = physical balance at the beginning of the balance period,

PI = physical inventory at the end of the balance period,

R = quantity of nuclear material received into the mass balance area.

S = quantity of nuclear material shipped out of the mass balance area.

2.11.3.2 *Safeguards techniques*

The basic requirement of international safeguards is for the Inspectorates to verify independently the plant operator's declared accountancy data. The operator is obliged to report to the Inspectorates on a monthly basis all nuclear material movements into and out of each mass balance area as well as any revised measurements. These data then permit the Inspectorates to maintain a book inventory which they can compare with the operator's nuclear materials working records and also with the physical inventory in order to satisfy themselves that the records and the reporting procedures are being adhered to and are functioning correctly, to evaluate the amount of nuclear material unaccounted for (MUF), to verify the adequacy of the operator's nuclear material control, and, finally, to draw conclusions regarding safeguards relating to any diversion of nuclear materials.

The Inspectorates, in performing their task, may use containment and surveillance techniques such as security seals and cameras in order to ease the demands on inspectors and to enable their inspections to be more effective as well as being less intrusive into the actual plant operations. There is an increasing interest in Near Real-Time Materials Accountancy using in-process measurements to provide physical inventories on a more frequent timescale for bulk handling facilities in situations where the verification of the detection criteria by existing means is difficult.

2.11.3.3 *Nuclear materials accountancy and safeguards provision*

To enable nuclear material balances to be achieved, measurement points need to be installed at specific places in the reprocessing plant. Normally the accountancy tank, into which the dissolver feed is taken, would be the first point at which the shipper's declaration would be measured. In a modern plant the formal accountancy reports are generated by means of a computerized system, with nuclear materials data input terminals allocated to suitable areas within the plant. The software and routines on the nuclear materials accountancy computer should be of a format suitable to both the operators and the Inspectorates.

2.11.4 Radiological monitoring

A comprehensive system of radiation monitors and air samplers are installed in a fuel reprocessing plant to provide assurance that the containment and ventilation systems are maintaining satisfactory operating conditions in areas occupied, or likely to be occupied, by plant personnel. Individual monitors for alpha, beta, and gamma radiation are located at strategic positions throughout the reprocessing complex. The detection of activity levels above the normal background level would cause the monitors to operate local audible and visual alarms. As a design principle these monitors are not used for any purpose other than for radiological or environmental monitoring, that is, they are not used for such functions as process control or to the activation of interlock devices. The individual monitors would be provided with self-checking facilities to indicate an instrument failure. Each type of monitor have, in addition, contact and activity rate outputs that are connected into a display system by means of plug and sockets. The use of these techniques means that individual monitors may be replaced or moved to other areas quickly and without disruption to the systems. The data obtained by these individual monitors within an area could be integrated by local computers which in turn are connected via a data link to the main control room. Thus, using this computer display system the plant operators and the health physics staff are able to assess and analyse the radiation conditions at specific points in the reprocessing complex and so permit early warning of potential alarm situations.

In addition to the radiation monitors, air samplers are provided in areas where there is a potential hazard from airborne contamination. These samplers rely on air being drawn through filter papers which are periodically collected and analysed in the laboratory by health physics staff.

2.11.4.1 *Location of monitors*

Monitors are provided in all areas of the fuel reprocessing complex where there may be the potential for contamination and where a breach in the containment could cause a hazard to personnel in that area. The use of these monitors, therefore, protect the operator at his or her place of work by indicating the integrated dose measured by the monitoring instrument. Then location would normally be arrived at by a consideration of the operational and health physics implications and the design of the plant ventilation system arrangements.

The choice of radiation monitors are determined by the potential hazard envisaged. In areas where, for example, plutonium would be handled the choice is easy and an alpha-in-air monitor provided. In other areas where irradiated fuel is handled then both alpha-in-air and beta-in-air monitors are installed. Gamma radiation monitors are provided in areas where there is a potential risk of shielding failure; for example, where there are manipulators or shield doors, or in areas which contain unshielded process lines which have the potential to contain active liquors.

In areas of the reprocessing complex where there are no direct sources of contamination, air sampling using passive samplers rather than air monitoring would probably be used.

2.11.4.2 *Personnel protection*

Personnel contamination control is carried out by using a combination of plant and personnel monitoring techniques. Personnel monitoring equipment is provided in the changerooms and would be so arranged that all personnel leaving the facility must pass through the equipment and be monitored.

2.11.4.3 *Discharge monitoring*

Another independent system of radiation detectors is installed in the reprocessing plant to provide duct and stack monitoring information. The system comprises both on-line monitors for the detection of alpha and beta particulates and iodine, and also proportional samplers. The data obtained from this equipment enable the plant operators and the heath physics staff to ensure compliance with the site operational procedures and with the site discharge limits determined by the regulatory agencies.

Proportional samplers, alpha monitors and on-line gamma spectrometry are used to monitor any liquid effluent discharges from the plant. Laboratory analysis of the samples prior to discharge are also performed and written records maintained in order to demonstrate compliance with the discharge levels set by the regulatory agencies.

Solid wastes would normally be stored either on site or in a depository after first having been monitored and classified.

2.11.4.4 *Criticality*

In all the areas within a fuel reprocessing plant which handles and stores fissile materials extensive measures should be taken to prevent the possibility of a criticality incident. These measures take the form of plant design based on the use of safe geometry backed up by managerial control on the movements of fissile material. Criticality detection systems are designed for the safety of personnel and to ensure their rapid evacuation in the event of a criticality incident.

2.11.4.5 *Criticality detection system*

The selection of a criticality detection system for a reprocessing plant depends upon the radiation environment envisaged and the physical constraints provided by the plant layout and shielding. Criticality detection systems may be based upon the measurement of gamma dose or upon neutron dose or rate of change of dose. In the United Kingdom, detection systems have been largely based upon gamma ray detection.

All the detectors used in the plant are of the type tested using radiation pulses from a pulsed reactor. In order to meet reliability requirements, three detector systems circuits are provided on each alarm system. These would be arranged to operate on a two-from-three majority voting basis to provide an acceptably low false alarm rate.

References

ADAR, M., HILDEBRANDT, R. A. and NAPLAN, L. (1956). *J. Inorg. Nucl. Chem.* **2**, 153.

AIKEN, A. M., BRUCE, T. and MOSS, M. (1951). *CEI*–18.

ALCOCK, K., BEST, G. F., HESFORD, E. and MACKAY, H. A. C. (1958). *J. Inorg. Nucl. Chem.* **6**, 328.

ALI, S. A. and EBERLE, S. H. (1969). *Radiochemica Acta* **11**, 149–53.

ALY, H. F., EL-DESSOUKY, M. M., KHALIFA, S. M., NAVRATIL, J. D., SHEHATA, F. A. (1985). *Solvent Extraction and Ion Exchange*, Vol. 3, pp. 867–77.

AWWAL, M. A. and CARSWELL, D. S. (1966). *Chem. Rev.* **66**, 279–95.

BAC, R. (1968). Kjeller Report KR–128.

BAC, R. (1968). *Radiochemica Acta* **9**(2–3), 130–9.

BALDWIN, W. H., HIGGINS, C. E. and RUTH, J. M. (1951). *ORNL–870.*

BALDWIN, W. H., HIGGINS, C. E. and RUTH, J. M. (1952). *USAEC Report ORNL–1338.*

BATHELLIER, A., CHESNE, A. and KOEHLY, G. (1963). *Nucl. Sci. Eng.* **17**, 557.

BAUMGARTNER, F., BECKER, R. and STIEGLITZ, L. (1976). *Report KFK 2304.*

BAUMGARTNER, F. and FINSTERWALDER, L. (1970). *J. Phys. Chem.* **74**, 108.

BAUMGARTNER, F., GOLDACKER, H., KOCH, G., OCHSENFELD, W. and SCHMIEDER, M. (1977). *Report KFK–2557.*

BAUTZ, H., BECKER, R., STIEGLITZ, L. and WUNSCHEL, A. (1978). *Report KFK 2613.*

BECKER, R. and STIEGLITZ, L. (1981). *Proceedings Conference on Reprocessing of Breeder Fuel, Obertraum, Austria*, p. 333.

BERSTEIN, G. J. (1973). *ANL–7969.*

BLAKE, C. A., DAVIS, W. and SCHMITT, J. M. (1963). *Nucl. Sci. Eng.* **17**, 626–37.

BLAKE, C. A. and SCHMIDT, I. M. (1965). *Solvent Extraction of Metals*, p. 161. Macmillan, London.

BLANCO, R. E., BRESEE, J. C. and BRUCE, F. R. (1959). *USAEC Report CF*–58–*11*–91.

BRADLEY, B. G. and STEVENSON, R. L. (1951). *HW–19170.*

BREESE, J. C. and SIMMS, R. (1959). *ORNL CF 59–10–136.*

BRESCHET, C. and MIQUEL, P. (1971). *ISEC71, The Hague*, Vol. 1, pp. 565–76.

BROWN, B. A. and THORNTON, J. D. (1966). UK Patent 1205562 (filed 1966).

BUCK, C., HOWELLS, G. R., PARRY, T. A., WARNER, B. F. and WILLIAMS, J. A. (1959). *Proceedings of the Second International Conference on the Peaceful Uses of Atomic Energy, Geneva*, Vol. 17, p. 25.

BURGER, L. L. (1955). *Report HW–40910.*

BURGER, L. L. (1958). *J. Phys. Chem.* **62**, 590.

BURGER, L. L. (1958). *Progress in Nuclear Energy, Series III* (Process Chemistry), Vol. 2, pp. 307–19.

BURNS, W. A., GROOT, C. and SLANSKY, C. M., (1949). USAEC Report HW-14728.

CAMPBELL, W. M. (1956). *Nucleonics* **14**(9), 92.

CASARCI, M., GASPARINI, G. M., GROSSI, G., MOCCIA, A. and PAGLIAI, V., (1986). *ISEC 86, Munich*.

CENTENO, J., DETILLEUX, E., ESCHRICH, H., VAN GEEL, J., JOSEPH, C. J. and SCHNEIDER, E., (1972). *Proceedings of the Fourth International Conference on the Peaceful Uses of Atomic Energy*, Vol. 8, pp. 349–66. United Nations, Geneva.

CENTENO, J., DETILLEUX, E., VAN GEEL, J., GUSTAFSSEN, B., HEINZ, W. and JOSEPH, C. (1971). *ISEC 71, The Hague*, Vol. I, pp. 577–92.

CHIARIZIA, R., DANESI, P. R., MEIDER-GORICAN, H. and SCIBONA, G. (1975). *J. Inorg. Nucl. Chem.* **37**, 1479.

CHIARIZIA, R. and HORWITZ, E. P. (1986). *Solv. Extr. Ion Exch.* **4**(4), 677–723.

CHMUTOVA, M. K., PALEI, P. N. and ZOLOTOV, U. A. (1968). *Zhurnal Analiticheskoi Khimi* **23**, 1476–80.

CLAY, P. G. and WITORT, M. (1974). *Radiochem. Radioanal. Lett.* **19**(2), 101–7.

CLELLAND, D. W. and HUGHES, T. G. (1963). OECD European Nuclear Energy Agency, *Brussels Symposium*, pp. 189–204.

CLEVELAND, J. M., (1979). American Nuclear Society Publication.

COOPER, V. R. and WALLING, M. T., Jr. (1959). *Proc. Second Int. Conf. Peaceful Uses of Atomic Energy*, Vol. 17, pp. 291–323. United Nations, Geneva.

CULLER, F. L. (1956). *Proc. First Int. Conf. Peaceful Uses of Atomic Energy*, Vol. 9, pp. 464–83. United Nations, Geneva.

DAVIDSON, J. K., HAAS, W. O. and SCHAFER, A. C. Jr (1957). *KAPL-1809*.

DAVIS, W. and KIBBEY, A. H. (1970). *ORNL-TM-2624*.

VAN DIJCK, W. J. D. (1935). US Patent 2011186.

DUBOZ, M. (1965). *Energy Nucleaire* **7**, 228.

DUKES, E. K. (1957). *DP-250*.

EGOROV, G. F., ILOZHEV, A. P., NIKIFOROV, A. S., SHEVCHENKO, V. B., SHMIDT, V. S. and SMELOV, V. S. (1979). *Atomnaya Energia* **47**(2), 75–79.

ELLERHORST, R. H., KISPERT, R. C., KLOPFENSTEIN, R. K., KREKELER, J. H. and NELLI, J. R. (1958). *NCLO-715*.

FAUGERAS, P. and TALMONT, X. (1967). *CEA-CONF-1265*.

FOREMAN, J. K., MARTIN, B. and OCKENDEN, D. W., (1961). *J. Inorg. Nucl. Chem.* **21**, 96–107.

FRENSDORFF, H. K. (1971). *J. Am. Chem. Soc.*, **93**, 4684.

FRYXELL, R. and KLEINSCHMIDT, R. (1945). *USAEC Report MDDC-461*.

GALLA, U., GOLDACKER, H., HEILGEIST, M., KLUTH M. and PETRICH, G. (1984). *Extraction 84, European Federation of Chemical Engineers, Publication Series No 43*, pp. 267–78.

GEIER, R. G. (1957). *HW-49542 A*.

GIEFER, D. L. and MERKELIN, J. F. (1975). *Nucl. Instr. Methods* **128**, 609–10.

GOLDSCHMIDT, B., PREVOT, I. and REGNAUT, P. (1956). *Proc. First Int. Conf. Peaceful Uses of Atomic Energy*, United Nations, Geneva, Vol. 9, p. 492.

GUILLAUME, B., HOEL, P., MORISSEAU, J. C. and MUSIKAS, C. (1987). *Extraction 87*, EFCE Publication Series 347, pp. 51–5.

GUREEV, E. S., KOSYAKOV, V. N. and YAKOVLEV, G. N. (1964). *Soviet Radiochemistry* **6**, 639.

HAHN, H. T. (1957). *J. Am. Chem. Soc.* **79**, 4625.
HEALY, T. V. and MCKAY, H. A. C. (1956). *Recuil.* **75**, 730.
HEALY, T. V. and PILBEAM, A. (1974). *ISEC 74*, Lyon, France, pp. 459–68.
HESFORD, E., JACKSON, E. E. and MCKAY, H. A. C. (1959). *J. Inorg. Nucl. Chem.* **9**, 279.
HESFORD, E. and MCKAY, H. A. C. (1958). *Trans. Faraday Soc.* **54**, 573.
HORWITZ, E. P. (1986). *ISEC 86*, Munich, West Germany.
HOWELLS, G. R., HUGHES, T. G., MACKEY, D. R. and SADDINGTON, K. (1959). *Proc. Second Int. Conf. Peaceful uses of Atomic Energy*, United Nations, Geneva, Vol. 107, pp. 3–24.
HOWELLS, G. R., HUGHES, T. G. and SADDINGTON, K. (1963). *Progress in Nuclear Energy* Series III (Process Chemistry), Vol. 3, pp. 151–87.
HUGGARD, A. J. and WARNER, B. F. (1963). *Nucl. Sci. Eng.* **17**, 638–50.
HUPPERT, K. L., ISSEL, W. and KNOCH, W. (1974). *ISEC 74, Lyon*, pp. 2063–74.
IRISH, E. A. (1959). *United States Atomic Energy Commission Report*, Vol. HW-60116.
IRISH, E. R. and REAS, W. H. (1957). *TID-7534*, pp. 83–101.
JONANNAUD, C. (1965). *Proc. Third Int. Conf. Peaceful Uses of Atomic Energy*, Vol. 10, pp. 215–21. United Nations, Geneva.
JOYCE, A. W., PERRY, L. C. and SHELDON, E. B. (1960). *Chemical Engineering Progress Symposium Series*, pp. 56, 28, 21.
KEDER, W. E., RYAN, J. L. and WILSON, A. S. (1961). *J. Inorg. Nucl. Chem.* **2**, 131–39.
KEDER, W. E., SHEPPARD, J. C. and WILSON, A. S. (1960). *J. Inorg. Nucl. Chem.* **12**, 327–35.
KENNEDY, J. and PERKINS, R. (1964). *J. Inorg. Nucl. Chem.* **26**, 1601.
KERTES, A. S. (1965). *Solvent Extraction Chemistry of Metals*, p. 377. Macmillan, London.
KERTES, A. S., MARCUS, Y. and YANIR, E. (1974). IUPAC additional publication to *Pure and Applied Chemistry*. Butterworths, London.
KOCH, G. (1965). *Kerntechnic* **7**(8–9), 394–99.
KOLARIK, Z. (1975). *Gmelins Handbuch der anorganischen Chemie*, Transurane D.
KUCA, L. (1962). *Collection of Czechoslovak Chemical Communications* **27**, 2372.
LANE, E. S. (1963). *Nucl. Sci. Eng.* **17**, 620–5.
LAWROSKI, S. and LEVENSON, M. (1957). *Symposium on the Reprocessing of Irradiated Fuels*, Brussels.
LILLYMAN, E. and MILLS, A. L. (1974). *ISEC74, Lyon*, Vol. 2, pp. 1499–517.
LOGSDAIL, D. H. and THORNTON, J. D. (1959). *J. Nucl. Energy* (B), 1, 15.
LOPEZ-MENCHERO, E. and GEHEM, L. (1963). *ENEA-Eurochemic Symposium on Aqueous Reprocessing Chemistry for Irradiated Fuels, Brussels*, p. 147.
LOPEZ-MENCHERO E. and SALOMON, L. (1967). *Eurochemic Report ETR-203*.
LOWES, L., TANNER, M. L. and WILLIAMS, J. A. (1958). *Trans. Inst. Chem. Eng.* **36**, 464–72.
MCKAY, H. A. C. (1979). *Proc. Symp. Fast Reactor Fuel Reprocessing*. Society of Chemistry and Industry, London.
MCKAY, H. A. C. and STREETON, R. J. W. (1965). *J. Inorg. Nucl. Chem.* **27**, 879.
MCKNIBBEN, B. M. and ORTH, D. A. (1969). *Trans. Am. Nucl. Soc.* **12**, 28.
MARSHALL, W. W., NAYLOR, A., SHORT, G. D. C. and WARNER, B. F. (1965).

Proc. Third Int. Conf. Peaceful Uses of Atomic Energy, Vol. 10, p. 224. United Nations, Geneva.

Merkelin, J. F. and Razvi, J. (1980). *Nuclear Instruments and Methods*, Vol. 169, pp. 223–6.

Michalik, J., Nowak, M., Nowak, Z. and Owczarczyk, A. (1972). *Radiochem. Radioanal. Lett.* **12**(2–3), 79–8.

Mills, A. L. (1979). *Proc. Symp. Fast Reactor Fuel Reprocessing*. Society of Chemistry and Industry, London.

Mills, A. L. (1984). *Institution of Chemical Engineers Symposium Series 88*, pp. 245–56.

Miquel, P., Patigny, P., Regnier, G. and Taillard, D. (1974). *ISEC 74, Lyon*, Vol. 3, pp. 2019–33.

Moccia, A. (1986). Pre-print of a paper presented at ICNR 86, Beijing, China, CNS-CAST.

Muller, W. (1967). *Actni. Rev.* **1**, 71–119.

Naito, K. (1960). *Bull. Chem. Soc. Jap.* **33**, 363.

Navratil, J. D. (1986). *ISEC 86, Munich*.

Navratil, J. D., Schultz, W. W. and Talbot, E. A. (1984). *Science and Technology of Tributylphosphate*, Vol. 1.

Nichols, C. M. and Spence, R. (1957) *Trans. Inst. Chem. Eng.* **35**(6), 380–93.

Nikipelov, B. V., Saprykina, E. I., Shevtsev, P. P. and Zakharkin, B. S. (1974). *Radiokhimiya* **16**, 449.

Nikitina, G. P., Pushlenkov, M. F. and Voden, V. G. (1959). *Radiokhimiya* 1, 2, 121.

Nikitina, G. P., Pushlenkov, M. F. and Voden, V. G. (1960). *Radiokhimiya* 2, 215.

Nowak, M. and Nowak, Z. (1973). *Radiochem. Radioanal. Lett.* **14**(3), 161–68.

Nowak, M. and Nowak, Z. (1979). *Radiochem. Radioanal. Lett.* **38**(5–6), 377–86.

Nowak, M., Nowak, Z. and Rochon, A. (1974). *Radiochem. Radioanal. Lett.* **20**(1), 47–58.

Nowak, M., Nowak, Z. and Seyel, A. (1979). *Radiochem. Radioanal. Lett.* **34**(5–6), 343–54.

Nowak, Z. (1977). *Nucleonika* **22** (2), 155–171.

Olcott, T. W. and Orth, D. A. (1963). *Nucl. Sci. Eng.* **17**, 593–612.

Orth, D. A. (1965). *Report DPSPU-64-30-28*. United States Atomic Energy Commission.

Pai, S. A., Shukla, J. P. and Subramanian, M. S. (1979) *Paper No. II-3, Symposium on Solvent Extraction of Metals, Bombay*.

Powell, R. (1958). *Analyst* **83**, 252.

Purex Technical Manual (1955). General Electric Company, Hanford, HW 31000 (Del.).

Pushlenkov, M. F., Shchepetilnikov, N. N., Shumkov, V. G., Shuvalov, O. N., Voden, V. G., Zemljanakhin, V. I. and Zilberman, B. Ja. (1965). *Proc. Third Int. Conf. Peaceful Uses of Atomic Energy*, Vol. 10, pp. 310–37. United Nations, Geneva.

Rais, J. and Selucky, P. (1971). *Radiochem. Radioanal. Lett.* **6**, 257.

Ransom, A. E. (1951). (ed.) *Redox Technical Manual*, HW-18700 DEL.

Schmieder, H. (1984). *Symposium on Fuel Reprocessing and Waste Management*, Jackson, Wyoming, USA, ANSI.

SEGE, G. and WOODFIELD, F. W. (1954). *Chem. Eng. Prog.* **50**(8), 396.
SHEPPARD, J. C. (1957). Vol. HW-51958.
SHEVCHENKO, V. B., TIMOSHEV, V. G. and VOLKOVA, A. A. (1960). *Sov. J. Atomic Energy.* **6**, 293.
SHUKLA, J. P. and SUBRAMANIAN, M. S. (1980). *Proceedings of the Nuclear Chemistry and Radiochemistry Symposium, Waltair, India.*
SIDALL, T. H. III (1960). *J. Phys. Chem.* **64**, 1863.
SPENCE, R. (1957). *J. Roy. Inst. Chem.*
STARR NICHOLLS, G. (1958). *DP-255.*
STEWART, G. and THORNTON, J. D. (1967). *Institution of Chemical Engineers Symposium Series* No. 26, 29, 37.
STIEGLITZ, L. I. (1971). *ISEC 71, The Hague*, pp. 155–9.
THORNTON, J. D. (1954a). *Chem. Eng. Progr.* Symposium Series **50**(13), 39.
THORNTON, J. D. (1954b). UK Patent 756049 (filed 1954).
THORNTON, J. D. (1955). UK Patent 774130 (filed 1955).
THORNTON, J. D. (1958). UK Patent 864677 (filed 1958).
THORNTON, J. D. (1976). *Birmingham Univ. Chem. Eng.* **27**(1), 6.
VASSALO, G. (1981). *EUR 6996.*
WAGNER, R. M. (1951). *Report HW-19959.*
WHATLEY, M. E. (1982). US/UK Exchange Meeting, Oak Ridge, USA, *CF-821056-7.*
WILKINSON, W. L. (1988). *Nucl. Eng. Int.* 32–6.
WILSON, A. S. (1959). *Proc. Second Int. Conf. Peaceful Uses of Atomic Energy*, Vol. 17, p. 557. United Nations, Geneva.
WINCHESTER, R. S. (1957). *USAEC Report LA-2170.*

3

LIQUID–LIQUID PARTITION IN THE PHARMACEUTICAL INDUSTRY

MICHAEL S. VERRALL

3.1 Introduction

Liquid–liquid partition is used by the pharmaceutical industry in a number of ways, some of which are entirely conventional while others have distinctive features. The former—for example the use of solvents to work up reaction mixtures—will not be discussed. This chapter will instead deal with the use of solvents in the isolation of products of biological origin, small-scale biochemical applications which have interesting partition systems and selected chemically assisted extractions.

Literature published in the area of pharmaceutical production processes is relatively sparse and much of it is in the form of patents which do not necessarily relate to actual practice. Thus it is not normally possible to provide detailed manufacturing procedures for specific medicinal products. The chapter will describe a number of liquid–liquid partition techniques which can be used for the isolation or purification of natural products together with examples where the techniques have actual or potential application. The methods described will be based mostly on bulk extraction and will not include liquid–liquid partition chromatography. An attempt will be made to give an indication of the general features of the feedstocks and the compounds they contain in the context of the solvent-extraction procedures employed.

In the course of drug evaluation a number of analytical or small-scale preparative techniques are used, some of which involve liquid–liquid partition. These will also be discussed as many could, in principle, be adapted for large-scale use.

The three main research approaches to the discovery of a novel biologically active substance are:

(1) a search in the natural environment for compounds with the required activity;

(2) to synthesize structures chemically similar to known active compounds;

(3) to design a suitable drug molecule to interact with a specific enzyme or other biological target.

The third approach has been pursued for a relatively short period of time, while the second partially relies on the first in that many chemically synthesized drugs have analogues in nature. Thus directly or indirectly we have been heavily indebted to natural products to supply pharmaceuticals in actuality or in concept. Over many millenia man has evaluated on a trial-and-error basis numerous biologically active materials, mostly from plant sources. Today the search for novel naturally occurring pharmaceuticals is concentrated mainly in the area of micro-organisms which can produce a wide variety of secondary metabolites (compounds having no obvious metabolic function, usually produced specifically by only a small group of organisms). The most widely investigated microbial products are antibiotics.

Thus generally speaking pharmaceutical compounds are biologically active, complex organic molecules often derived directly or indirectly from natural sources, especially micro-organisms. The range of compounds involved is very diverse with many different types of structure represented and with relative molecular mass from less than one hundred to more than one million. Many compounds, in order to be effective drugs, are reactive compounds in aqueous solution and in consequence suffer from problems of stability during isolation. Additionally, in crude or partially purified extracts stability may be variable depending on the nature of co-occurring impurities.

The discussion will not be restricted to established industrial processes and commercial compounds but will also include applications which have so far been used only only on a small scale or experimental basis.

The pharmaceutical industry, in common with all chemical industry, must conform to the appropriate national legislature with respect to health, safety, and environmental aspects of all manufacturing operations. In particular the use of flammable solvents requires special facilities and operational procedures.

The industry has additional constraints in that all operations must be conducted in accordance with good pharmaceutical manufacturing practice using only procedures approved by the various national regulatory bodies. Good manufacturing practice cannot be defined as a set procedure; rather it reflects the general philosophy which governs the manner in which the drug substance is developed and produced. It is inevitable that the final product will contain impurities, some of which may be unidentified, especially with natural products. The impurities and their concentration can often only be defined by the manufacturing process itself. Thus detailed standard operating procedures must be rigorously followed and documented. Any development of the process (for example, a new strain of producing organism, raw material, or procedure) cannot be put into operation until the product so produced has been thoroughly tested and the revised procedure approved by the regulatory authority.

Since the final products are usually destined for internal administration, toxic impurities must be at a very low level. An example of the demanding criteria for purity is provided by one potential impurity, pyrogen. This material (a complex bacterial lipopolysaccharide) elicits a febrile (fever producing) response at a dose of 0.5 ng/kg body weight. Thus a 70 kg man will react to the administration of 35 ng pyrogen and it follows that a 1 g dose of drug must contain less than 35 parts per billion of this impurity.

3.2 Microbial products

3.2.1 General aspects

The production of antibiotics is probably the largest single application of solvent extraction in the pharmaceutical industry. World production of benzylpenicillin (penicillin G) alone was 11 100 tons in 1985 (Anon 1986).

The first commercial antibiotic, penicillin (Fig. 3.1) was initially produced by growing the producing organism on the surface of a nutrient medium but, due to the severe limitations of this method, submerged fermentation techniques were soon developed. The term 'fermentation' is used in its general sense to cover the cultivation of micro-organisms in a (usually aerated) liquid medium. Anaerobic cultivation of yeast to produce alcoholic beverages is only one specialized application.

In general, the production process for a secondary metabolite involves the growth of a selected strain of micro-organism in submerged (i.e. suspension) culture through several 'seed stages' of increasing volume, culminating in the final production stage which lasts from a few hours to over 7 days, depending on the product. During the latter stage there may be continuous or intermittent addition of extra nutrients and/or chemical precursors. The fermentation medium usually consists of a carefully optimized mixture containing (in solution or suspension) carbohydrate and proteinaceous material supplemented with various inorganic compounds, for example phosphate, sulphate, calcium, and ammonium ions together with lower concentrations of metal ions such as iron, zinc, cobalt, or copper. Details of the genetic regulation and biosynthetic pathways involved are given in a number of review volumes, for example by Vining (1983) or Vandamme (1984).

The principal nutrients are often residues from other industries, for example cotton seed meal, corn steep liquor, or molasses. The fermentation

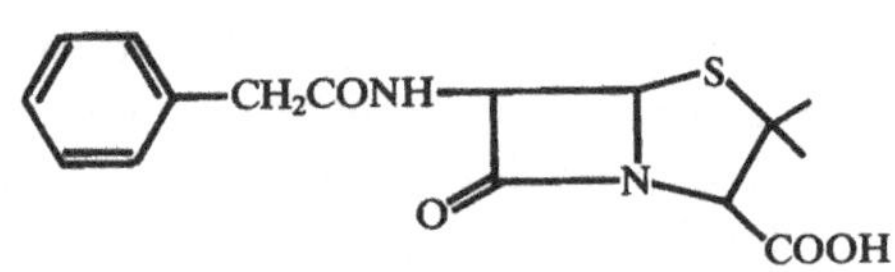

Fig. 3.1. Structure of penicillin G.

process is usually operated batchwise rather than in continuous mode because production of secondary metabolites is not normally associated with the growth phase of the organism and also because many high-yielding strains have evolved over a long period of time using selection based on batch culture. Thus the final fermentation stage may consist of two distinct phases: initial growth of the organism, followed by a 'bioreactor' phase when the accumulated biomass synthesizes product from substrates existing in the fermenter or added during this stage.

The organisms used are usually bacteria or fungi, although in principle other types of cell (for example algae, plant cells, or animal cells) can be grown in suspension culture to produce pharmaceutical products.

Bacteria (except for actinomycetes) grow in suspension culture as individual cells which may be spherical or cylindrical in shape. Their minor axis typically measures 0.2–2 μm. The Actinomycetales (usually referred to as actinomycetes) comprise an order of bacteria which possess the distinctive morphological feature of forming branching filaments. Their individual cells are of a similar size to those of other bacteria. They are notable for the enormous variety of secondary metabolites they produce. An estimate of the number of known antibiotic compounds from all micro-organisms (in 1984) was given as 7150 (Bérdy 1985), of which 64 per cent were from actinomycetes.

Microfungi grow as a mass of intertwining, branching filaments (hyphae) of diameter 5–50 μm. The entire mass of such hyphae is termed mycelium. During an extended fermentation a proportion of the producing organism may lyse (that is die and disintegrate) and some mycelial masses may form pellets greater than 1 mm in diameter. Thus the feedstock produced for downstream processing can have particulate impurities of a very wide size range.

During the final fermentation stage the product is biosynthesized within the cells and, depending on its properties and the physiology of the cell, will be either retained within the cell or excreted into the culture fluid. Compounds retained within the cell are mostly lipophilic molecules or macromolecules. Some lipophilic compounds may also be found in the culture filtrate at a higher concentration than their nominal aqueous solubility due to conjugation with carrier molecules, adsorption on to colloidal particles, or micelle formation.

At the end of the production phase the fermenter contains 'whole broth' consisting of biomass suspended in a dilute aqueous solution of unused nutrients and precursors, undefined impurities (unwanted metabolites), and surfactants (antifoam agents), plus, of course, the required product. As already mentioned, this may be in solution or partially or wholly associated with the microbial cells. In appearance whole broth of a fungal culture may resemble a 'porridge', with the biomass representing 4–5 per cent of the dry weight, but its hydrated volume may be over 30 per cent. A highly developed fermentation process, such as that for penicillin G, can produce 1 g of product per gram dry weight of cells (Cooney 1983).

The whole broth may be variable from batch to batch with respect to rheology, particle size distribution, or impurity levels. An additional point is that the fermentation technologist is constantly modifying the producing culture or other parameters with a view to improving productivity. Thus the downstream processing system cannot be too highly tuned because it must be able to process with consistent efficiency variable batches as they are produced. There is usually no opportunity for reprocessing primary material.

Direct solvent extraction (that is where extraction is based only on solvation of the solute) is obviously not appropriate for all products. Most macromolecular products are not amenable to this technique (but see section 3.7), nor are very hydrophilic compounds such as the aminoglycoside family (for example streptomycin, Fig. 3.2). Neutral compounds with at least a portion of the molecule having lipophilic properties, or ionic compounds where ionization can be suppressed by pH adjustment or other means, are the best candidates.

The original philosophy of product recovery was to separate biomass from the culture fluid before proceeding to subsequent operations. This has usually been achieved using rotary vacuum filters for filamentous organisms (fungi and actinomycetes) or centrifuges for bacteria. However, there is a growing tendency to bypass the broth clarification stage and extract whole broth instead. In either case mixtures of clarified or whole broth and a water-immiscible solvent emulsify very readily and gravity separation is usually slow if not impossible. The problem is due partially to the presence of proteinaceous impurities in solution or suspension and partially to other entities with detergent-like properties, which may even be the product itself. Proteins have low water solubility at, or near, their isoelectric point. If, during the change in pH of a crude protein solution (that is protein impurities) the value passes through the isoelectric point much protein will

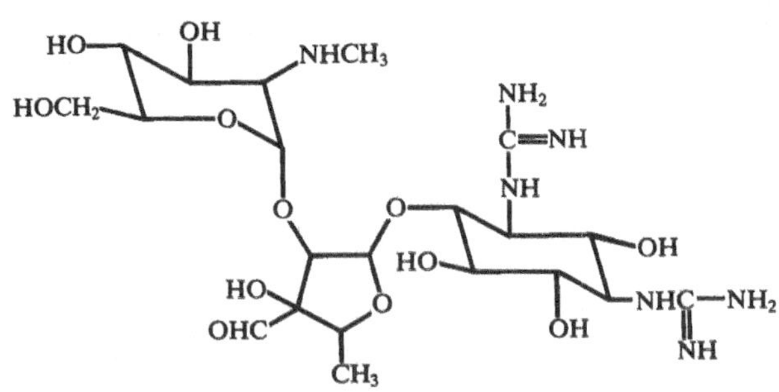

Fig. 3.2. Structure of streptomycin.

precipitate, some of it irreversibly. That which is resolubilized as the pH is adjusted away from the isoelectric point will usually redissolve quite slowly. This fact can have two possible consequences. Either prompt clarification can be carried out to give a cleaner solution or, if it is imperative to immediately proceed to solvent extraction, addition of surfactant will be necessary to minimize emulsification.

The choice of solvent and pH value is chosen to:

(1) maximize partition coefficient of the desired substance;

(2) minimize extraction of co-occurring impurities especially any which are difficult to remove during subsequent downstream processing operations.

Sometimes the choice will be influenced by the availability of existing equipment and solvent recovery facilities.

Chemically assisted solvent extraction based on ion exchange or chelation is a well known technique in hydrometallurgy but is less frequently used for pharmaceuticals. In general, if direct solvent extraction is inapplicable, some form of adsorption procedure has been used. Some exceptions to this statement are mentioned in section 3.4. In many cases solvent extraction may be only one of a number of viable alternatives. Some of the many factors to be considered have been mentioned in the context of cephalosporin C production (Wildfeuer 1985):

(1) capital costs;

(2) processing costs;

(3) throughput requirements;

(4) yield potential;

(5) product quality;

(6) technical expertise available;

(7) conformance to regulatory requirements;

(8) waste treatment considerations;

(9) requirement for continuous or batch processing;

(10) potential for automation;

(11) health and safety aspects.

To enable a rational choice to be made, chemical process design flowsheeting techniques have been adapted to the analysis and design of downstream biotechnological processes (Jackson and de Silva 1986).

Having decided on the overall concept of the isolation process, detailed optimization can then be carried out. An algorithm for the overall optimization of a fermentation and isolation process incorporating three-stage solvent extraction has been described by Okabe and Aiba (1975).

The foregoing discussion is centred on the use of solvent extraction as a primary isolation/purification stage for fermentation products. There are instances where the technique may be used further downstream but in such cases conditions will more closely resemble those in other chemical processes and these will not be discussed in detail.

3.2.2 Applications

In the following examples general features of the solvent extraction stages used in the isolation of microbial secondary metabolites will be illustrated with reference to two extracellular products, penicillin and erythromycin, and three intracellular products, a polyene macrolide, an anthracycline, and avermectin.

3.2.2.1 *Penicillins*

Penicillin was the first commercial antibiotic and its production is the best documented. Many of the considerations mentioned in the context of penicillins are also applicable to other antibiotics.

Penicillin G (benzylpenicillin) is produced for use as an antibiotic *per se* and also as an intermediate in the preparation of semisynthetic penicillins and cephalosporins. Penicillin V (phenoxymethylpenicillin) is also produced by fermentation but in smaller quantities. The most significant difference between the compounds with respect to isolation is their stability in dilute acid, the latter compound being relatively stable at low pH values. Penicillins have the structure indicated in Fig. 3.3. For convenience the 1-thia-4-aza-7-oxo-[3,2,0]-bicycloheptane ring system has been given the trivial name 'penam' (Brown 1982). Penicillins may also be described as penicillanic acids. Using this nomenclature, the parent compound of all natural penicillins is 6-amino penicillanic acid. A large number of semisynthetic penicillins are available, based on a common penam structure but with different acyl substituents on the 6-amino group. An example is amoxycillin. Several other families of antibiotics also have the 4-membered β-lactam (azetidinone) ring structure, in some groups fused to a 5- or 6-membered ring, and in other groups free.

The essential features of penicillin G to be considered when isolating it from fermentation broth are:

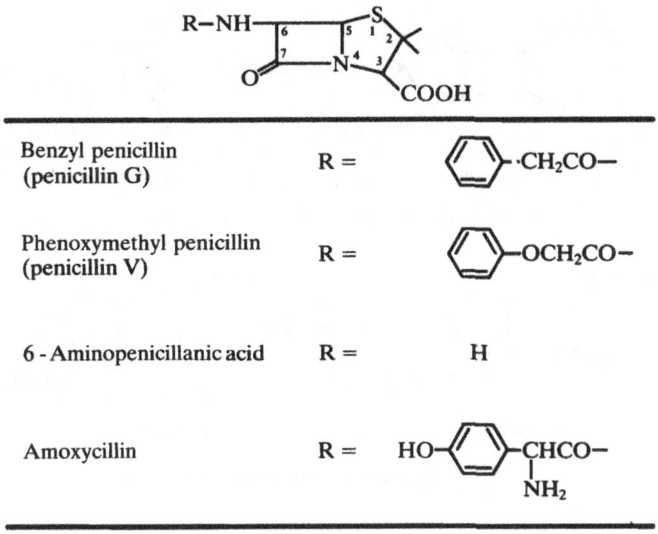

Fig. 3.3. Structure of natural and semisynthetic penicillins.

1. The compound is a monocarboxylic acid of pK_a 2.75 (Reschke and Schügerl 1984a).

2. It is very unstable as the free acid in aqueous solution and of limited solubility. Acid hydrolysis gives the penillic acid (Fig. 3.4). The reaction involves intramolecular attack of the side chain amide carbonyl on the β-lactam moiety. Thus penicillins with electron withdrawing groups in the side chain such as penicillin V are less susceptible to acid hydrolysis. Certain ions have a catalytic effect on the acid hydrolysis of penicillin G (Finholt *et al.* 1965); they reported that dihydrogen citrate, monohydrogen phosphate, and borate catalyse degradation of penicillanate whereas monohydrogen citrate, dihydrogen phosphate and acetate were almost without effect.

3. Dilute alkali (pH > 9) hydrolyzes the compound to the corresponding penicilloic acid (Fig. 3.5). The enzyme β-lactamase found in many bacteria catalyzes the same reaction.

Both acid and alkaline hydrolysis follow first order kinetics. The stability of penicillin G at intermediate pH values is depicted in Table 3.1 (Reschke and Schügerl 1984a). The same authors have plotted the data in a different way to indicate the time for 5 per cent decomposition at 25°C (Fig. 3.6). These curves show the importance of co-occurring ions on the stability of penicillin G. Phosphate, although a widely used biochemical buffer, promotes degradation of the compound (see also Section 3.4.2.2).

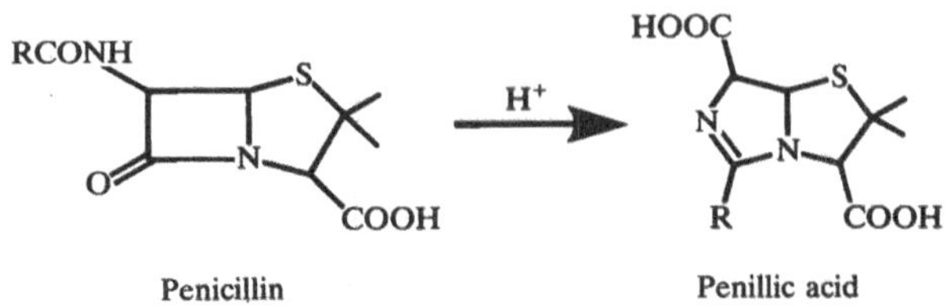

Fig. 3.4. Acid hydrolysis of penicillins.

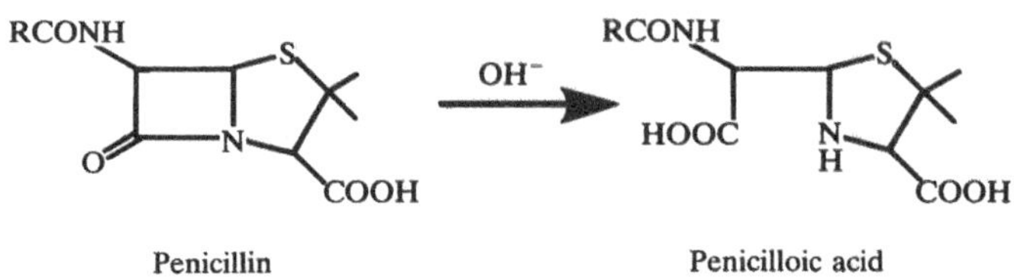

Fig. 3.5. Alkaline hydrolysis of penicillins.

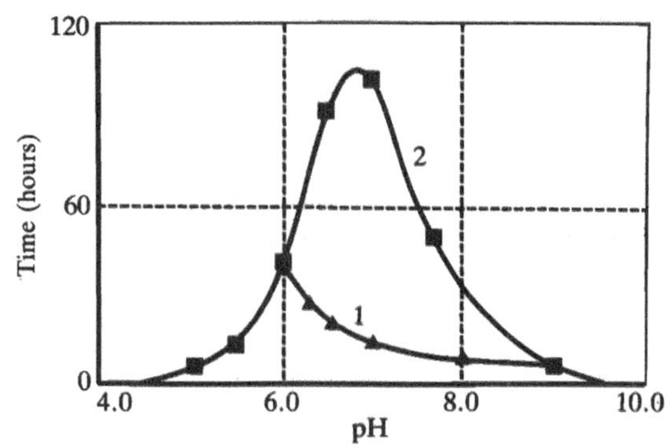

Fig. 3.6. Decomposition times for penicillin G giving 5 per cent loss in activity. Curve 1: phosphate buffer. Curve 2: citrate buffer. (From Reschke and Schügerl 1984a.)

The above stability data impose severe constraints on the operational parameters which may be employed for the isolation of penicillin G and the quality of process control must be of a very high order.

The history of penicillin discovery and development has been well documented (see for example Abraham 1974 or Wilson 1976). Before its chemical nature was appreciated carbon absorption was used for isolation but, as soon as it was realized that penicillin is a labile carboxylic acid, solvent extraction techniques were investigated.

Initial commercial development used fermenters of 2m^3 capacity (current production vessels have up to 100 times this capacity). It was established that the properties of the product (in the early days a mixture of compounds) and the matrix (fermentation broth) rendered most currently available solvent

TABLE 3.1 *Decomposition of penicillin G in aqueous solution*

pH	$k(h^{-1})$
4.0	4.40×10^{-2}
5.0	1.08×10^{-2}
6.0	1.16×10^{-3}
6.25	2.02×10^{-3}
6.5	2.60×10^{-3}
7.0	3.29×10^{-3}
8.0	6.23×10^{-3}
9.0	1.06×10^{-2}

extraction equipment inapplicable. The Podbielniak extractor had been described in 1935 (Podbielniak 1935) but few applications had thus far been found. However, it was soon appreciated that it was suitable for penicillin extraction and it was no doubt a key contributor to the rapid development of a practical solvent extraction process.

The basic isolation process for penicillin G has changed very little since the 1940s. Various aspects of the procedure will be described in detail as many features of the process and general considerations apply to some extent to other antibiotics.

In summary, filtered fermentation broth is acidified to pH 2.0–2.5, extracted with a polar solvent, such as 4-methylpentan-2-one or butyl acetate, and a stable salt of penicillin recovered by precipitation.

Since the manufacture of penicillin G (and indeed all antibiotics) is highly competitive each producer will have developed his own variant of the basic method incorporating hard-won and confidential knowledge. For this reason discussion of the process must necessarily be in general terms.

At the end of the fermentation stage the producing organism, *Penicillium chrysogenum*, will have converted most of the initial nutrients and added precursors into penicillin G, which will be present as the anion (penicillanate) in a neutral aqueous solution containing various dissolved and suspended impurities.

A detailed isolation procedure is illustrated in Fig. 3.7 and is reported to be that used by Gist-Brocades in Holland (Hersbach *et al.* 1984). It is used as the basis of the following discussion.

Mycelium is separated from fermentation broth using a rotary vacuum filter. If a precoat filter is used the filtrate will be of good clarity, but this feature is not crucial as subsequent acidification will precipitate proteinaceous and colloidal impurities. However, it may be advantageous to add a flocculating agent to the broth prior to filtration to reduce the colloidal impurity content.

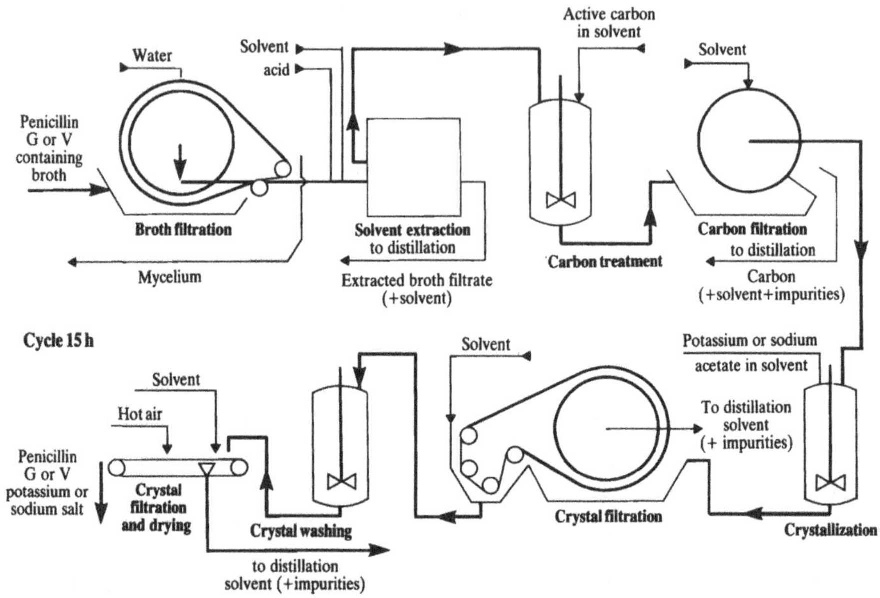

Fig. 3.7. Penicillin purification process of Gist-Brocades. (From Hersbach *et al.* 1984.)

Broth is acidified to convert the extremely water-soluble penicillanate anion to largely undissociated free acid. The precise pH value chosen will depend on finding a compromise between stability and extraction efficiency. The data in Fig. 3.8 (Reuben and Sjöberg 1981) show that the partition coefficient increases steadily with increase in acidity of the aqueous phase. Unfortunately this is accompanied by a concomitant rapid decrease in product stability. A second factor influencing choice of extraction pH is selectivity. This factor has become relatively less important as product titres have increased and the fermentation is operated in a manner which ensures minimal phenylacetate (precursor) concentration at harvest. It is reported that penicillin titres have increased from 1.8×10^{-4} moles per litre in the 1940s to 0.06–0.1 moles per litre in 1979 (Queener and Schwartz 1979).

Acidification may be performed with any convenient strong acid. In addition to converting the penicillanate anion to the free acid the acid must also be sufficient to overcome the strong buffering effect of impurities. In practice sulphuric acid is normally used. Hydrochloric acid is not used to avoid corrosion in the mainly stainless steel process plant and centrifugal extraction equipment used further downstream. If a single stage extraction system is used it is preferable to add the mineral acid after the solvent, as the penicillanic acid can then immediately transfer into the organic phase in which it is more stable. Additionally, with high titre fermentation broth, some free penicillanic acid will precipitate when in contact with mineral acid.

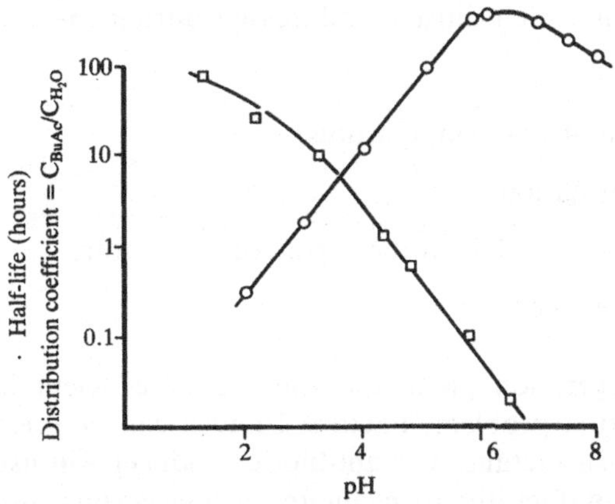

Fig. 3.8. Extraction of penicillin G with butyl acetate. pH dependence of distribution coefficient and stability. (From Reuben and Sjöberg 1981.)

A suitable in-line mixing device is used to ensure rapid and thorough mixing of the broth/solvent/acid mixture. Countercurrent extractors for antibiotic production are designed with internal mixing devices and operate with high flowrate and minimal residence time to overcome these problems. The extraction efficiency will be maximized if the extractor can be operated with the solvent as predominantly the continuous phase.

The process streams are chilled throughout, but particular attention must be paid to efficient cooling at the acidification stage as this is the point where the product is of minimum stability. In an aqueous solution of pH 2 and 25°C a 1 per cent loss of penicillin G is experienced every 10 seconds (Reschke and Schügerl 1984a). The penicillanic acid is transferred into a polar organic solvent such as 4-methylpentan-2-one, butyl acetate, or amyl acetate. In the early days of penicillin production some halogenated solvents and cyclic ketones were recommended on the basis of their selectivity but, as mentioned above, this factor is of minimal importance now that the ratio of product to impurities is two to three orders of magnitude greater.

After transfer into solvent the product is more stable, being removed from the bulk of the hydrolytic acidic aqueous phase, but in order to obtain a high yield it is still essential to isolate a stable salt as quickly as possible. It must of course be remembered that the rich organic phase will certainly contain dissolved, and may contain entrained, an acidified aqueous phase which will adversely affect stability.

The extraction is preferably carried out in countercurrent mode using devices such as the Podbielniak extractor (favoured in North America) or a vertical bowl machine (for example, Alfa-Laval ABE-216). The latter type of machine is much used in Europe.

Four problems which must be addressed during the solvent extraction stage are:

(1) maintenance of low temperatures;

(2) emulsion formation;

(3) accumulation of solids in the extractor/separator;

(4) overall process control.

A suitable refrigeration plant and the use of efficient heat exchangers overcomes the first problem. Factors 2 and 3 can be greatly ameliorated using a suitable surfactant. The antibiotic producer will usually liaise with a surfactant manufacturer to evaluate various agents and, if necessary, develop a new formulation. The approach is largely empirical and a surfactant devised for one manufacturer's penicillin process will not necessarily work for another. Typical surfactants used are quaternary ammonium compounds or esters of phosphoric acid (Reschke and Schügerl 1984a). The surfactant must also be chosen with due regard to its effect on penicillin stability. It is generally found that anionic surfactants (for example sodium dodecyl sulphate) enhance acid-catalyzed degradation whereas stability is increased in the presence of cationic and non-ionic surfactants. The rate constants of hydrolysis initially decrease with increase in concentration of the latter surfactants, then approach a constant value above the critical micelle concentration of surfactant, suggesting that penicillin–micelle complexes are formed (Tsuji *et al.* 1978). For high yields a sophisticated process control system is essential. The outline of that installed in the Glaxo plant at Cambois is described by Berkovitch (1976).

After separation the spent aqueous phase is stripped to recover dissolved solvent, while the organic phase may be treated with carbon to remove various pigments and impurities. In some plants the primary solvent extract is contacted with aqueous buffer from which the penicillanic acid is re-extracted into solvent to increase concentration and purity. With penicillin titres now in the order of 0.1 moles per litre this secondary concentration stage is often no longer necessary.

The required salt of penicillin G may be precipitated from the organic phase by the addition of a suitable reagent such as sodium or potassium acetate. In some cases an organic base salt is prepared using, for example, N-ethyl piperidine. The product may be filtered, washed, and dried at this stage but if it is required as pharmaceutical grade further recrystallization or other processing may be needed. The penicillin salt crystals are washed with an anhydrous volatile solvent such as propan-2-ol. Should it be necessary to remove residual coloured impurities at this stage, butan-1-ol is reported to be a particularly effective wash fluid (Schwartz 1985).

Approximately 85 per cent of penicillin produced is used as an intermediate in the production of 6-aminopenicillanic acid or cephalosporin compounds (Anon 1986) and for these applications the initial salt preparation is normally of adequate purity.

Stage yields for the above type of procedure are given by Schwartz (1985) and are listed in Table 3.2. These yields are probably rather conservative as another report (Hersbach *et al.* 1984) quote an overall yield of 90 per cent.

The foregoing account is based on the extraction of filtered broth. There have been periodic attempts to avoid broth filtration prior to solvent extraction. Although it provides a cleaner feedstock for extraction, broth filtration has a number of disadvantages. It is usually necessary to use a diatomaceous earth filter aid to obtain satisfactory filtration. It is essential to wash the filter cake to maximize yield in the filtrate but, as shown in Table 3.2, product losses still occur. The wash water dilutes the filtrate, which means that a higher volumetric throughput extraction train is required.

These disadvantages can be overcome by extracting whole (unfiltered) broth. However, other factors must now be considered. Entrainment of solvent in the spent aqueous phase becomes more of a problem and it will be necessary to re-evaluate the choice of surfactant and its concentration. A higher level will almost certainly be required.

Workers at Parke Davis (Anderson and Lau 1955), having had a fair measure of success using a Podbielniak 9500 extractor for chloromycetin (chloramphenicol) recovery from whole broth, attempted to use the machine in a similar manner for penicillin G but it was found that extraction efficiency was only about 65 per cent. The countercurrent extractor was operating at only one half of a theoretical stage.

Whole broth extraction of penicillin G was reported by Podbielniak *et al.* (1970), but there is little indication that it was widely adopted at that time. However, Queener and Schwartz (1979) mention that at least one major

TABLE 3.2 *Penicillin production: typical stage yields*

Stage	Yield %
Broth holding	95
Filtration	90–95
Solvent extraction	
single stage	80–90
two stage	92–96
Back extraction	95–97
Crystallization	95
Drying	95
Overall	approx 78

manufacturer was using whole broth extraction for penicillin G production.

To prevent accumulation of solids the extractors must be operated at a high throughput, giving rise to 10 per cent entrained solvent in the waste aqueous stream. This contains negligible penicillin and the solvent is recovered by distillation. The rich organic extract is free of suspended solids and entrained aqueous droplets. The surfactant required to assist phase separation is dosed at three to four times the amount required for filtered broth. Some solvent is premixed with the whole broth before the first extractor. The solvent tends to adhere to the mycelial and other solids, adding buoyancy which minimizes settling under gravity and consequent deposition in the extractor. Using this procedure the extractors can operate for 16 to 60 hours without cleaning.

There has recently been an increased interest in whole broth extraction and a centrifugal extractor has been produced specifically for this purpose (Brunner 1985). The machine was developed from a scroll decanter centrifuge which was adapted to operate with two immiscible liquid phases. It is described in more detail in section 3.9.3.2.

A computer-based flowsheet simulator has been used to model three approaches to penicillin G isolation, that is a single extractor process, a two extractor process incorporating back extraction into aqueous buffer, and whole broth extraction (Cooney *et al.* 1988). The simulation covered energy and material balance for the process and also an estimate of capital and running costs.

As mentioned before, acid stability of penicillins can be increased by introducing electron-withdrawing groups into the 6-side chain. Thus the other major penicillin produced by fermentation, penicillin V (phenoxymethyl penicillin), is considerably more stable to acid than penicillin G. Although a penicillin G type of process can be used it is also possible to recover the free acid in stable form by precipitation from aqueous solution. A procedure has been patented (Edmundowicz and Allanson 1982) for the direct precipitation of penicillin V free acid at pH 2.0–2.5 from fermentation broth which has previously been filtered at pH 4.0–4.3. It should be noted that this compound is also highly unstable in alkaline solution and susceptible to β-lactamase hydrolysis.

3.2.2.2 *Macrolide antibiotics*

The term 'macrolide' was proposed by Woodward (1957) to describe the family of antibiotics consisting of 12-, 14-, or 16-membered macrocyclic lactones substituted mainly with alkyl groups and sugar residues. The first member of the group to be isolated was pikromycin (Brockman and Henkel 1950), while the first commercially produced (in 1952) was erythromycin (Fig. 3.9).

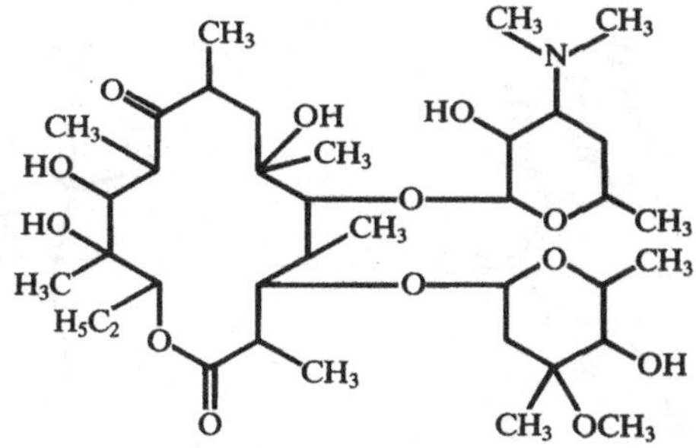

Fig. 3.9. Structure of erythromycin.

More than 90 compounds of this type are now known (Omura and Tanaka 1984). They are usually produced by actinomycetes, although there is one report of erythromycin being produced by a bacterium (*Arthrobacter* sp.) (French *et al.* 1970). Macrolides are particularly useful against gram positive bacteria including *Legionella*.

Macrolides, in common with many microbial products, are produced by the wild strain of the organism as a mixture of closely related products. Culture modification and fermentation development progressively increase the proportion of the desired component. Macrolides possessing an amino sugar substituent have weakly basic properties (erythromycin has pK_a 8.6) and solvent extraction is usually carried out on an aqueous phase of pH 8.5–10.0. A solvent of moderate polarity such as ethyl acetate, butyl acetate, or dichloromethane is used.

Under the conditions used for solvent extraction macrolides are of adequate stability but precautions must be taken to avoid excessively low or high pH values. Acidic conditions promote nucleophilic attack of the 6-hydroxyl on the C-9 carbonyl which is followed by dehydration to give initially 8,9-anhydroerythromycin-6,9-hemiketal which may degrade further, as indicated in Fig. 3.10 (Kurath *et al.* 1971). Alkaline hydrolysis will cleave the lactone. Erythromycin is stable for weeks in neutral solution at 5°C but degrades in a few days at room temperature (Nakayama 1984).

The desired product is either a salt or ester of the parent compound. A suitable salt is readily obtained by extracting the organic phase with a dilute solution of the appropriate acid or by direct precipitation.

Erythromycin is produced by *Streptomyces erythraeus* (recently redesignated *Saccharopolyspora erythraea* (McAlpine *et al.* 1987)) which, like most members of this genus, produces mycelium which is relatively difficult to remove by filtration or centrifugation. Filtration is best performed at pH 4 in order to maximize solubility of the product and also to precipitate some proteinaceous impurities.

In one application filtered broth is adjusted to pH 10 and extracted with

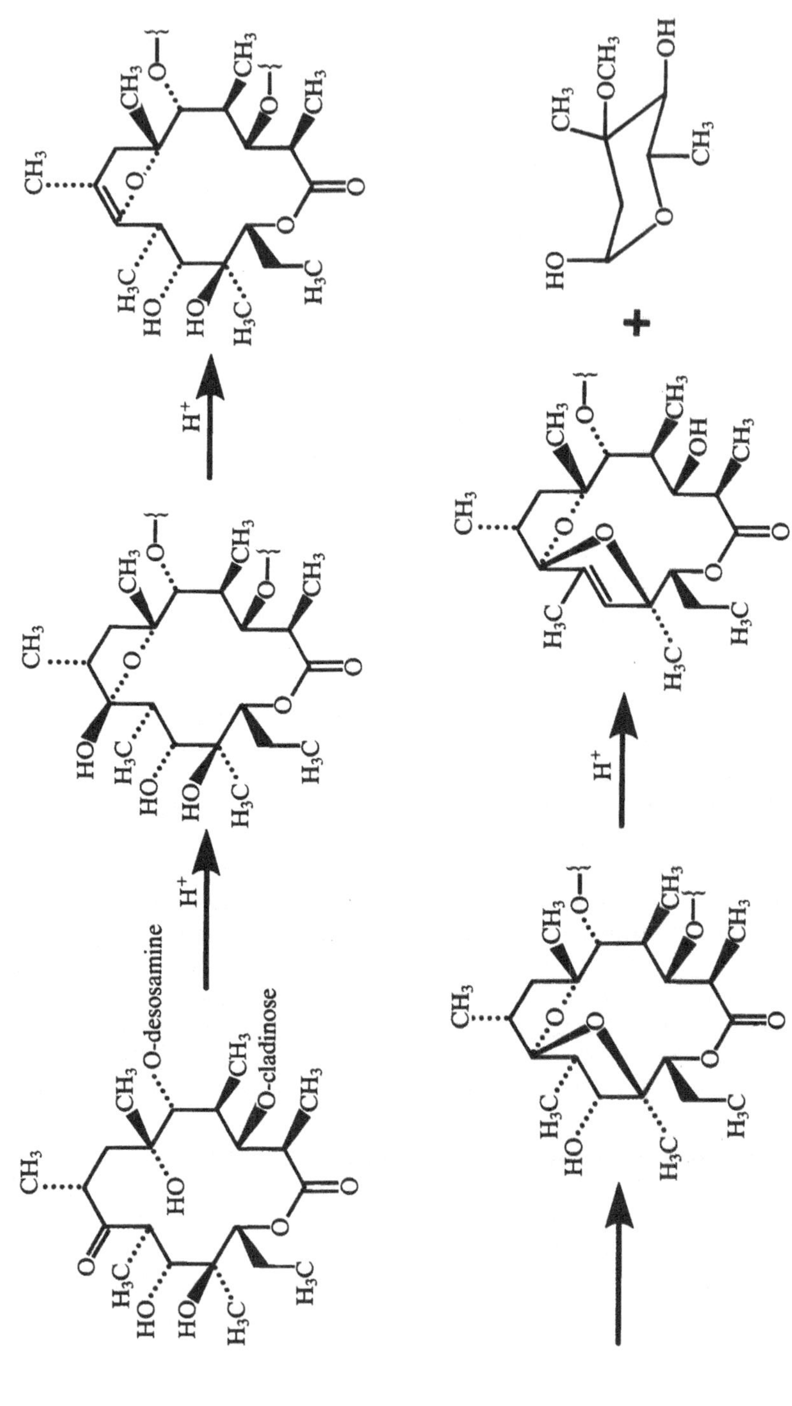

Fig. 3.10. Acid hydrolysis of erythromycin.

butyl or amyl acetate at 30°C (Baker Perkins 1987). A broth to solvent ratio of 7 : 1 to 3 : 1 can be used giving extraction efficiencies of 95–98 per cent. A continuous back extraction is carried out using acetate or phosphate buffers of pH 4.1–5.4. A third stage extraction, transferring product back into solvent, may be employed, as depicted in Fig. 3.11. This figure gives an indication of some typical broth pretreatment stages which may be required to obtain satisfactory operation of the solvent-extraction equipment. In this process crude product is precipitated from the solvent as the thiocyanate salt for further processing.

Calcium carbonate is often present in the fermentation medium to assist in pH control. Calcium ions present at the pH 10 extraction stage are likely to give rise to deposits in the extractor. The problem can be alleviated by the addition of chelating agents or an occasional wash with dilute nitric acid.

As with penicillins, whole broth extraction has been considered by various manufacturers. One is reported to have decided against the method due to the cost of necessary broth-treating reagents and has completely eliminated the need for such reagents by employing ultrafiltration prior to extraction (Baker Perkins 1987). Another manufacturer uses whole broth from which readily sedimentable solids have been removed using a centrifugal decanter. The broth containing the majority of the mycelial solids is passed through two Podbielniak extractors in series.

The recently developed decanting extractor has been successfully used for erythromycin isolation (Brunner 1985). Using butyl acetate as extractant, a 10–15 per cent overall yield increase was obtained compared with the two discrete operations of filtration and solvent extraction. As with the majority of antibiotic extractions, a suitable surfactant had to be found to aid phase separation. In this application Arquad (Akzo Chemie) was found effective.

Thus the unit operations required for the liquid–liquid extraction stages of erythromycin production are essentially the same as for penicillins. The main difference is that the product is essentially stable under normal extraction conditions.

3.2.2.3 *Polyenes*

The polyene antifungal antibiotics are an example of a group of antibiotics which are found largely as intracellular products. They represent another type of macrocyclic lactone but differ from the erythromycin type in three significant ways. Firstly, they have pronounced antifungal activity especially against yeasts. Secondly, they are based on a much larger ring system (20 to 44 ring atoms) and thirdly, as their name suggests, they contain a set of conjugated double bonds (usually 4 to 7).

The first member of the family to be discovered (Hazen and Brown 1950) was nystatin (Fig. 3.12), produced by *Streptomyces noursei*, and this

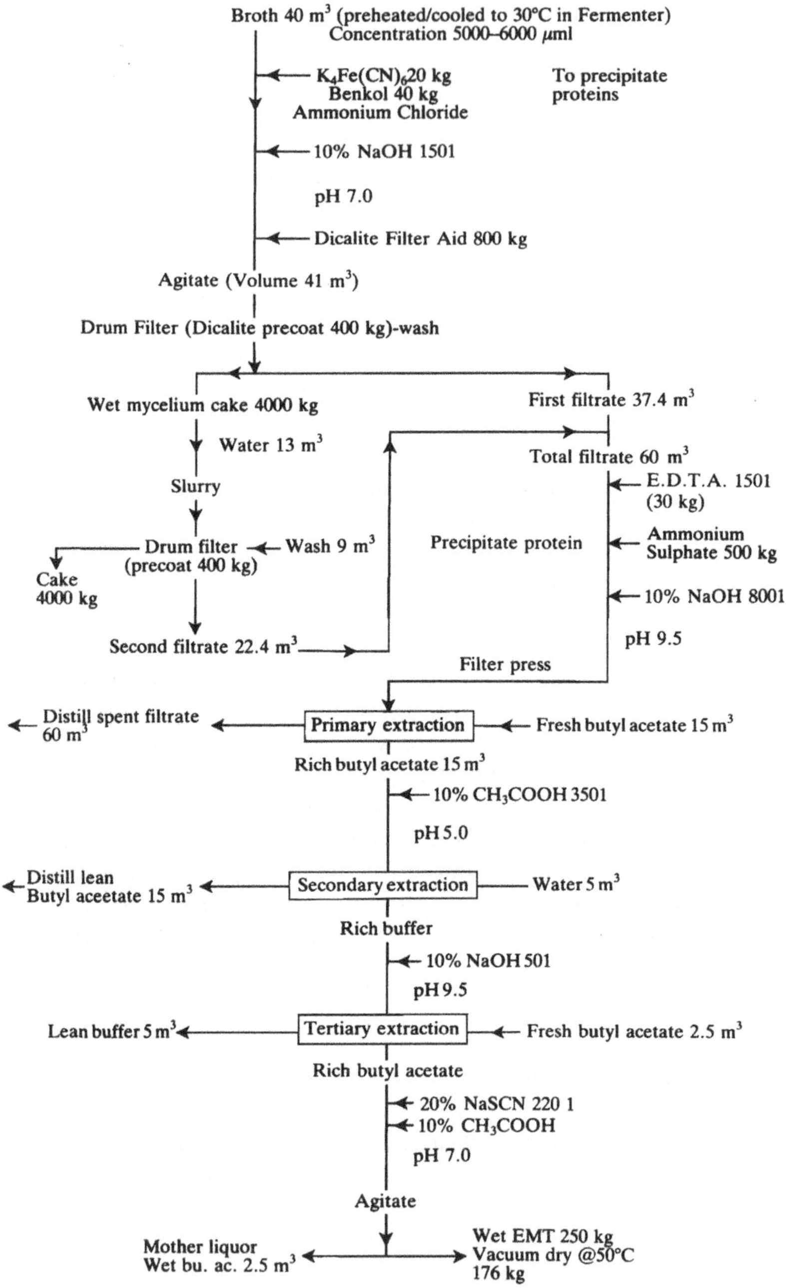

Fig. 3.11. Extraction procedure for erythromycin (Baker–Perkins technical literature).

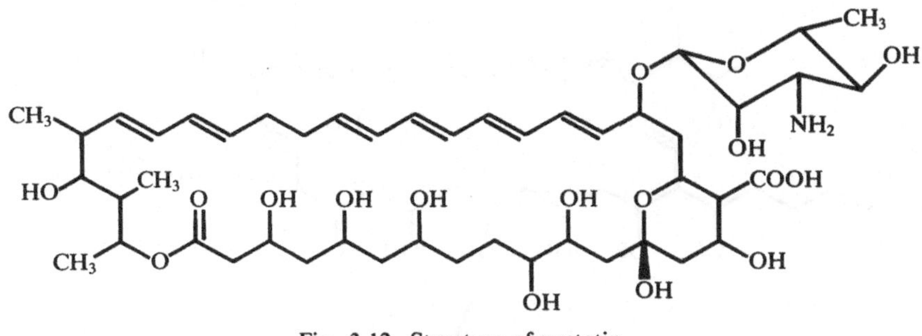

Fig. 3.12. Structure of nystatin.

compound is still widely used in antifungal therapy. Another early polyene, amphotericin B (Fig. 3.13), is still the drug of choice for many serious fungal infections. Since nystatin over 200 polyene antifungal compounds have been reported but some of these have subsequently been found to be mixtures. Indeed it is quite usual to find more than one polyene produced by a single species of micro-organism. In many cases it is necessary to use sophisticated separation methods such as countercurrent distribution or high resolution chromatography to obtain individual components.

Many polyenes are amphoteric compounds. The acidic group is a carboxyl group attached to the ring and the basic function is represented by an amino sugar. A smaller number are acidic or neutral. The molecules are amphipathic, having a rigid planar lipophilic portion and a flexible hydrophilic polyhydroxylated region. The conjugated double bond system gives rise to a characteristic ultraviolet or visible absorption spectrum and is also responsible for their sensitivity to oxidation, photolability, and low water solubility. The compounds are also unstable under acidic or basic conditions. At pH 7 amphotericin B degrades by 30 per cent in 20 hours at 20°C. The corresponding figures for pH 2 and pH 11 are 50 per cent in 5 hours and 50 per cent in 1.5 hours.

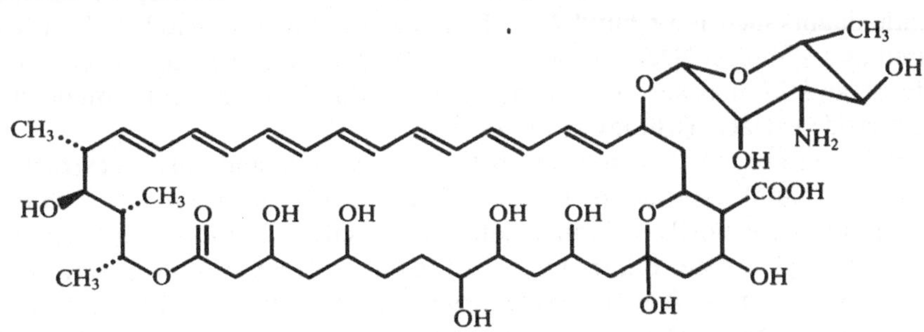

Fig. 3.13. Structure of amphotericin B.

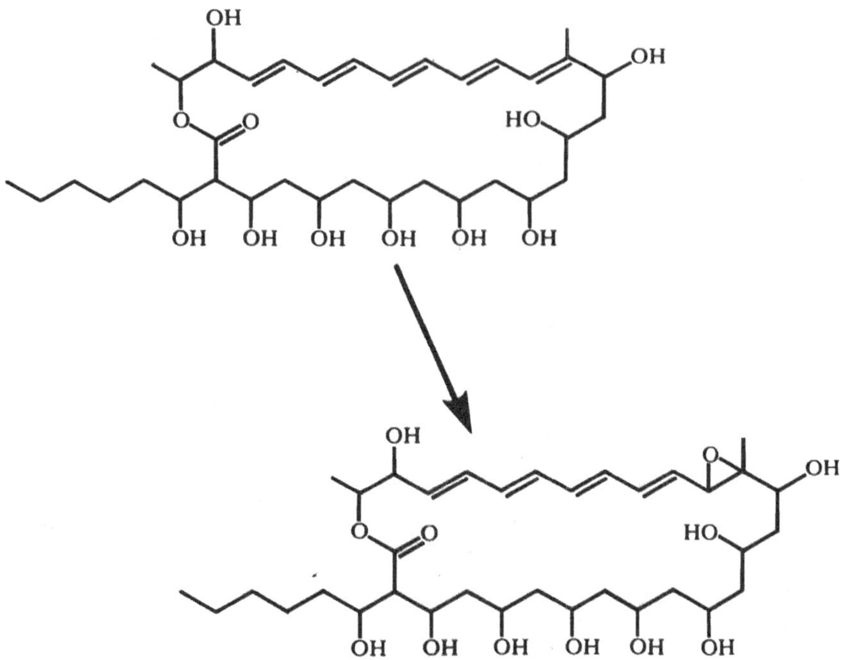

Fig. 3.14. Oxidation of filipin.

Oxidation usually results in the initial formation of a monoepoxide followed by higher oxidation products (Fig. 3.14) (Rickards *et al.* 1970). In some cases polyenes exist in aqueous 'solution' as micelles which can exert a self-protecting effect with respect to various degradative influences. This is stated to be the case with amphotericin B (Bonner *et al.* 1975). Possibly conflicting evidence is given by Rickards *et al.* (1970) who reported that filipin is stable in contact with air and light as a 0.001 per cent methanolic solution but more concentrated solutions (0.8 per cent) are unstable, even in the dark, at 4°C. The latter could be stabilized considerably by adding antioxidants such as 3-t-butyl-4-methoxyphenol. Ascorbic acid has also been used as a stabilizer (Dekker and Ark 1959). Polyenes form complexes with sterols and this may account for their greater stability in fermentation broths and crude extracts (Klimov *et al.* 1971).

Polyenes are mostly associated with the mycelium and may be extracted using methanol, acetone, or butan-1-ol. Polar solvents are required as the compounds are mostly of limited solubility in solvents of moderate polarity such as butyl acetate. Butanol may be the only solvent which is practical to use on whole broth. The ionizable groups are fairly weak and extraction is not greatly influenced by pH. It appears to be preferable to choose the pH value for greatest stability rather than for maximal partition coefficient.

Having obtained a crude extract, the product may be purified by selective precipitation methods. To obtain concentrated solutions of polyenes very polar solvents such as formamide, dimethylformamide, or dimethylsulphoxide are required. Polyenes are readily precipitated from such solutions by the addition of lower polarity solvents.

It is usually impractical to fractionate on a commercial scale any mixtures of closely related polyene compounds and production of a single product is best achieved by manipulation of the producing organism and the fermentation procedure.

In summary, the main requirements for successful extraction of polyene macrolides are the use of conditions which minimize oxidation and photolysis. It is preferable to maintain pH near neutrality.

3.2.2.4 *Anthracyclines*

The anthracycline anticancer agents consist of a family of compounds which are based on an anthraquinone nucleus attached to one or more sugar residues. They are also largely intracellular compounds and are included here as an example of metabolites which may require some simple chemical manipulation before, or during, isolation. Due to their high biological activity they also require special containment procedures.

The first clinically effective anthracycline was isolated independently by three groups. One group (Grein *et al.* 1963) named the compound 'daunomycin', the second, 'rubidomycin' (Dubost *et al.* 1963) and the third group chose 'rubomycin' (Brezhnikova *et al.* 1966). The hybrid name 'daunorubicin' has now been adopted. Compounds of this class exert their action by intercalation of the planar ring system with DNA thus inhibiting its replication.

Daunorubicin is a monoglycoside produced by *Streptomyces peucetius* (Fig. 3.15) but much of the product in the fermentation broth is in the form of higher glycosides which are subjected to carefully controlled hydrolysis

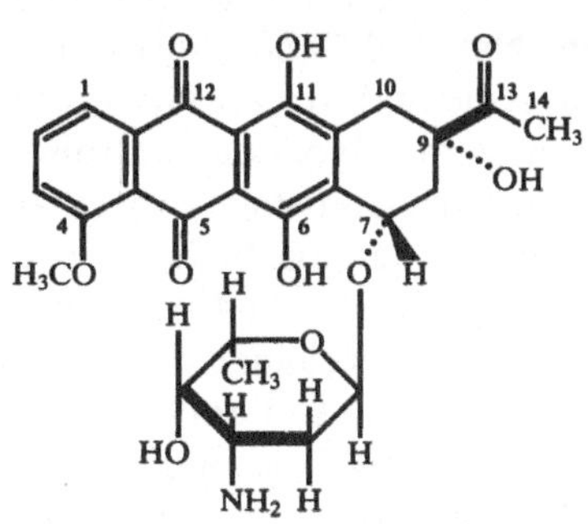

Fig. 3.15. Structure of daunorubicin.

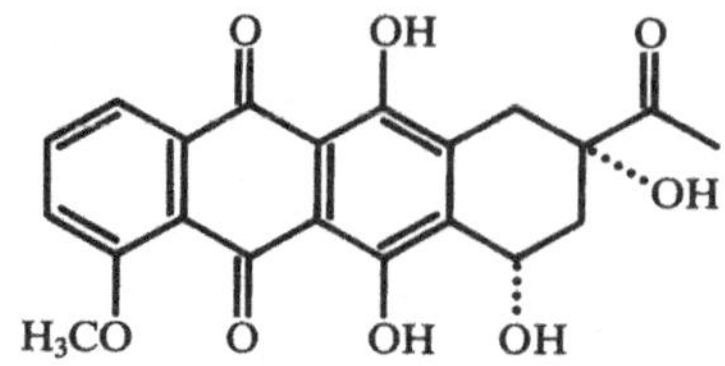

Fig. 3.16. Structure of daunorubicinone.

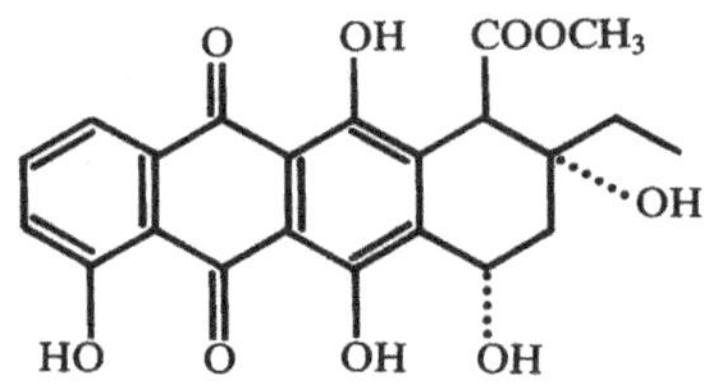

Fig. 3.17. Structure of ϵ-rhodomycinone.

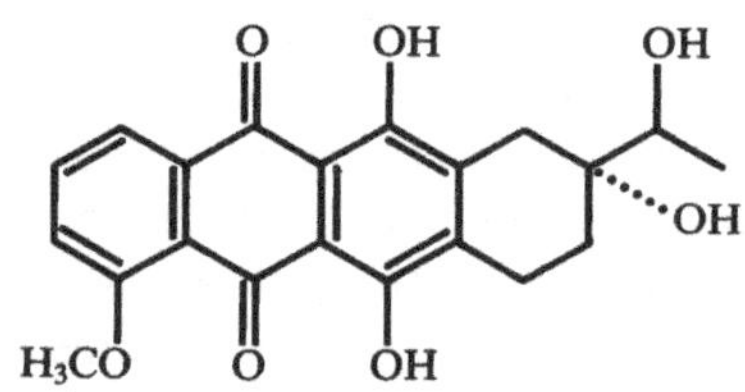

Fig. 3.18. Structure of 7-deoxydihydrorubromycinone.

prior to, or during, extraction. Excess hydrolysis must be avoided to minimize generation of the aglycone daunorubicinone (Fig. 3.16), some of which will already be present in the broth together with ϵ-rhodomycinone (Fig. 3.17). If fermentation broth is left unaerated for a few hours much product is converted to 7-deoxydihydrodaunorubicinone (Fig. 3.18). Daunorubicin is degraded in alkaline solution.

The process operated by Rhône-Poulenc uses hydrolysis with oxalic acid at 50°C while that of the National Cancer Institute operates at pH 1.5 and ambient temperature. Both methods release anthracyclines from the cells but the latter does not hydrolyse the glycosides. In this case they are extracted as a mixture then hydrolysed at a later stage with ethanolic hydrochloric acid at 45°C for 10 minutes to give daunorubicin hydrochloride.

The Rhône-Poulenc method effects the first purification stage by absorbing the product on to the cation exchange resin Amberlite IRC-50 (Rohm and Haas) followed by elution with sodium chloride in aqueous methanol. This is followed by extraction of the base into chloroform and precipitation with hexane. Further precipitation and acid treatment stages are used to give daunorubicin hydrochloride.

The National Cancer Institute procedure uses a number of partition stages alternating between the free base in organic solution and the sulphate or hydrochloride in aqueous solution. It is not clear which procedure gives the highest yield. The two procedures are summarized below (Flickinger 1985).

Rhône-Poulenc method

1. Add oxalic acid to whole broth, heat at 50°C for 1 hour.
2. Filter, wash filter cake, and discard.
3. Adjust filtrate to pH 4.5, percolate through Amberlite IRA 458.
4. Wash column with water, 50 per cent methanol, 90 per cent methanol.
5. Elute product with sodium chloride in aqueous methanol.
6. Concentrate solution, adjust to pH 7.5, extract with chloroform.
7. Concentrate solution, precipitate daunorubicin base with hexane.
8. Dissolve in butanol-water-hydrochloric acid 94 : 5 : 1, precipitate with acetone-hexane 1 : 1, filter.
9. Dissolve in butanol–water, 95 : 5, precipitate with acetone, filter.
10. Dissolve in dioxane-water 1 : 1, add dioxane, filter, wash, dry.

National Cancer Institute method

1. Acidify whole broth to pH 1.5, filter.
2. Adjust filtrate to pH 8.6, extract with butanol.
3. Adjust pH to 4.0, concentrate.
4. Precipitate product with heptane, dissolve in dilute sulphuric acid (pH 2).
5. Adjust to pH 8.6, extract with dichloromethane.
6. Extract with dilute sulphuric acid (pH 2).
7. Adjust to pH 8.6, extract with butanol.
8. Adjust to pH 4.0, concentrate.
9. Add ethanolic hydrochloric acid to pH 1.5, heat at 47°C for 10 minutes.
10. Cool, filter, recrystallize from chloroform–methanol–water.

The development and simplification of the procedure used at NCI is discussed by Pandey *et al.* (1979). The large number of stages were originally required to separate product from the numerous impurities present in fermentation broth. The simplified method uses solvent mixtures which are more selective for the impurities. A 30 per cent acetone in toluene wash of the crude aqueous solution was particularly effective. The modified method is summarized below.

Modified NCI method

1. Adjust whole broth to pH 1.5, stir 45 minutes, centrifuge.
2. Slurry mycelium in acidic water, centrifuge, combine acidic extracts.
3. Wash twice with 30 per cent acetone in toluene or ethyl acetate.
4. Adjust pH to 8.2, extract twice with butanol.
5. Wash with water, adjust to pH 5.3, concentrate.
6. As stages (8) and (9) above.

As mentioned above, due to the high biological activity (cytotoxicity) of the anthracyclines the isolation process must be performed in specially controlled containment facilities. In some respects the requirements are similar to those of a radiochemical facility. The general design concepts for a suitable plant, which have been given by Flickinger and Sansone (1984), are:

1. The area must be of controlled access.
2. Working areas should be under negative air pressure.
3. All air exhausted to the environment must be passed through HEPA (high efficiency particulate air) filters.
4. Any equipment which is likely to generate aerosols is protected by additional containment.
5. Operating personnel must wear appropriate protection.
6. All process wastes must be treated to render them free of toxic substances.
7. All personnel potentially at risk are subjected to routine medical monitoring.
8. There should be adequate monotoring of the immediate environment.

3.2.2.5 *Avermectins*

The avermectins are a group of macrolide compounds produced by *Strep-*

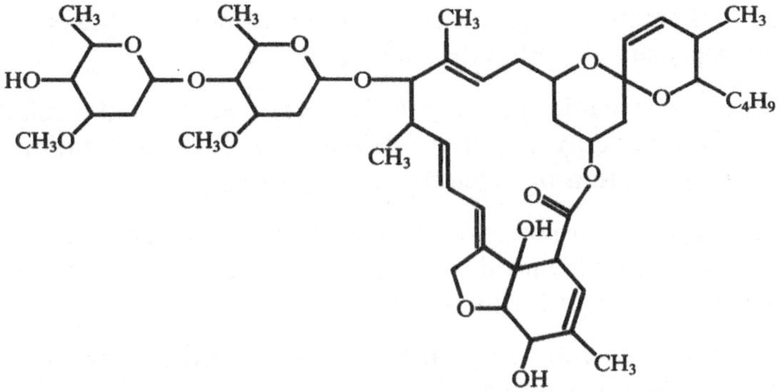

Fig. 3.19. Structure of avermectin B1a.

tomyces avermitilis which are highly active against veterinary parasites. A commercial product, Ivermectin, is produced by controlled hydrogenation of avermectin B1 (Fig. 3.19) (Chabala *et al.* 1980).

Avermectins were initially isolated by acetone extraction of the microbial cells (Cole and Goegelman 1979). This method involves many material handling stages and it would appear that whole broth extraction should provide a simpler alternative. The compound is of low water solubility but is soluble in a wide range of organic solvents. However, it is found that whole broth extraction with most water-immiscible solvents is very low yielding. Butan-1-ol is effective but suffers from the disadvantage of high water solubility and low selectivity. A patented process (Begner and Wildman 1983) which overcomes these problems exploits the compound's unusual acid stability. Whole broth is heated to 80°C at pH 2.5 for 2 hours to permeabilize the cells, then one fifth volume of toluene is added to leach out the product. The latter procedure uses only liquid phase handling equipment which is easier to operate and control. It is also simpler to detoxify all residues using alkaline hydrolysis.

3.2.3 Summary

The factors which distinguish microbial metabolite isolation from most other solvent extraction procedures are:

1. The product occurs as a dilute aqueous solution or within the microbial cells (or both).
2. The product is frequently chemically unstable.
3. It is often a minor component of the total solute mixture.

4. Particulate impurities have a large size range but are always very light, having a density only slightly above unity.
5. The aqueous/organic system emulsifies very readily and a suitable surfactant is invariably required both to aid phase separation and to minimize deposition of precipitated impurities.
6. It may be necessary to perform solvent extraction under conditions where the product is highly unstable. Centrifugal extractors or separators with sophisticated control facilities are essential.
7. Where products are of very high biological activity, appropriate containment facilities are required.

3.3 Plant products

Plants are the oldest source of medicinal products and are still widely harvested for this purpose. Some 25 per cent of purified drug substances, especially steroids and alkaloids, are derived from plant sources (Bu'Lock and Kristiansen 1987).

Two main features emerge when comparing the isolation of pharmaceutical products from plants with fermentation products. Firstly, the compounds in plants have been formed slowly over a long period of time and consequently are expected to have a higher biochemical stability. Secondly, with the exception of products extracted from juice or sap, it is necessary for the product to be released or extracted from a rigid matrix.

It has been shown that some products can be produced by growing plant cells in suspension culture. In principle plant cell cultures can be established for any plant species (Berlin 1984). Unfortunately, current product yields from these cultures is generally low but there are exceptions. Some cell cultures can produce metabolites at an equivalent level to the natural plant. The main problem is that most plant products of interest are accumulated only in specialized cells. However, there are instances where potentially useful substances have been detected in a cell culture but not in the plant from which it was derived. The undifferentiated cells of suspension culture are less productive and have low growth rates, requiring a long residence time in the fermenter. Nevertheless at least one product (a shikonin derivative) is reported to be in development for commercial production (Fujita *et al.* 1982). Once conditions are established there seems no problem in increasing the scale of suspension culture. Conventional fermentation vessels have been used for plant cell culture up to 20 m^3 in volume, but in general plant cell culture is best performed in specially designed vessels. As with secondary metabolites produced by micro-organisms, those from plant cells are usually produced in the idiophase (that is after cell growth has ceased) and there is scope for the development of immobilized cell systems. It is then necessary for the product to be excreted from the cells. Where this does occur it is often unclear whether it is the result of active release or of cell lysis. In many instances product is retained within the cell and must be released by cell disruption or leaching. Both operations are more readily performed on plant cells than on micro-organisms due to the relatively greater size and fragility of the former.

Should an immobilized cell reactor be used the above methods are inapplicable and *in situ* release methods must be employed. The main approach is permeabilization of the cell by physical or biological means. There appears to be no generally applicable method. Cells have been permeabilized without destroying viability by treatment with 5 per cent dimethylsulphoxide for 30 minutes. After recovering the product and washing out excess dimethylsulphoxide, the cells can continue biosynthesis for a further period (Brodelius and Nilsson 1983). In another instance a glucoside was released by treatment with cetyltrimethylammonium bromide. In some cases a simple pH change will effect the required release. Cells can also be permeabilized by infecting with a bacterium such as *Agrobacterium rhizogenes*. Instead of investigating ways of releasing product from the cell an alternative approach would be to develop a cell line with the required release characteristics. Where the product is released from the producing cell its continuous removal from the liquid phase may increase the biosynthetic capacity (Robins and Rhodes 1986; Mavituna *et al.* 1987).

In general the extraction requirements for plant cell cultures are similar to those of other fermentation products. The main difference is that plant cells are larger and more fragile than micro-organisms.

The extraction procedure for products from normal plant material obviously depends on the type of tissue to be processed and any limitations imposed by the product. If it is not possible to finely comminute some types of tissue (for example, root) prolonged extraction may be necessary. Certain products may occur in the lipid phase within the plant (especially the seed) and are found associated with the oil produced therefrom; for example, steroids in soya bean oil (Blunden *et al.* 1975).

Where plant secondary metabolites are extracted by solvent extraction it appears that simple batch processes are normally used. One notable exception is the extraction of caffeine from coffee using supercritical carbon dioxide, but this example belongs in the chapter on food processes. As noted later (section 3.5), supercritical extraction has potential for the extraction of other drug substances and it should be particularly appropriate where valuable plant products must be extracted from a rigid matrix.

An example of the isolation of a plant product using modern equipment is given by Brunner (1984) (Fig. 3.20). The product (morphine) is extracted from milled poppy plants by countercurrent extraction in a multistage

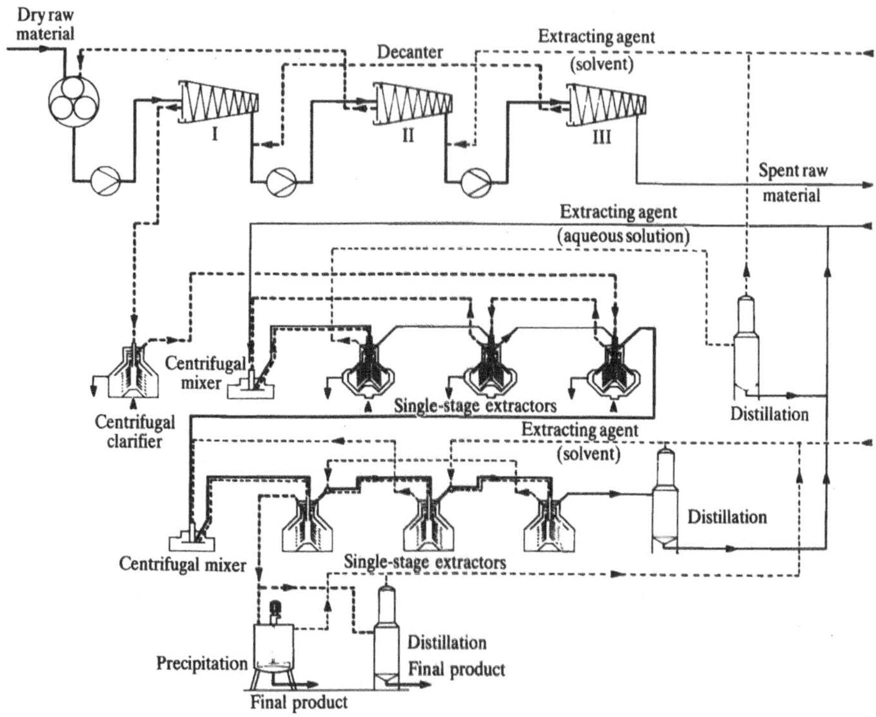

Fig. 3.20. Extraction procedure for morphine. (From Brunner 1984.)

decanter system (see section 3.9.3.2). The solution of morphine base so obtained is contacted with an acidified aqueous phase, then the free base is reextracted into solvent which is concentrated and crystallized. The secondary extraction stages use disc-stack centrifuges for separation.

3.4 Chemically assisted extraction

3.4.1 General aspects

Chemically assisted solvent extraction is widely used in hydrometallurgy. Both chelating agents and liquid ion exchangers find numerous large-scale applications. The structure of the chelate or ion pair extracted is usually relatively simple and well defined. It is thus possible to design a selective reagent for a particular purpose.

The situation is rather different with pharmaceuticals where the solutes of interest are complex organic molecules. It is much more difficult to design reagents for precise physicochemical interaction. It should be noted however that in recent years the selectivity of nature has been harnessed for downstream processing using immunoaffinity adsorbents and such ligands can be attached to the polymers used in two-phase aqueous systems (see section 3.7.2.1.2). Nevertheless the majority of chemically assisted solvent extraction processes investigated by the pharmaceutical industry depend on ion-pair or ion exchange interactions with the reagents being selected empirically.

In simple terms a hydrophilic organic anion (for example, penicillanate and a relatively hydrophobic cation such as an alkyl substituted ammonium salt) tends to exist in aqueous solution in equilibrium with the neutral ion pair which may partition favourably into a water-immiscible organic solvent (Fig. 3.21). A similar net result may be obtained by extracting the aqueous solution with a solvent solution of the hydrophobic cation. In this case the procedure is usually termed 'liquid ion exchange'. The term 'ion-pair extraction' should be reserved for cases where the equilibrium show in Fig. 3.21 can be shown to exist in the aqueous phase. If the extractant is an amine dissolved in organic solvent instead of a quaternary salt it is valid to designate the process ion pair extraction as a pair of ions, that is anion and proton are extracted.

$$A_{org} + P^-_{aq} + H^+_{aq} = AHP_{org}$$

Naturally the ions can be reversed; a hydrophobic anion can be used to extract a water-soluble cation. In the literature the terms ion-pair extraction and liquid ion exchange are often used interchangeably. When used in the context of chemical synthesis the process is usually termed 'phase transfer catalysis'.

Ion-pair interactions have been studied for analytical applications particularly where low concentrations of drug substance must be extracted from a biological matrix. One of the first reports was the colorimetric determination of quaternary ammonium salts using bromophenol blue in dichloroethane as extractant (Auerbach 1943). Ion-pair extraction has been studied extensively by Schill and co-workers at Uppsala, again from an analytical viewpoint (Schill 1974; Schill *et al.* 1977). They investigated a wide range of pharmaceutical substances and appropriate extractants. Unlike hydrometallurgical applications of liquid ion exchange which often operate with a hydrocarbon such as kerosene as diluent, organic applications usually require a more polar diluent. Halogenated solvents such as dichloromethane

$$R_4N^+ + P^- \rightleftharpoons R_4NP_{aq} \rightleftharpoons R_4NP_{org}$$

Aqueous phase | Organic phase

Fig. 3.21. Ion-pair formation: simplified equilibria R_4N^+ = quaternary ammonium cation. P^- = organic anion, for example penicillanate.

or chloroform are widely used. However, it is reported that relatively hydrophilic ion pairs can be accommodated in a lipophilic solvent such as cyclohexane if the solvent contains only a small amount of dipolar solvating agent (Higuchi *et al.* 1967). In cases where the cation is the large lipophilic member of the pair the most effective solvating agents appear to be those with relatively acidic protons, for example chloroform, alcohols, and phenols. In the reverse situation solvating species with nucleophilic sites, for example ethers, ketones, amides, and phosphate, esters are most effective. Strongly solvating solvents such as higher alcohols should, if possible, be avoided. Although they give good extraction they are of low selectivity and many inorganic ions may also be extracted in the presence of excess ion-pair reagent (Schill 1972).

Other reports of enhancement of ion-pair extraction efficiency using adduct-forming reagents include those of Schröder-Nielsen (1976) who used trioctylphosphine oxide for this purpose. Modin and Schröder-Nielsen (1971), investigating the extraction of penicillin with tetrabutylammonium hydrogen sulphate in chloroform, found that extraction was enhanced by the addition of 1-dodecylamine as adduct-forming reagent. Some studies have shown that the ion-pair complex is stoichiometrically solvated (Higuchi *et al.* 1967).

Ion-pair extraction studies have been extended to zwitterions. Nordgren and Modin (1975) extracted amino acids with tetrapentylammonium hydroxide in dichloromethane. The method is also applicable to certain macromolecules. Egan *et al.* (1970) extracted transfer ribonucleic acid into a commercial solvent/reagent mixture consisting of Adogen 464® (a methyltrialkyl ammonium chloride, Ashland Chemical Company) in Freon 214 (tetrafluorotetrachloropropane, DuPont).

It is emphasized that the applications reported above are essentially analytical in scale and most of the literature reports concentrate on the forward extraction (transfer of solute from aqueous to organic phase). The references given provide an overview of some practical and theoretical considerations of ion-pair extraction.

Stripping has not been investigated to the same extent. It is usually implied that washing the organic phase with a suitable strong salt solution readily dissociates the complex and returns the solute to an aqueous phase. The studies reported in section 3.4.2.2 show that this is largely true.

Selective stripping can sometimes be achieved by the sequential use of different aqueous back extractants or, alternatively, the same extractant employed at progressively increasing concentration. Thus the species with the highest distribution ratio will be retained in the organic phase and those with lower ratios will transfer back to the aqueous phase.

Little has been reported on the recycling of reagents but some aspects are mentioned by Reuben and Sjöberg (1981).

A recent comprehensive review discussing general aspects of extraction equilibria for carboxylic acids produced by fermentation includes extraction with single and mixed solvents and also with organic phosphates and aliphatic amines (Kertez and King 1986).

There are relatively few reports of ion pair extraction being used preparatively in the pharmaceutical industry and even fewer discuss the problems of recycling or regenerating the reagents.

Ion-pair extraction relies on ionic effects and hydrogen bonding for its success and the interactions are readily reversed using a slight change in conditions. In some cases it may be desirable to extract a stable derivative of the antibiotic and here the technique of extractive esterification can prove useful. A dilute aqueous solution of a carboxylate (for example fermentation broth or crude extract) is contacted with a substituted diazomethane in a suitable water-immiscible solvent. The carboxyl group is esterified and transfers to the organic phase leaving non-reacting impurities in the aqueous phase. The technique has been successfully used for cephalosporins and is further discussed in section 3.4.2.4. The following areas of application give an indication of the scope and potential of the methods described.

3.4.2 Applications

3.4.2.1 *Liquid cation exchange*

Some strongly basic hydrophilic compounds such the aminoglycoside antibiotics are not amenable to direct solvent extraction as even in their free base form they remain very hydrophilic and they are normally isolated using ion exchange resin technology. However, from time to time process patents for liquid cation exchange of these compounds are published. One patent (Gallaher *et al.* 1964) describes extractions using alkyl acid phosphates such as di (2-ethylhexyl) hydrogen phosphate, isooctyl dihydrogen phosphate, and octyl phenyl hydrogen phosphate. It was claimed that a wide range of diluents could be used. Hexane, toluene, dichloromethane or 4-methylpentan-2-one were all effective. Generally extraction was performed at pH 5–9 with back extraction into dilute sulphuric acid to give a pH value of 1–3, thus generating the antibiotic sulphate. The method was shown to be applicable to neomycin, bacitracin, streptomycin, viomycin, and polymyxin (Fig. 3.22a & b).

Another patented procedure (Jahnke 1967) uses aromatic/alkyl sulphonic acids. Salts of dinonylnaphthalene sulphonic acid are favoured. Shorter alkyl chain substituents tended to promote emulsification. The process is claimed to be useful for a wide range of nitrogen basic antibiotics although the examples concentrate on lincomycin (Fig. 3.23).

A novel feature of the process is that the products are effectively back extracted using amines, amine salts, or quaternary ammonium compounds.

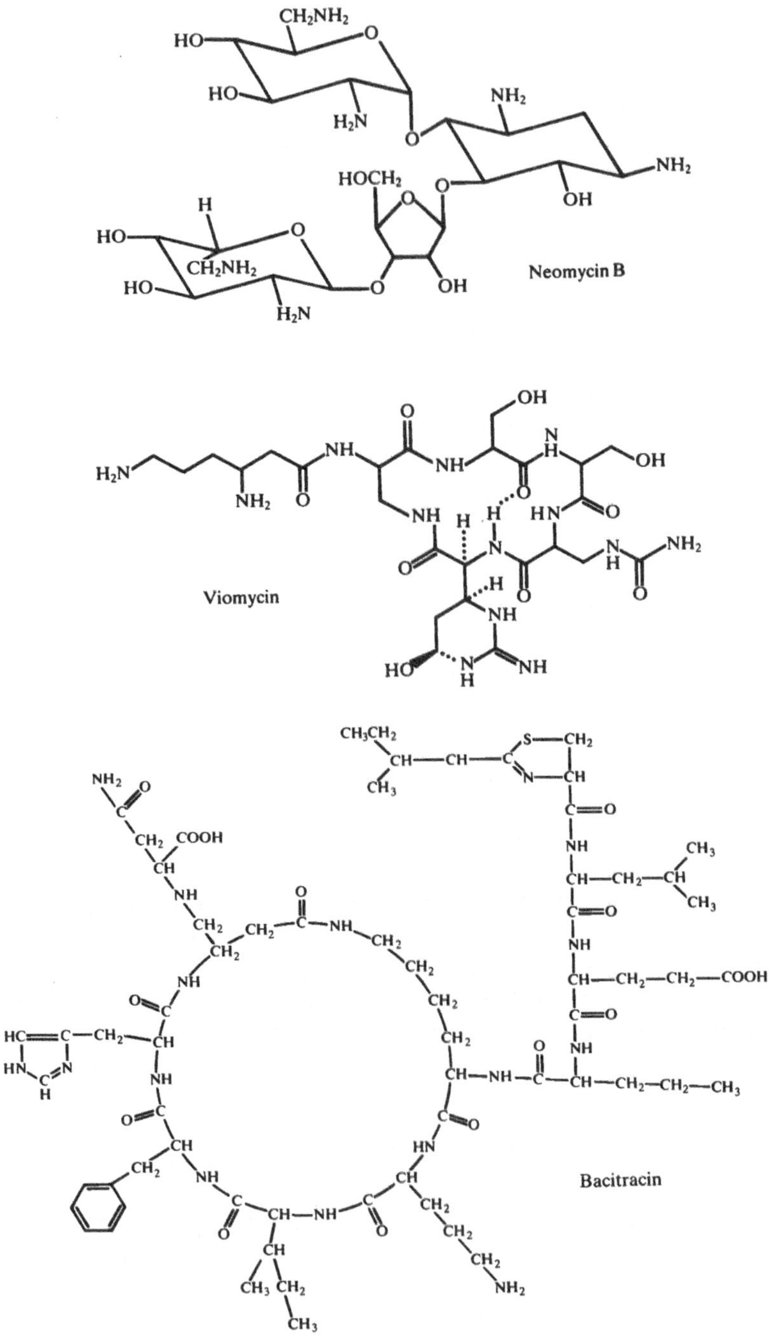

Fig. 3.22. Structures of neomycin, bacitracin, viomycin.

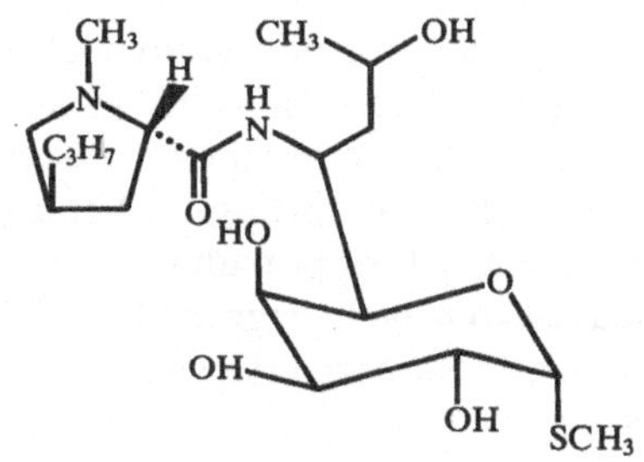

Fig. 3.23. Structure of lincomycin.

In general the stripping agent should be fairly lipophilic and should be able to transfer into the organic phase to interact with the loaded exchanger. Alternatively, if the stripping agent is essentially insoluble in water it may be added directly to the loaded cation exchanger and the displaced antibiotic extracted with water. Amines of low molecular weight and greater water solubility were less effective but economically this could be balanced by their lower equivalent weight.

The patent is exemplified using the extraction of lincomycin from filtered fermentation broth ($1.6\,m^3$) containing $795\,\mu g/cm^3$ antibiotic. The broth was adjusted to pH 8 and contacted in a Podbielniak countercurrent extractor with 2 per cent dinonylnaphthalene sulphonic acid in dichloromethane ($350\,dm^3$). Extraction efficiency was in excess of 98 per cent. The organic phase was evaporated to $30\,dm^3$ then octylamine ($2.2\,dm^3$) and water ($5\,dm^3$) was added. The mixture was adjusted to pH 3 with hydrochloric acid, agitated for five minutes and the phases were separated. The organic phase was washed with four portions of water. Lincomycin in the combined aqueous extracts was further purified by converting to the free base (at pH 10), extracting into dichloromethane and finally crystallizing the hydrochloride: overall yield was 83 per cent.

3.4.2.2 *Liquid anion exchange*

A series of papers on the ion-pair extraction of penicillin G discuss in detail the influence of extractant, carrier solvent, pH, and other parameters which must be considered when developing an ion pair extraction procedure (Reschke and Schügerl 1984a, b, c). Direct solvent extraction of penicillins via the free acid has been developed over the years into a highly efficient process and chemically assisted extraction is unlikely to be a serious competitor. However, due to the fact that penicillin G is a well-known product and can now be reliably produced in high yield it is often used as a model for other fermentation products.

It is often found that microbial metabolites are more stable in organic solvent solutions as they are then removed from hydrolytic conditions and enzymic degradation. In the case of compounds such as penicillin G the main incentive for using ion-pair extraction is to avoid the acidic conditions (which may degrade the product) necessary for direct solvent extraction. If a single extraction or a single pass through a countercurrent system gives inadequate extraction there is little opportunity for re-extracting the solution as the product would have extensively degraded although ion-pair extraction was suggested by Reuben and Sjöberg (1981) as a means of recovering penicillin G from waste streams.

Comparative extraction curves for penicillin G in butyl acetate and in the same solvent containing 100 mM Amberlite LA2 (a commercial secondary amine extractant; Rohm and Haas) are presented in Fig. 3.24. In the presence of the amine, good extraction can be achieved at pH values where the penicillin is of acceptable stability.

A number of amine reagents (primary, secondary, and tertiary) were investigated (Reschke and Schügerl 1984a). With some reagents penicillin stability was greater than in pure organic solvent while certain other reagents decreased stability. They found that the picture is rather complex. At pH 5 Amberlite LA2 and Adogen 283 have a stabilizing effect, whereas dioctylamine promotes degradation. At pH 7 Amberlite LA2 still inhibits degradation, whereas dioctylamine slightly inhibits the degradation and Adogen 283 slightly accelerates decomposition. The effect of these reagents on stability may to some extent be due to the fact that they are able to coextract certain ions other than penicillanate which may participate in degradation. However, stability in all the organic phases was higher than in the aqueous phase with which it was in contact.

As many other studies have shown, the choice of carrier solvent is critical, with halogenated solvents being better than esters; hydrocarbons are very poor (Fig. 3.25).

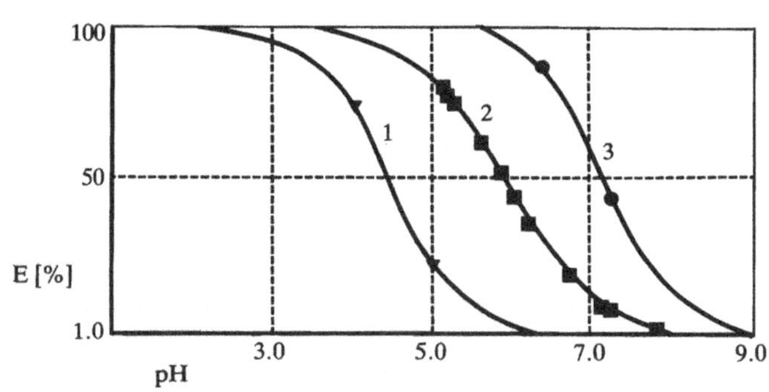

Fig. 3.24. Extraction of penicillin G with butyl acetate or Amberlite LA-2 in butyl acetate: degree of extraction as a function of pH. 1. Butyl acetate. 2. Butyl acetate + LA-2 (equimolar). 3. Butyl acetate + LA-2 (10 fold excess).

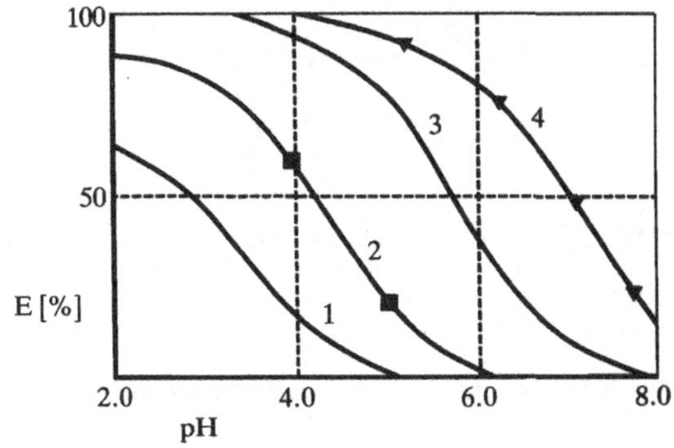

Fig. 3.25. Extraction of penicillin G with LA-2: effect of carrier solvent. 1. Kerosene. 2. Xylene. 3. N-Butyl acetate. 4. Chloroform. All curves obtained using both penicillin and LA-2 at 10 mM.

Quaternary ammonium extractants gave good forward extraction and were largely unaffected by pH but were not investigated in depth as aqueous back extraction was difficult, with a 200-fold excess of sodium chloride being required.

The same group continued their studies by showing that simple extraction equipment such as a Karr column or a four-stage mixer-settler could be used to give greater than 90 per cent extraction of penicillin from culture filtrate (Reschke and Schügerl 1986; Lidikis and Schügerl 1987). Regeneration and recycle of the extractant have not yet been studied.

Some other aspects of liquid anion exchange were investigated during pilot scale studies of carbepenem antibiotics. The pharmaceutical industry investigates many novel compounds but only a small proportion become commercial products. Some are secondary metabolites, initially produced by the wild strain of the micro-organism in very low concentrations, and a limited amount of process development is usually required in order to obtain sufficient material for evaluation. This process development work, which will often include solvent extraction, is rarely published. However, in the case of the carbapenem antibiotics a certain amount of information is available (Butterworth *et al.* 1983). Liquid anion exchange procedures were found very effective as an initial purification stage for these compounds.

Carbapenem antibiotics are based on the 7-oxo-1-azabicyclo-[3,2,0]-hept-2-ene nucleus (Fig. 3.26). The trivial name of carbapenem indicates their structural relationship with penicillins (cf. p. 201). A considerable number have been reported during the past 8 years since the initial report of thienamycin (Kahan *et al.* 1979). They are mostly produced by *Streptomyces* but some have been reported from other bacteria. The producing organisms have worldwide distribution but all appear to produce very low levels of antibiotic and this, combined with the poor stability of the compounds, has

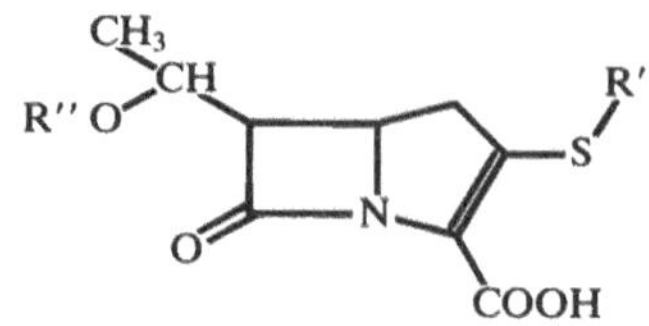

MM13902:	R' = $-CH{=}CHNHCOCH_3$;	R'' = $-OSO_3H$
MM22380:	R' = $-CH_2CH_2NHCOCH_3$;	R'' = $-H$
MM22382:	R' = $-CH{=}CHNHCOCH_3$	R'' = $-H$
Thienamycin:	R' = $-CH_2CH_2NH_2$	R'' = $-H$

Fig. 3.26. Structures of some carbapenem compounds.

Probably been the reason for their being overlooked in the past.

Carbapenems are all carboxylic acids (with the exception of the thienamycins which are amphoteric). Some members of the family have a sulphate ester or sulphonic acid as an additional acidic function.

Stability is often more of a problem with these compounds than with penicillins. In aqueous solution the compounds are relatively stable only over a narrow pH range near neutrality and at low temperatures, preferably below ambient. Stability can also be affected by concentration, choice of buffering agent, and purity. The last factor is important as stability may be variable and less than expected in solutions such as fermentation broth and partially purified extracts containing unidentified impurities. Some of these effects are illustrated in Table 3.3. It is seen that buffering agents with any nucleophilic properties such as phosphate or tris(hydroxymethyl)amino methane are detrimental to stability and the compounds are also less stable with increasing concentration. It has been found that transition metal ions generally promote degradation and many workers have incorporated ethylenediaminetetraacetic acid into reagents used in purification processes.

All carbapenems are released into the culture fluid and no significant amount accumulates in the cells of the producing organism.

The compounds are unstable in aqueous solution at the low pH values where they would be undissociated and liquid–liquid extraction of the free acid does not appear very attractive. However, liquid ion exchange (ion-pair extraction) was found useful in isolating the carbapenem compounds MM4550, MM13902 and MM17880 (Hood *et al.* 1979). This group of compounds has two acidic functions, a carboxyl group and a sulphate ester. It was found that they could be extracted from culture filtrate into certain water-immiscible solvents, especially halogenated solvents such as dichloromethane, containing a hydrophobic quaternary ammonium compound. The data in Table 3.4 show that hydrophobicity and structure of the quaternary ion is crucial to effective extraction. Under the conditions used the tetrabutylammonium ion was ineffective, but dimethyldioctylammonium with only

TABLE 3.3 *Stability of carbapenems in aqueous solution*

Compound	Temp (°C)	Concentration (mM)	Buffer	pH	t_{50} (h)
Thienamycin	25	0.1	0.03 M phosphate	2.0	0.11
	25	0.1	0.03 M phosphate	3.0	0.50
	25	0.1	0.03 M phosphate	4.0	5.8
	25	0.1	0.03 M phosphate	7.0	94.0
	30	0.1	0.02 M borate	8.2	0.17
	30	0.1	0.02 M borate	8.8	0.095
	30	0.1	0.01 M NaOH	12.0	0.035
	37	0.1	0.2 M TRIS[a]	7.5	0.55
	23	0.05	1.0 M MES[b]	7.0	210
	25	60	1.0 M MES	7.0	13.3[e]
	25	140	1.0 M MES	7.0	7.3[e]
	25	210	1.0 M MES	7.0	4.3[e]
	25	620	1.0 M MES	7.0	3.1[e]
MM13902	25	0.137	McIlvane's[c]	2.2	0.042
	25	0.137	McIlvane's	4.0	0.125
	25	0.137	McIlvane's	6.0	4.0
	25	0.137	McIlvane's	7.0	12.0
	25	0.137	McIlvane's	7.8	27.0
	25	0.137	carbonate	10.2	4.0
	28	0.229	0.05 M MES	6.0	30.0
	28	0.229	0.05 M MOPS[d]	7.0	60.0
	28	0.229	0.05 M MOPS	8.0	67.0

[a] TRIS: tris (hydroxymethyl) aminomethane. [b] MES: 2-(N-morpholino) ethane sulphonate.
[c] McIlvane's phosphate-citrate buffer. [d] MOPS: 3-(N-morpholino) propane sulphonate.
[e] t_{30} values: time for 30 per cent degradation.

TABLE 3.4 *Efficacy of quaternary ammonium ions*

Quaternary ion	Extraction efficiency (%)
tetra-*n*-butylammonium	negligible
dimethyldioctylammonium	41
dimethyldidecylammonium	73
benzyldimethyl-*n*-hexadecylammonium	65
methyltrioctylammonium	73

Culture filtrate was extracted with one half volume 5 mM quaternary salt in dichloromethane.

two more carbon atoms extracted 41 per cent of the MM13902 present. A further four carbon atoms in the quaternary salt (dimethyldidecylammonium) increased the recovery to 73 per cent.

Thus the presence of some long and some short alkyl chains in the quaternary ion appears to be important but the quaternary salt must not be too hydrophobic otherwise it is totally excluded from the aqueous phase and is unable to pull out the required anion (Brändström 1977). In practice most work was done using a commercial product consisting mainly of trioctylmethylammonium chloride ('Aliquat 336', Henkel). It was rather interesting to find that another commercial compound of similar specification ('Adogen 464', Ashland Chemical Co.) was only 86 per cent as effective as Aliquat 336. Similar differences were observed between the two compounds when they were used as phase transfer catalysts (Reuben and Sjöberg 1981). It is possible that the compounds possess some undisclosed difference in isomeric composition. Thus the carbon content of the extractant may not be the only criterion for successful extraction and steric effects connected with formation or solvation of the complex may be important.

Carbapenems lacking the sulphate ester group, for example MM22382, are less readily extracted as in this case there is only the carboxyl group to interact with the extracting cation (Harada *et al.* 1983). When extracting MM22382 from culture filtrate, it was found that one quarter volume of 200 mM Aliquat 336 in dichloromethane was required to achieve the same extraction efficiency obtained using one tenth volume 5 mM Aliquat 336 on MM13902. Other structural features of the compound to be extracted are also important. Some carbapenems containing a C-3 saturated side chain are less readily extracted than those containing a side chain double bond. Thus conditions which extracted 80 per cent MM22382 gave only 21 per cent recovery of MM22380, the dihydro analogue.

The extracted complex can be dissociated and the product transferred back to aqueous phase by extracting with an aqueous solution of a suitable inorganic salt. In practice nitrate and iodide have been found suitable and chloride is less effective. In general it was discovered that good dissociating agents give quantitative back extraction when used at one molar equivalent with respect to the quaternary ion. If a dilute quaternary extractant is used a useful volume reduction can be obtained employing a concentrated salt solution as back extractant. A variety of inorganic cations may be used as counterion and the procedure can provide access to a range of salts, for example sodium, potassium, calcium, etc.

The results discussed above for both forward and back extraction were obtained using batch methods but they apply also to continuous flow procedures as equilibrium appears to be rapidly attained. A flow system outlined in Fig. 3.27 consisting of a metering pump for each phase, a static mixer and a disc-stack centrifuge with an overall contact time of 6 seconds gave

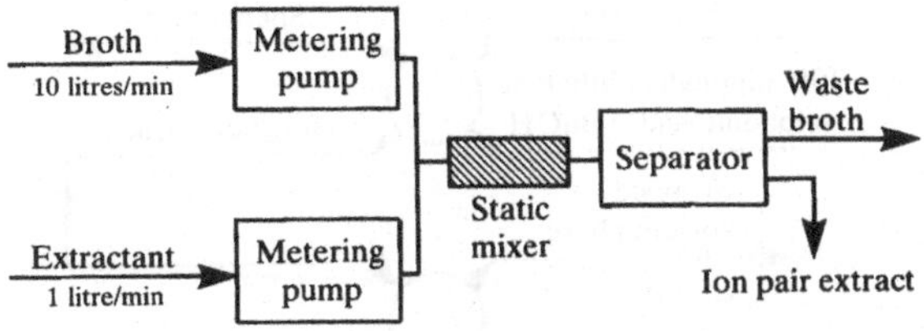

Fig. 3.27. Continuous flow extraction system for MM 13902. Extractant: 0.08 M Aliquat 336 in dichloromethane. Separator: Westfalia TA7-01-506.

results similar to those obtained using batch methods when used for the isolation of MM13902.

Liquid ion exchange can give a more selective extraction of some compounds from culture filtrate than liquid–solid ion exchange. In a comparative experiment, MM13902 was absorbed from filtered broth on to the ion exchange resin Amberlite IRA 458 (Rohm and Haas) and eluted with sodium chloride. After desalting, the product was 6.8 per cent pure. The same broth was subjected to ion pair extraction using Aliquat 336 with sodium nitrate back extraction. The resulting solution was desalted in the same way and gave product 18.8 per cent pure (further purification stages followed).

3.4.2.3 *Liquid ion exchange: amphoteric compounds*

The group of carbapenem compounds provides an example of an amphoteric compound which has been extracted using liquid ion exchange. The principal amphoteric carbapenem is thienamycin. Unlike most other related compounds, the amino group on the C-3 side chain is unsubstituted and is thus able to form a zwitterion with the carboxy group (Fig. 3.26). Three isolation procedures have been published from the laboratories of Merck, Sharpe and Dohm, two using liquid–solid and one using liquid–liquid ion exchange methods. The initial procedure (absorption on to a cation exchange resin, followed by elution with aqueous pyridine) gave a very low yield (Kahan *et al.* 1979). The liquid ion exchange method gave greatly improved recovery and was based on a two-stage procedure using successively the cationic and anionic forms of the compound (Datta and Wildman 1979). The procedure, employing continuous flow techniques, is outlined in Fig. 3.28.

Filtered broth was acidified to pH 3, contacted with 10 per cent dinonylnaphthalene sulphonic acid in butan-1-ol, and back extracted with an aqueous pyridine/phosphate solution. This was immediately adjusted to pH 11 with sodium hydroxide and contacted with 30 per cent Aliquat 336 in butanol. Back extraction was with aqueous potassium acetate. The cation exchange stage yield was 93 per cent and 80–85 per cent yield was obtained from the anion exchange stage, giving an overall recovery of 74–79 per cent. This compares very favourably with the overall yield of 10 per cent for the

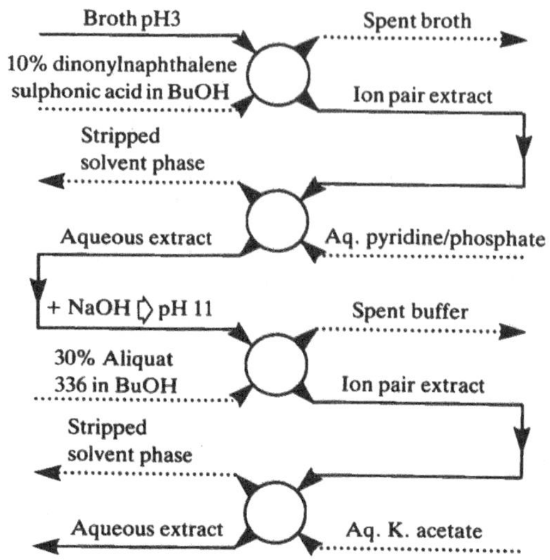

Fig. 3.28. Continuous flow liquid ion exchange extraction of thienamycin. ------ process streams; ...... waste streams. (Adapted from Datta and Wildman 1978.)

first two ion exchange stages using the original resin procedure. The success of the liquid ion exchange procedure was undoubtedly due to the use of continuous flow extractors (Podbielniak). In fact the procedure requires minimal hold-up times (measured in seconds) because thienamycin is very unstable under broth acidic and basic conditions (see data in Table 3.3).

3.4.2.4 *Extractive esterification*

Cephalosporin antibiotics are chemically related to penicillins and much of their biosynthetic pathway is similar. However, in contrast to penicillin G, which is acidic, cephalosporin C, the only compound of the family produced commercially by fermentation, is amphoteric (Fig. 3.29). Thus ionization cannot be suppressed in the same way as with penicillins to render the compound solvent extractable. Cephalosporins share with penicillins a marked instability to dilute alkali and to certain β-lactamases. The hydrolysis products are the compounds analogous to penicilloic acids, the cephalosporanoic acids. In neutral aqueous solution at 37° 2-(D-4-amino-4-carboxybutyl) thiazole-4-carboxylic acid is slowly produced (Jeffery *et al.* 1960). In other respects the cephalosporins are more stable than penicillin G; the β-lactam ring is less sensitive to acid hydrolysis due to decreased nucleophilicity of the nitrogen in the dihydrothiazine ring compared with the thiazolidine ring of penicillins (Abraham 1968).

Most early processes for the isolation of cephalosporin C used carbon absorption and ion exchange resins for isolation. More recently, non-

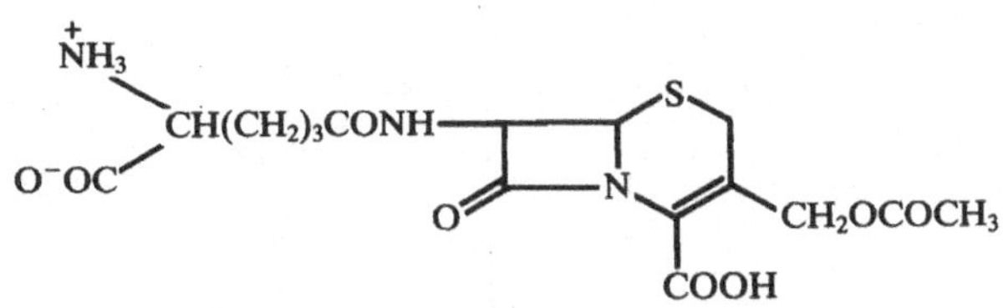

Fig. 3.29. Structure of cephalosporin C.

functional macroporous styrene–divinylbenzene adsorbent polymers have been used. It is possible to isolate cephalosporin C in a number of different ways as has been discussed by Wildfeuer (1985). Since cephalosporin C is used solely as an intermediate in the manufacture of other cephalosporins it could be advantageous to transfer the compound to an organic solvent at an early stage. This can be achieved by blocking the 5′-amino group with a suitable reagent to give an acidic compound which is then amenable to solvent extraction procedures. A number of different reagents have been proposed for this purpose and are the subject of various patents. Acetyl analogues can be prepared using chloroacetyl chloride or diketene to give the N-chloroacetyl (Hayes and Huff 1971) or N-acetoacetyl derivative (Buitar *et al.* 1973). Oxycarbonyl derivatives have been prepared using reagents such as isobutylchloroformate (Johnson *et al.* 1970). Urea derivatives have also been prepared from the amino compound using n-butylisocyanate (Johnson *et al.* 1970) or phenyl isocyanate (Morimoto *et al.* 1972).

Regardless of the reagent chosen, a large excess will be required as it will react with many other amino-containing impurities and some may be hydrolysed by the aqueous environment. A more economical use of acylating agent can be achieved if the product is first partially purified, for example by absorbing from the fermentation broth on to a polystyrene–divinylbenzene resin and eluting with an aqueous alcohol. Alternatively, instead of using a chemical reagent to block the amino group it can be removed by enzymic deamination (Suziki *et al.* 1976).

It is thus possible to extract the derivative from acidified aqueous solution into an immiscible organic solvent. Unfortunately, another problem now arises. The acetoxy group on the 3-side chain is rather labile and readily undergoes hydrolysis to give the 3-hydroxymethyl compound which under acidic conditions forms the lactone (Fig. 3.30). In any case the fermentation broth will contain some desacetyl cephalosporin C as a result of chemical or biochemical hydrolysis and it would be useful to recover this as well as the cephalosporin C to maximize yield of the cephalosporin nucleus. Once the lactone has been formed it is difficult to cleanly and economically regenerate the parent hydroxyacid.

In order to eliminate lactonization during solvent extraction the concept of extractive esterification has been used. The amino group is first blocked

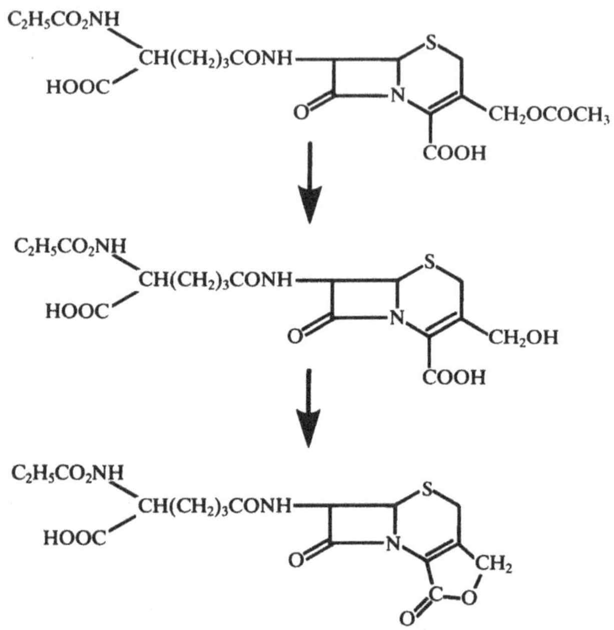

Fig. 3.30. Deacetylation and lactonization of N-ethoxycarbonyl cephalosporin C.

using an acid chloride or a chloroformate ester at 0–5°C and pH 8.0. The latter reagents are preferred as they are less readily hydrolysed by the aqueous medium. Ethyl chloroformate is favoured for a commercial process although trichloroethylchloroformate gives a derivative which is more readily extractively esterified (Bywood *et al.* 1977).

The aqueous solution is then mixed with a solution of a substituted diazomethane (diphenyldiazomethane appears to be favoured) in dichloromethane and the aqueous phase is acidified to pH 2.0–3.5. This stage is performed at 20°C. The bis(diphenylmethyl) ester transfers in high yield to the organic phase and can then be subjected to further reactions. The 5′-amino-5′-carboxypentanamido-side chain is not required in any commercial cephalosporin and it may conveniently be removed at this stage. The N-derivatized ester is treated with phosphorus pentachloride in the presence of a weak base and aqueous methanol to give 7-aminocephalosporanic ester as the hydrochloride which can be crystallized from the mixture. The ester group is removed using formic acid to give 7-aminocephalosporanic acid, the intermediate for many semisynthetic cephalosporin antibiotics.

For certain purposes the 3-hydroxymethyl or 3-chloromethyl cephalosporin derivatives are preferred. Initial treatment of the cephalosporin C and desacetylcephalosporin C mixture with an enzyme from the yeast *Rhodotorula rubra* effectively removes the acetyl group from the 3-position side chain. Protection of the 5′-amino group and extractive esterification can be carried out as before. Treatment of this derivative with phosphorus pentachloride under appropriate conditions gives the diphenylmethyl ester of 7-amino-3-chloromethylceph-3-em-4-carboxylic acid (Bywood *et al.* 1977).

The choice of diazoalkane as extractive esterifying agent is subject to a number of constraints, some general and some specific for these applications. The general constraints are those of cost and stability. Each application will have its own requirements regarding either regeneration of the parent acid or other use of the ester. For cephalosporins, diazodiphenylmethane satisfies all requirements. It is prepared by the reaction of benzophenone hydrazone with peracetic acid in a water–dichloromethane mixture. Iodine is used to catalyse the reaction and a quaternary ammonium salt acts as phase transfer catalyst. Sufficient inorganic base is added to neutralize the acid produced. After the cephalosporin ester has been hydrolysed with formic acid the resulting diphenylmethyl formate may be oxidized to benzophenone and recycled via the hydrazone to diazodiphenylmethane (Bywood *et al.* 1977).

The chemistry of cephalosporin extraction has been discussed at some length as it illustrates some of the considerations required before solvent extraction is chosen as the preferred route.

The use of extractive esterification requires that the two-phase reaction mixture be held for a suitable time (about 1 hour) for reaction to occur before separating the phases. Therefore the solvent extraction equipment will be based on single stage separators.

3.4.2.5 *Extractive hydrolysis*

Extractive esterification was discussed in the context of cephalosporin production. The process for another β-lactam antibiotic, imipenem, uses extractive hydrolysis for one stage of its synthesis. Although originally derived from a fermentation product (thienamycin) this compound is now produced entirely by chemical synthesis. The development of one stage of the process is described by King *et al.* (1985).

It involves the hydrolysis of an O-formyl group to the corresponding alcohol, as indicated in Fig. 3.31. The starting material was in dichloromethane solution and also contained reagents from earlier stages (triphenylphosphine oxide, a hydrazide and formic acid). Originally the hydrolysis was performed with acid in the presence of sufficient methanol to give a monophasic system. The procedure suffered from problems of excessive solvent usage and required large amounts of equipment for operation and solvent recovery.

It was noted that most of the impurities were insoluble in water so separation of the product from the co-occurring impurities without the use of addi-

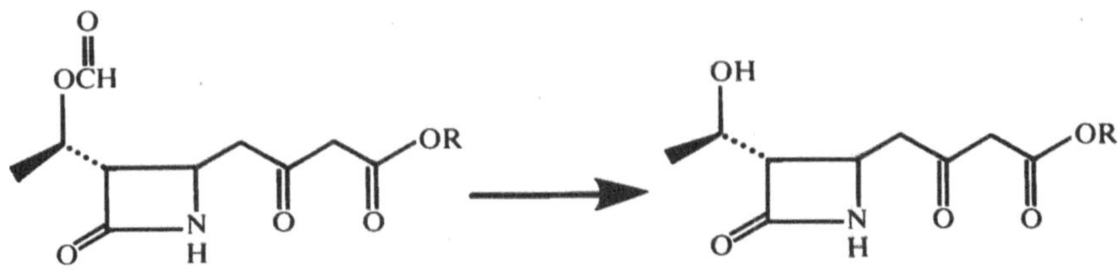

Compound I — Compound II

MeCl, solvent
$(C_6H_5)_3$ P-O
triphenylphosphine oxide
O H H O
|| | | ||
$(CH_3)_2$CH-O-C-N-N-C-OCH $(CH_3)_2$
N, N'-di isopropoxycarbonyl hydrazide
HCOOH
formic acid

>90% pure

Fig. 3.31. Hydrolysis of formyl intermediate of imipenem.

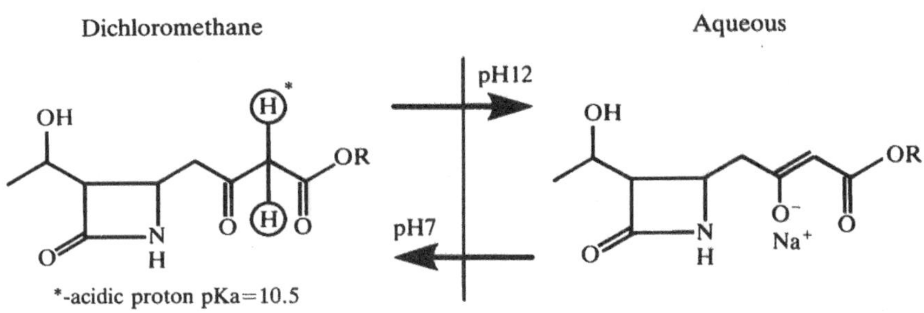

Fig. 3.32. Extraction of imipenem intermediate enolate salt.

tional solvents could be achieved by forming the water-soluble enolate salt. It was found that hydrolysis with aqueous alkali was rapid but both reactant and product were very unstable in alkaline solution, decomposing at 0.3 per cent per minute at 0° C and 10 per cent per minute at 20°C. Using a Podbielniak extractor to operate the process indicated in Fig. 3.32, the reaction time can be equated to the residence time in the extractor. Thus the reactant is rapidly hydrolysed and simultaneously tautomerizes to the enolate salt which transfers to the aqueous phase. This is neutralized immediately after phase separation precipitating the required hydroxy compound. Initially the extractor was used in its normal mode as depicted in Fig. 3.33. Yields were excellent but product purity was less than desired. It was realized that some impurities had a greater water solubility than the product and the extractor was in effect too efficient.

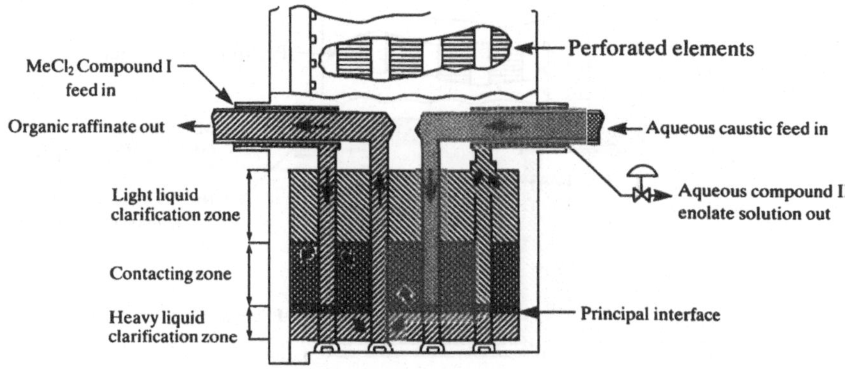

Fig. 3.33. Podbielniak extraction of enolate: normal extraction mode.

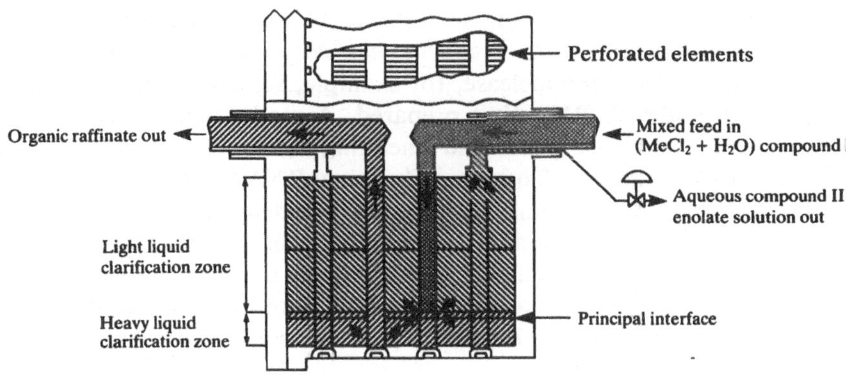

Fig. 3.34. Podbielniak extraction of enolate: single stage extraction mode.

The extractor operation was modified so that the premixed hydrolysis mixture was fed to the light phase inlet. This had the effect of converting the countercurrent extractor to a single stage separator (Fig. 3.34). A further refinement incorporated a backwash stage using fresh dichloromethane, as indicated in Fig. 3.35.

3.4.2.6 *Extractions using phenols*

Phenol is one of the few water-immiscible solvents in which proteins are soluble. Although a solid at ambient temperature, it may be used as 90 per cent aqueous solution or admixed with m-cresol. On passing into the phenol phase with its combination of strong hydrogen bonding properties and π-π interactions, proteins lose, often irreversibly, their natural conformation. Thus phenol extraction is used for removing protein from other biological macromolecules such as nucleic acids, polysaccharides, and lipids. It is

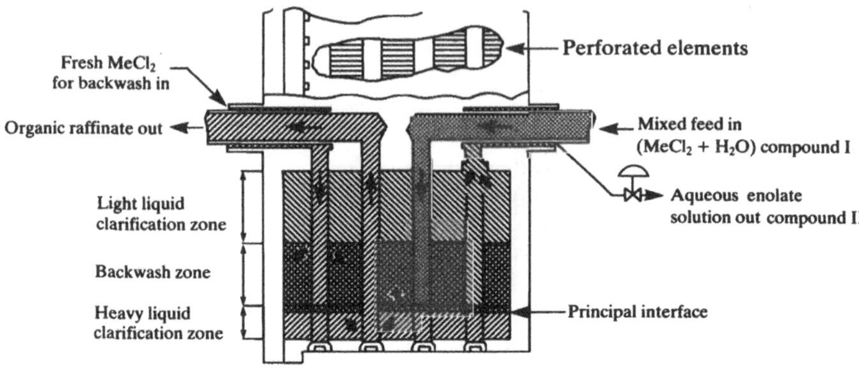

Fig. 3.35. Podbielniak extraction of enolate: single stage with backwash mode.

especially useful when isolating such products from degradative enzymes, but one cannot assume that the enzyme has been completely inactivated. It has been found that ribonuclease, for example, survives phenol extraction and may degrade the RNA thus prepared. A modification introduced by Kirby (1965) used detergents and chelating agents to affect additional ribonuclease degradation. To isolate ribosomal RNA from rat liver the fresh tissue was dropped into liquid nitrogen and blended with a 1 : 1 mixture of 0.5 per cent aqueous disodium naphthalene disulphonate and phenol/cresol/8-hydroxyquinoline mixture. The separated aqueous phase was treated with 5 per cent triisopropylnaphthalene sulphonate and extracted again with the phenol/cresol mixture. Further separation stages gave largely undergraded RNA.

A simple phenol extraction system can be used to deproteinize virus particles prior to purification of nucleic acid from the core.

Two-phase systems containing phenol have been employed extract lipopolysaccharide for biochemical studies from bacterial cell walls.

Lipopolysaccharide forms a major part of the outer membrane of gram negative bacteria. It is a complex molecule consisting of lipid A, oligosaccharide and O-antigenic side chain. Lipid A is always present and usually consists of diglucosamine phosphate substituted with a variety of 2- and 3-hydroxy fatty acids unique to this material. The oligosaccharide consists of ten to twelve sugar units, largely 2-keto-3-deoxyoctonate and mannoheptose derivatives. The O-antigenic side chain, which varies with the bacterial serotype, consists of repeating units of tri- to pentasaccharide units.

The most hydrophilic types of lipopolysaccharide (from smooth type bacteria) can be extracted using phenol–water 45 : 55 v/v at 69°C. This mixture is monophasic above 68°C. On cooling it forms two phases, with the aqueous phase containing the lipopolysaccharide and nucleic acid. With certain strains of bacteria which do not possess an O-antigenic side chain and possibly no heptose either, the above method is unsatisfactory as the lipopolysaccharide, being in this case more hydrophobic, partitions into the phenol phase and is thus not separated from protein. A modified solvent system devised by Galanos *et al.* (1969) consisting of 90 per cent phenol, chloroform, and petroleum ether (b.p. 40°–60°) 2 : 5 : 8 was found to be effective, giving a product free of protein and nucleic acid.

A more recent method which extracts lipopolysaccharide from both rough and smooth type bacteria uses enzymes and detergents to remove contaminating materials (Darveau and Hancock 1983).

3.4.3 Summary

It appears that chemically assisted extraction is not widely practised in the pharmaceutical industry but, as always, the proviso must be made that many secret processes are operated which appear in the literature only after they have been superseded by improved procedures or have eventually become common knowledge. The foregoing examples indicate that chemically assisted extraction has been used or considered for the purification of several different types of compound. The examples give an indication of some procedures which are feasible and various practical points are highlighted.

Phenol extraction is normally used only for the preparation of small-scale experimental batches of material.

3.5 Supercritical extraction

[See also Hamm Chap. 4, Weatherley Chap. 5]

Extraction with supercritical fluids has been a known technique for about 100 years but it is only in the last 25 years or so that the method has been seriously studied. The main areas of present application are the food and petrochemical industries. Supercritical chromatography has also been under investigation for some time and commercial analytical scale equipment is now available.

The most convenient and mostly widely used supercritical fluid used is carbon dioxide, which, compared with conventional liquid/liquid extractants, has the following advantages:

(1) solvent residues are non toxic;

(2) the solvent is not flammable;

(3) it is inexpensive;

(4) the solvent is volatile and readily removed from the product;

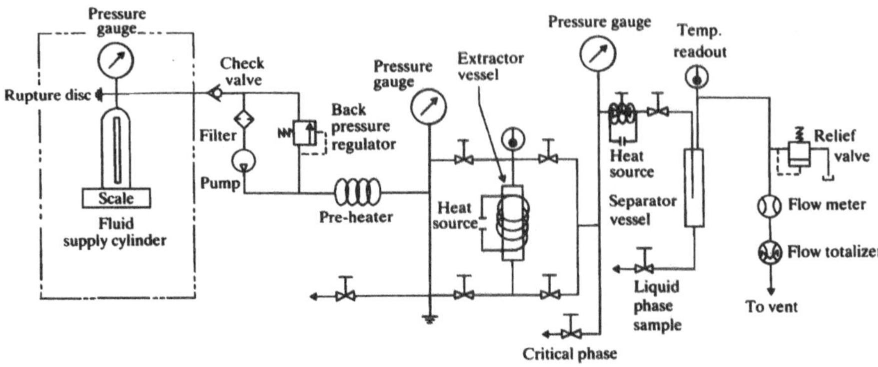

Fig. 3.36. Test apparatus for supercritical extraction.

(5) it has low viscosity and surface tension and can thus easily penetrate into solid matrices;

(6) solvent power can be readily altered by changing the pressure.

The one significant disadvantage is the necessity for high pressure equipment, giving rise to high capital costs. Another disadvantage of carbon dioxide as an extractant of general applicability is its relative low polarity. Various alternative fluids, for example sulphur dioxide, ammonia, or hydrocarbons, can be used for supercritical extraction but they all possess various practical disadvantages such as toxicity or flammability. Polarity modifiers ('entrainers') have been investigated. It has been found that supercritical extraction of nicotine from tobacco with carbon dioxide requires some water to be present in the system (Roselius *et al.* 1979). Large-scale plant is in operation for applications such as the removal of caffeine from coffee and the extraction of hop oils. A general discussion of the pertinent features of supercritical extraction as applied to natural products is given by Steytler *et al.* (1986) and a comprehensive treatise on the technique has been written by McHugh and Krukonis (1986).

For evaluation of the potential of supercritical fluid extraction small-scale equipment is available. An example is illustrated in Fig. 3.36. This device is manufactured by Autoclave Engineers (Erie, Pennsylvania, USA). The extractor vessel is loaded with the material to be extracted. A liquid carbon dioxide supply is connected to the pump (a pump designed for high-performance liquid chromatography is suitable) and the pump outlet is connected to a heat exchanger where the fluid is converted to the supercritical state. The fluid passes through the extractor then to the separator vessel where the pressure is lower and the fluid is no longer supercritical. Solutes extracted from the sample now precipitate and are deposited. In the experimental apparatus the gas is vented to atmosphere. Using this equipment, extractability of a compound can be measured gravimetrically either by weighing the product collected or by weight loss from the sample. The latter is preferable because some product may collect in the lines connecting the extractor to the collecting vessel.

Alternatively, the potential for supercritical extraction can be assessed by microscale extraction (2 cm^3 extracting chamber), collecting the product on the surface of a thin layer chromatography plate for subsequent analysis (Stahl *et al.* 1978). They found, using carbon dioxide at 40°C, that substances containing polar functional groups are difficult to extract, as would be expected from the low polarity of the extracting fluid. Sugars and amino acids are not extractable; even at 2000 bar their solubility is less than 1 : 200 000. The results obtained from a series of hydroxy acids is illustrative of the effect of polar substituents (Table 3.5).

A wide range of plant products were found to be extractable at 80–90 bar, for example cannabinol, colchicine, codeine, atropine, vincristine (Figs. 3.37–3.41). For certain compounds supercritical nitrous oxide was found to be a better extractant but there were also examples where the reverse was true. Nitrous oxide having a dipole moment (unlike carbon dioxide) is better able to interact with certain polar groups.

Some additional work on pharmaceutical compounds has been reported by Larsen and King (1986). They determined the solubility of several pharmaceutical compounds in supercritical carbon dioxide, obtaining the data summarized in Table 3.6 (p. 262).

It will be noted that the zwitterionic compound imipenem (Fig. 3.42) is insoluble in supercritical carbon dioxide. Mevinolin (Fig. 3.43), which is moderately polar and has low solubility. It is interesting to note that the considerably more complex molecule efrotomycin (Fig. 3.44) is of similar solubility. It was confirmed that solubility can be enhanced by the addition of a small proportion of a polar solvent (or 'entrained solvent'). For example, the addition of 5 per cent methanol increased mevinolin solubility by an

TABLE 3.5 *Effect of polar substituents on supercritical extraction with carbon dioxide*

Compound	Extraction starts at (bar)
Benzoic acid	80
Salicylic acid (2-hydroxy benzoic)	85
p-Hydroxy benzoic acid (4-hydroxy benzoic acid)	120
Genistic acid (2,5-dihydroxy benzoic acid)	120
Gallic acid (3,4,5-trihydroxy benzoic acid)	not extractable

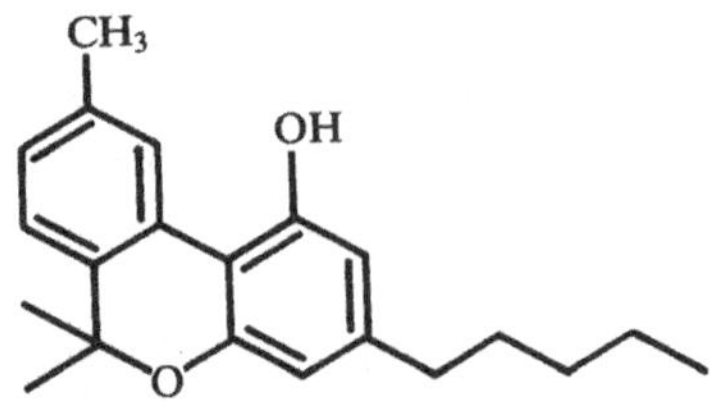

Fig. 3.37. Structure of cannabinol.

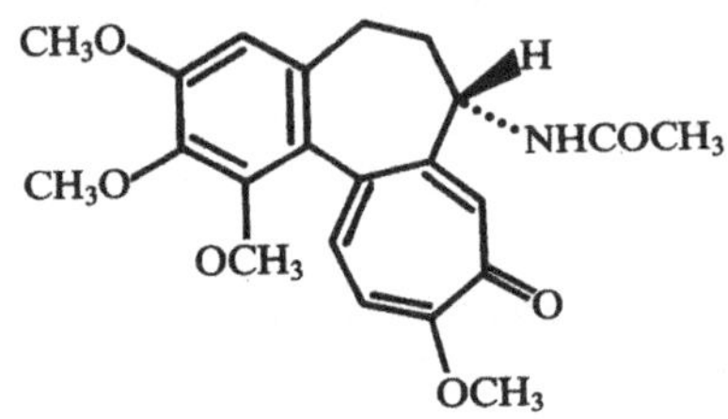

Fig. 3.38. Structure of colchicine.

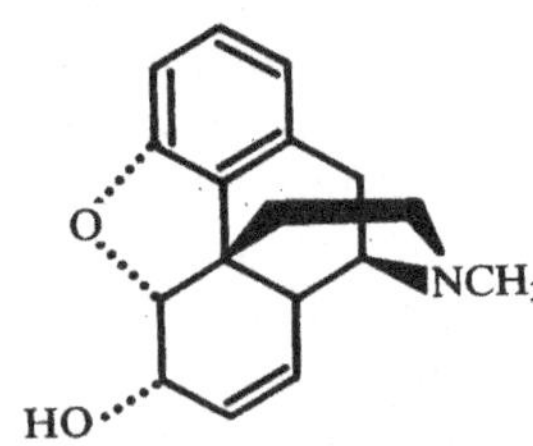

Fig. 3.39. Structure of codeine.

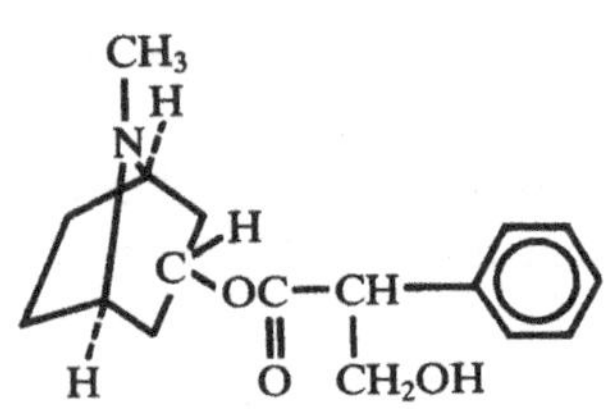

Fig. 3.40. Structure of atropine.

order of magnitude. This compound exists in many systems as a mixture of the lactone and hydroxyacid structures, the latter predominating under neutral and basic conditions. It was found that during the extraction procedure, using 3 per cent methanol in carbon dioxide, the proportion of the lactone increased. It was ascertained that this was due to conversion and not to selective extraction. Should the hydroxy acid be the preferred product this could be obtained by replacing the methanol with 3 per cent *t*-butylamine in

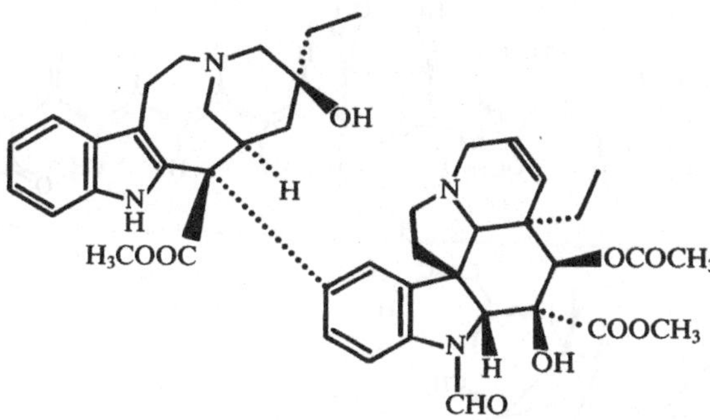

Fig. 3.41. Structure of vincristine.

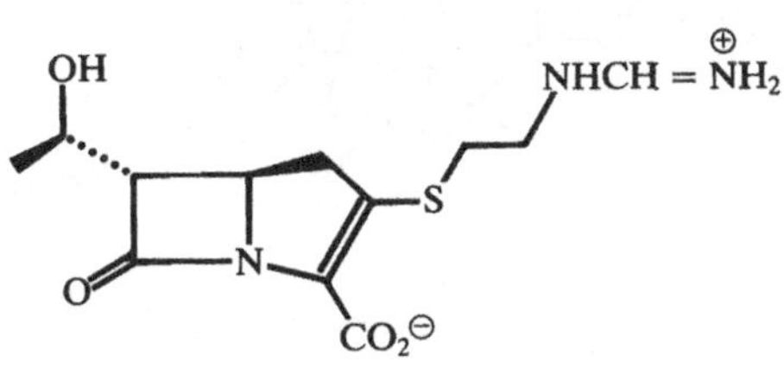

Fig. 3.42. Structure of imipenem.

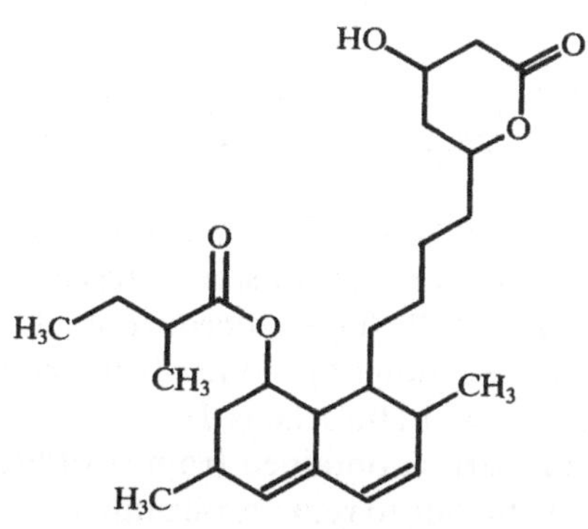

Fig. 3.43. Structure of mevinolin.

the solvent. However, even with the addition of 5 per cent methanol, acetone or water, imipenem remained insoluble. This confirms the studies of Stahl *et al.* (1978), who reported that very polar compounds are insoluble in carbon dioxide even at pressures of 2000 bar.

It was observed that some products recovered from supercritical extraction were in the form of microfine crystals (10–50 μm). X-ray diffraction studies indicated that the particles had the same crystal structure as the starting material prior to supercritical extraction. The procedure was

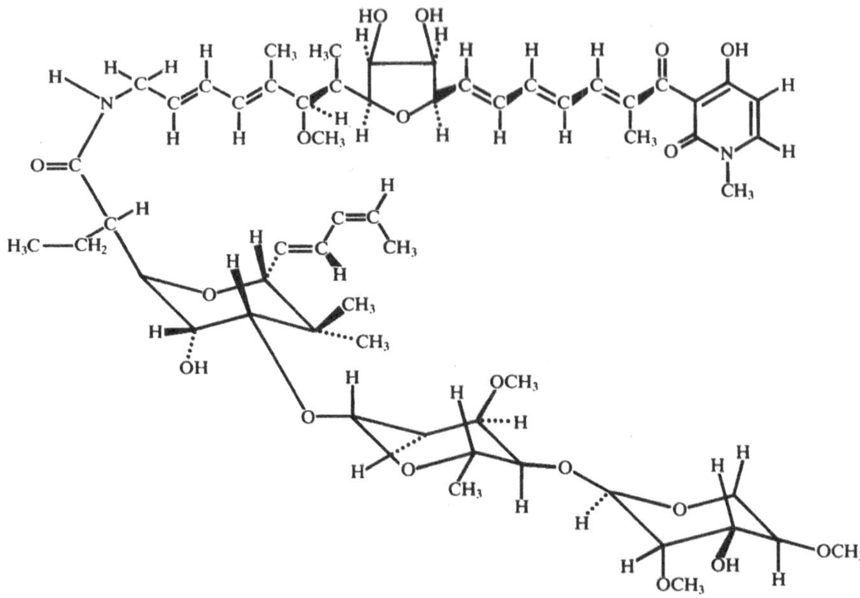

Fig. 3.44. Structure of efrotomycin.

suggested as an alternative to milling, having the advantage of avoiding some conditions (high temperature, shear etc.) which are detrimental to certain compounds.

It has been reported that supercritical extraction can kill microbial cells and it has been suggested as a means of sterilizing certain heat-sensitive products (Taniguchi *et al.* 1987). The procedure could have potential as a preliminary stage in the isolation of products from pathogenic organisms. A patented process (Heigel and Hueschens 1983) for the production of pure lecithin from the crude extract obtained from soya beans describes the use of supercritical carbon dioxide to remove oily impurities. The elimination of solvent residues compared with the established process (which uses acetone for de-oiling) is an obvious advantage. It would thus seem that there is a niche for supercritical extraction in the pharmaceutical industry.

It has been noted that solvent residues are less important in a primary extraction stage as the compound will usually pass through several more purification procedures before pure product is obtained. However, a final stage supercritical extraction has been suggested as a means of eliminating solvent residues derived from preceding stages (Kamihira *et al.* 1987). Figure 3.45 shows the rate of extraction of various solvents from potassium penicillin G. It is seen that supercritical carbon dioxide is a poor extractant for water. Even so it was found that 0.1 per cent entrained water in the fluid

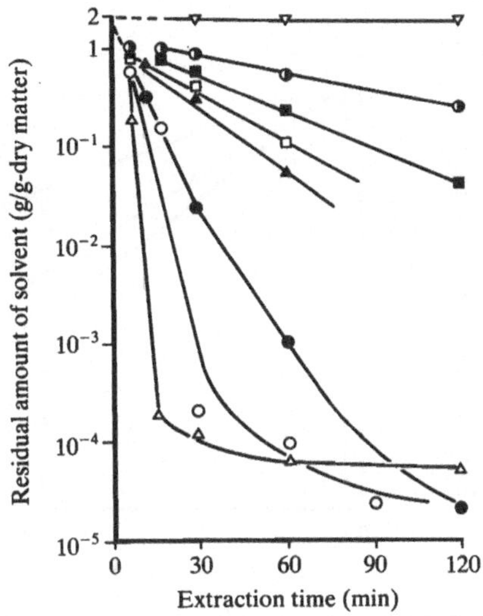

Fig. 3.45. Extraction of solvents from penicillin G using supercritical carbon dioxide. △ acetone; ○ methanol; ● ethanol; ▲ 2-propanol; ▫ ethyl acetate; ▪ n-butyl acetate; ◑ 1-butanol; ▼ water. (From Kahimura *et al.* 1987.)

increased the rate of extraction of butanol and butyl acetate which were relatively slowly extracted with carbon dioxide alone (Fig. 3.46).

Supercritical fluid extraction appears to be an economic alternative to conventional solvent extraction for several relatively inexpensive products in other industries and it must be concluded that the main factor retarding its adoption in the pharmaceutical industry is its solvent power. Many pharmaceutical compounds are very polar compounds which are not amenable to carbon dioxide extraction. The use of supercritical fluids with greater solvent power which are usually toxic or flammable outweighs the advantages of the technique.

3.6 Enzymes in non-aqueous systems

3.6.1 General considerations

Enzymes are proteins consisting of mainly linear polymers of amino acids. The polymer conformation is determined predominantly by hydrogen bonding and hydrophobic interactions between various amino acids in the molecule. Covalent crosslinks, usually disulphide bonds, and chelated metal atoms are often present as well but only in relatively small numbers compared with the number of other interacting groups. The net result is that an enzyme molecule has a very precise arrangement of hydrophobic and hydrophilic domains, the former usually within the molecule and only accessible

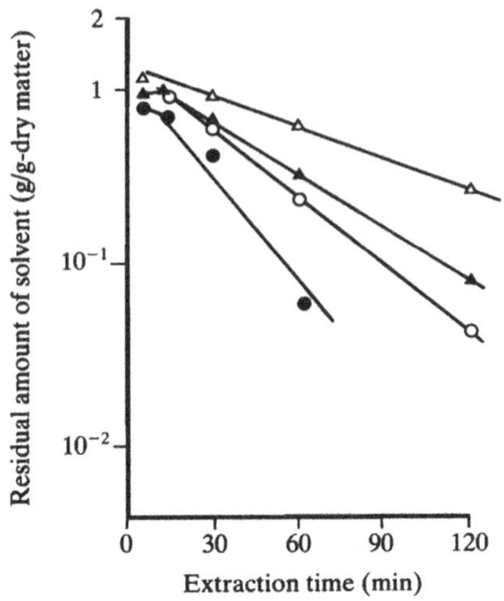

Fig. 3.46. Extraction of solvents from penicillin G using supercritical carbon dioxide: the effect of 0.1 per cent water as entrainer. ○ *n*-butyl acetate; △ 1-butanol. Closed symbols: with water; open symbols: without water. (From Kahimura *et al.* 1987.)

to specific substrates. If the medium surrounding the molecule is changed from water to a less polar solvent the overall conformation may be changed, causing the hydrophobic clefts to open up and embrace the new solvent. Enzyme activity is thus destroyed. However, if the enzyme can be maintained in its 'native', that is hydrated, conformation there is no reason why it should not be used to catalyse a reaction in an organic solvent.

The brief description of enzyme molecules given above indicates that some water must be present in the system but in some cases it appears that, provided there is sufficient water intimately associated with the enzyme molecule itself, it may be possible for it to function with very little additional water in the system.

Enzymes used in organic solvents have a number of advantages and applications.

1. It is possible to solubilize otherwise insoluble substrates or products.
2. It may be possible to carry out new reactions which kinetic or thermodynamic restrictions make impractical in water. For example, certain transesterifications may be more efficient in the absence of water.
3. Product recovery may be facilitated from an organic solvent solution.
4. The insolubility of enzymes in organic media makes recovery and reuse a simple matter.
5. Microbial contamination is minimized.

Two possible disadvantages of aqueous/organic systems are:

(1) enzyme denaturation at the interface;

(2) enzyme denaturation in the aqueous phase due to dissolution of the organic solvent.

However, organic systems are less suitable for cofactor dependant reactions because common cofactors are insoluble in these media (but see below).

The best solvents for enzyme stability appear to be non-polar liquids of relatively high boiling point, for example higher alkanes (Laane *et al.* 1985). Under certain conditions enzyme activity may be retained in more polar solvents. Zaks and Klibanov (1984) have reported that porcine pancreatic lipase is stable and catalytically active at 100°C in 2 M heptanol in tributyrin. It was found essential that the water content of the system be very low. With 0.015 per cent water the half life of the enzyme was over 12 hours at 100°C, whereas it was reduced to 15 minutes in the presence of 0.8 per cent water. The extra water increased conformational mobility of the protein, resulting in irreversible inactivation. It was shown that the dry lipase had a narrower substrate specificity, suggesting that it had a more rigid conformation, thus accounting for the greater thermostability.

Many enzymes are used as cell suspensions and in this case it is more difficult to optimize conditions for non-aqueous operation. One way to overcome the problem is to entrap the cells in a hydrophilic gel such as calcium alginate. With a hydrophobic substrate it may be necessary to employ a polar solvent to promote penetration of the substrate into the gel. It will be necessary to find a compromise between a suitable partition between the phases and inactivation of the entrapped enzyme.

Some enzymes can be used directly in aqueous/organic solvent systems in either a single phase or two immiscible phases. The former, using solvents such as lower alcohols or dimethylformamide, do not involve phase transfer and are outside the scope of this chapter.

3.6.2 Applications

The majority of pharmaceutical compounds rely on precise stereochemistry for their activity and often one enantiomer lacks biological activity or may even exhibit undesirable properties. There is therefore an increasing interest in the use of enzymes in enantioselective organic synthesis (Butt and Roberts 1987). The scope can be greatly enhanced by the use of non-aqueous or aqueous/organic reaction systems as illustrated by the following examples.

3.6.2.1 *Enzymic reactions*

(1) *Reactions inhibited by water.* Transesterification reactions catalysed by various lipases require minimal amounts of water. Indeed in the presence of excess water the normal hydrolytic activity of lipase would predominate. Zaks and Klibanov (1985) reacted lipases from three different sources (mammalian, yeast, and fungal) with glyceryl tributyrate and heptanol in pentan-2-one. The enzymes were added as 'dry' powder containing 0.5, 6.1, and 4.8 per cent water respectively. An additional 1 per cent water (with respect to the reaction medium) increased the rate of reaction of the mammalian and yeast enzymes but did not affect the fungal lipase. Thus an enzymic reaction is apparently occurring in a 99 per cent organic phase. However, as far as the enzyme molecule is concerned it still thinks it is in an aqueous system as all it can 'see' is a surrounding layer of water a few molecules thick. Therefore in the context of liquid–liquid partition the water layer is extracting the reactants, in this case the ester and the alcohol, transporting them to the active site of the enzyme where the ester is hydrolysed and the acyl group so formed is used to re-esterify any suitable hydroxy group. Since the alcohol is the predominant hydroxyl-containing compound in this case it is acylated to the appropriate ester. Thus in the above studies heptyl butyrate was formed (Fig. 3.47). This reaction would not occur in the presence of excess water. The product of transesterification is then exported via the aqueous layer surrounding the enzyme molecule back to the organic phase. Similarly, direct esterification of butyric acid with heptanol each at 0.3 M catalysed by porcine pancreatic lipase in water reached equilibrium at less than 0.1 per cent conversion whereas in *n*-hexane the reaction exceeded 90 per cent conversion after 2 hours.

Conventional enzymology states that an enzyme has a pH optimum. How does this apply to a non-aqueous system? Zaks and Klibanov (1985) found that the 'pH history' of the enzyme preparation affected the rate of reaction. Thus the enzyme powders used above were precipitated from aqueous solutions of various pH values and it appeared that the enzyme 'remembered' its

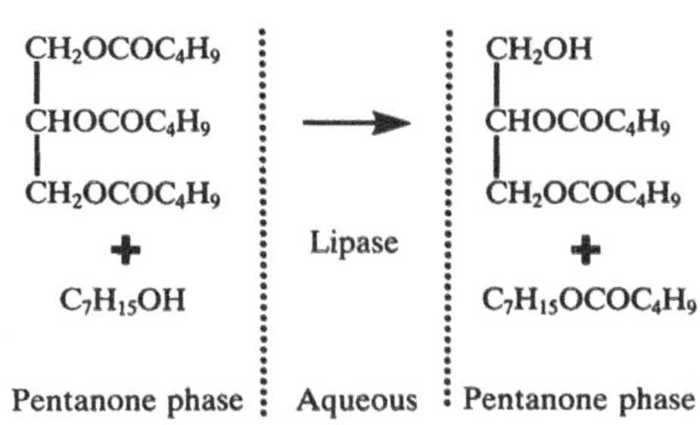

Fig. 3.47. Transesterification of glyceryl tributyrate and 1-heptanol.

previous environment, that is the charges and ionization states acquired by the enzyme molecule in water were retained in the solid state.

An example of a lipase-catalysed regioselective transesterification is reported by Therisod and Klibanov (1986). D-glucose, 2,2,2-trichloroethyl butyrate, and lipase were reacted in anhydrous pyridine for two days at 45 °C. Analytically pure 6-O-butyryl glucose was isolated and represented 82 per cent of all reaction products. It appeared from this, and other studies, that the reaction is very specific for primary hydroxyl groups.

(2) *Reactants solubilized by organic solvents.* Enzymic transformation reactions are widely used in the manufacture of steroidal drugs but they are usually performed using whole cells in aqueous suspension with subsequent product recovery by solvent extraction, sometimes on 'whole broth'. Some reports have shown the feasibility and some advantages of performing enzyme catalyzed steroid transformations in a two-phase aqueous/organic mixture. As judged by more recent literature, the methods have not been followed up, although if they have found industrial application the information may have been 'classified'.

It was found that *Nocardia rhodocrous* cells entrapped in urethane-based polymers will catalyse the dehydrogenation of various steroids in a variety of solvent systems. The choice of entrapment polymer and solvent depends mainly on the hydrophobicity of the substrate. This can be illustrated by oxidation of the very hydrophobic compound cholesterol to cholestenone (Fig. 3.48). A gel which is too hydrophilic impedes access of the substrate to the enzyme. A solvent that gives a high partition coefficient will not allow sufficient substrate to pass into the aqueous phase. Thus the rate limiting factor appears to be the partition of the substrate between the solvent and the entrapment medium (Omata *et al.* 1979). An organic phase consisting of benzene : *n*-heptane 1 : 1 gave poor results, whereas chloroform : *n*-heptane 1 : 1 was more successful. The more polar solvents must be used with caution as there is the possibility of inactivating the enzyme.

It was found that the same reaction can also be performed using a free suspension of cells in an aqueous/organic mixture (Buckland *et al.* 1975). A number of solvents (for example toluene, hexadecane, carbon tetrachloride) were effective but the latter was chosen on the basis of its non-flammability.

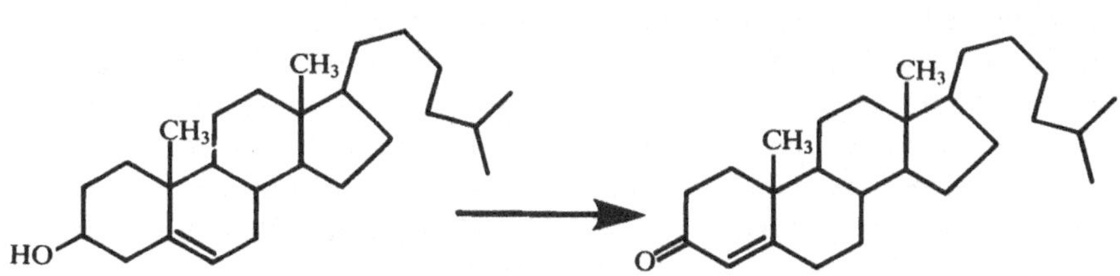

Fig. 3.48. Oxidation of cholesterol to cholestenone by *Nocardia rhodocrous*.

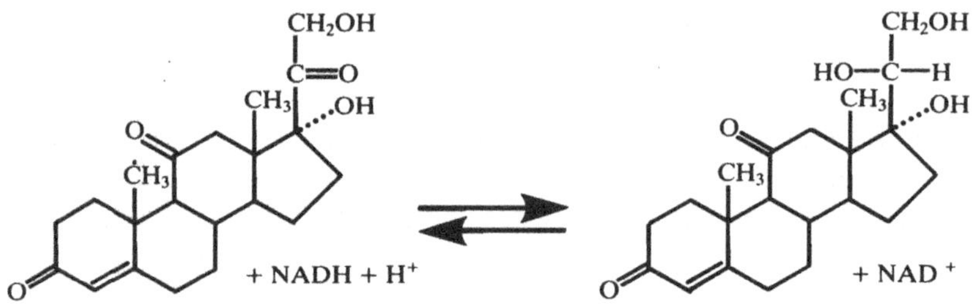

$$C_2H_5OH + NAD^+ \underset{}{\overset{ADH}{\rightleftharpoons}} CH_3CHO + NADH + H^+$$

$$CH_3CHO + NH_2NH\overset{O}{\overset{\|}{C}}NH_2 \longrightarrow CH_3CH = N\,NH\overset{O}{\overset{\|}{C}}NH_2$$

Fig. 3.49. Reduction of cortisone using a coupled enzyme system.

The cell suspension used seven times still retained 50 per cent of its activity. It appears that the degree of agitation requires careful optimization in two-phase aqueous/organic enzymic reactions especially where the dissolution of oxygen is a rate-determining factor.

Another example from the steroid area is provided by the reduction of 20-ketosteroids such as cortisone to pregn-4-en-17α, 20β, 21-triol-3, 11-dione (Cremonesi *et al.* 1975). The cofactor was regenerated by coupling the reaction to the alcohol dehydrogenase catalysed oxidation of ethanol to acetaldehyde. This equilibrium reaction is driven to the right by the addition of semicarbazide to react with the aldehyde (Fig. 3.49). It was necessary to find a compromise between enzyme activity/stability and solvent polarity. As already mentioned, enzymes are at their most stable in the presence of non-polar organic solvents but in this case it was necessary to use an ester solvent for reasons of solubility.

A further example of the advantage of two-phase aqueous/organic systems is given by the enantioselective hydrolysis of (±)-menthyl succinate to (−)-menthol using gel-entrapped cells of the yeast *Rhodotorula minuta* var. *texensis*. The succinate is water soluble, but using the yeast cells in aqueous suspension the product, (−)-menthol, being insoluble, formed an impervious coating on the cells and impeded further reaction. The use of gel-entrapped cells in *n*-heptane gave good conversion and also increased the stability of the enzyme from a half life of 2 days in free cells to 63 days in the gel (Omata *et al.* 1981). It should be noted however that complete elimination of the aqueous phase confers further advantages to the system. Using a (±)-menthyl laurate/isobutanol transesterification system, it was noted that the optical purity of the (−)-menthol produced was higher using only heptane as the reaction medium compared to an emulsified heptane/aqueous buffer (Langrand *et al.* 1986).

(3) *Cofactor-dependent reactions.* It was mentioned earlier that cofactor-requiring enzymic reactions could be problematical in non-aqueous systems. One method which has been proposed by Grunwald *et al.* (1986) involves the use of an enzyme–cofactor mixture adsorbed on a solid carrier (glass beads). Alcohol dehydrogenase was used to effect chiral oxido-reduction reactions such as the oxidation of (±)-*trans*-3-methylcyclohexanol to (−)-3-methylcyclohexanone or the reduction of (±)-2-phenylpropion-aldehyde to (−)-2-phenylpropanol. The philosophy was that since the enzyme and cofactor are both insoluble in organic media and therefore inactive they could be intimately mixed and would remain 'dormant' until activated by the presence of a suitable substrate with just sufficient water for transfer of the reactants to the enzyme–cofactor mixture. Reactions were performed in ethyl acetate or isopropyl ether presaturated with aqueous buffer. Continuous regeneration of the cofactor was achieved using a suitable alcohol or aldehyde, as indicated in Fig. 3.50. The system was stated to have the dual advantage of increased solubility of certain reactants and higher cofactor turnover numbers. However, under the conditions used rates of reaction were rather slow.

(4) *Gaseous rectants.* Molecular oxygen is about ten times more soluble in hydrocarbons than in water and thus enzymic reactions involving this reagent may be carried out advantageously using non-aqueous systems. A model system studied by Brink and Tramper (1987) was the oxidation of propene to propene oxide with oxygen using as catalyst the enzyme present in a strain of *Mycobacterium*. The bacterial cells were immobilized using calcium alginate, dispersed in glass beads and packed in a column. Using *n*-hexadecane as solvent, the two reactants could be passed through the column to give good conversions. In aqueous solution oxygen depletion was a problem due to its lower concentration in the inlet stream and a gas/liquid/solid trickle bed would be necessary to provide an adequate supply of

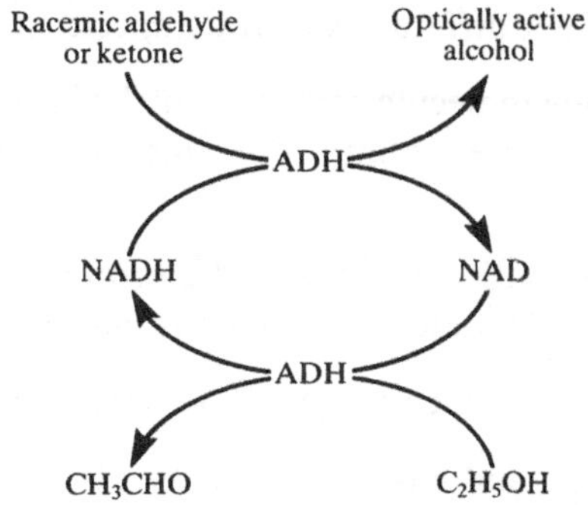

Fig. 3.50. Enantioselective reduction of carbonyl compounds to alcohols in non-aqueous solvents. Enzyme and cofactor immobilized on glass.

oxygen. In the organic system it was found that the static water film surrounding the immobilized cells was not a rate-limiting factor.

(5) *Enzymes in supercritical fluids*. It has been shown that enzymic reactions can occur in supercritical carbon dioxide (Randolph *et al.* 1985). Using alkaline phosphatase, a 71 per cent conversion of *p*-nitrophenyl phosphate to *p*-nitrophenol was obtained after 8 hours at 35°C and 100 bar pressure.

(6) *Biosensors*. An interesting application of a non-aqueous enzymic reaction is provided by the development of a 'temperature abuse' sensor. When marketing potentially labile products it is important to be able to document the 'temperature history' of the product.

Horseradish peroxidase catalyses the reaction between *p*-anisidine and hydrogen peroxide to give an intensely yellow-coloured product. Under appropriate conditions the reaction can occur in liquid hydrocarbons but not in the solid phase. By mixing various alkanes, for example tetradecane, pentadecane and hexadecane, a mixture of defined melting point around ambient temperature is produced. This can be used as the medium for the above enzymic reaction. Providing the system remains in the solid state no reaction occurs. If the temperature rises to a point where the hydrocarbon melts rapid reaction occurs to give a yellow colour which is retained on cooling and solidifying (Boeriu *et al.* 1986). The intensity of the colour is proportional to the time the mixture is above the melting point and can therefore be used to ascertain the time that the system was above the predefined temperature.

In summary it has been shown that enzymic reactions apparently occurring in non-aqueous systems are actually proceding via an organic/aqueous phase transfer and *vice-versa*. Such systems have a number of potential advantages:

(1) organic substrates are solubilized and reaction rates are enhanced;

(2) enzymic reactions which are impractical in water can be carried out. The enzyme may have a different substrate specificity;

(3) reactions involving molecular oxygen may be more rapid, as the gas has up to ten times greater solubility in organic solvents compared with water;

(4) microbial contamination leading to the production of extraneous proteases which may degrade the enzyme is eliminated;

(5) thermal stability of some enzymes is greater in non-aqueous systems;

(6) the regioselectivity and enantioselectivity of enzymes can be extended to a wider range of applications;

(7) novel applications of enzymes can be devised.

3.6.2.2 *Use of reverse micelles*

Reverse micelles consist of aqueous droplets enclosed in a liquid membrane of surfactant molecules suspended in a non-aqueous continuous phase. The hydrophilic 'head' of the surfactant is directed towards the centre of the micelle and the hydrophobic 'tail' interacts with the organic liquid. The aqueous core contains a certain number of water molecules tightly bound to the head groups of the surfactant. In the case of AOT (sodium 1,2-bis (2-ethylhexyloxycarbonyl)-1-ethane sulphonate) studied by Luisi (1985) this number is 6 to 8. Only when it is exceeded does 'free' water exist within the micelle. The aqueous core can contain dissolved enzyme at high concentration, and due to the high surface area available, rapid rates of reaction are theoretically possible. Inactivation of the enzyme by the surfactant is a potential problem. This may be the reason for the 30 per cent loss of activity reported by Dekker *et al.* (1986) when purifying α-amylase via reverse micelles. The enzyme was transferred from aqueous phase at pH 9.85 to an isooctane phase in the form of reversed micelles with trioctylmethylammonium chloride and was transferred back into aqueous phase by contacting with a buffer of pH 6.9 resulting in an eightfold concentration of enzyme. The enzyme has a pI of 5.2 and will therefore have a net negative charge at pH 9.85. The quaternary ammonium surfactant has a net positive charge at this pH value. It is stated by Dekker *et al.* (1986) that transfer of the enzyme from aqueous solution to the reversed micelles occurred when the charges on the enzyme and surfactant are equal and opposite. Moving away from this condition, to pH 6.9, thus means that the reversed micelle can no longer retain the enzyme which is transferred back to aqueous solution.

The use of reverse micelles for the isolation of enzymes is reviewed by Kadam (1986).

Having encapsulated the enzyme in a reverse micelle, how can it be used? Applications have not yet been developed but the use of micellar enzymes could have certain advantages such as:

1. Reaction rates of lipophilic substrates could be enhanced due to the high surface area of the micelles.

2. A reverse micelle-containing phase can coexist with with an aqueous phase. This provides the unusual situation of a two-phase aqueous/organic system with the enzyme residing in the organic phase. A possible use would be to remove water soluble products from a lipophilic substrate.

3. Enzyme activity may be increased and specificity modified (Luisi 1985).

The study of reverse micelles is still in its early stages and further developments are awaited with interest.

3.7 Two-phase aqueous systems [See also Hamm Chap. 4]

Two-phase aqueous systems produced as a result of mixing two polymer solutions, or a polymer and a salt, are most appropriate for partitioning biological macromolecules and can be used in a number of ways which are mostly analogous to the application of aqueous/organic systems. Thus it is possible to operate a simple partition system, chemically assisted extraction or extractive reaction. A number of interesting analytical scale biochemical applications have been reported but in biotechnology the systems have been most widely investigated for the isolation and purification of enzymes.

3.7.1 Properties of the systems

Phase separation as a result of mixing two polymer solutions is a fairly common occurrence and was first reported by Beijerinck (1896) as a result of mixing agar with gelatine or starch. One can visualize in physical terms a homogeneous polymer existing in solution in the form of interlocking coils. A polymer molecule of different size and shape may not be able to penetrate the first polymer and would thus be unable to disrupt the existing homogeneous structure. Therefore, unless the polymer molecules are very similar, they prefer to be surrounded by their own kind instead of being mixed with those of a different polymer. The system will consequently be in the energetically most favoured state when the two polymers are separated. The polymer solutions must be above a certain minimum concentration for phase separation to occur as indicated in a typical phase diagram in Fig. 3.51.

Higher molecular weight polymers require lower concentrations to form two phases. Referring to Fig. 3.51, a mixture of total composition *S* will separate into two phases, the composition of the upper phase being given by point *U* and that of the lower phase by point *L*. The length of the line UL

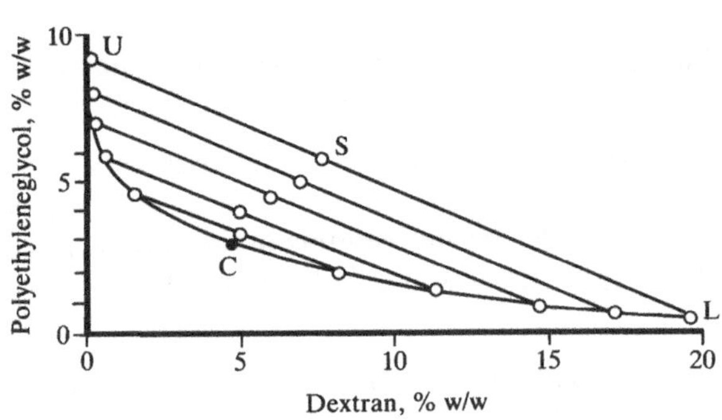

Fig. 3.51. Phase diagram of the system dextran D48 – polyethylene glycol 6000 at 4°C. (From Albertsson 1986, p 279.)

(the tie line) affects the separation rate of the phases as it defines density difference. However, with increasing tie line length there is also a concomitant increase in lower phase viscosity (especially when this is dextran) which partially counteracts the effect of increased density difference. Increased tie line length also indicates increased interfacial tension. At the critical point *C* the two phases merge into a single phase and close to this point, when two phases have just formed, the interfacial tension can be below 10^{-8}N/cm. The value increases roughly logarithmically with increasing tie line length. The rate of separation under gravity depends on the complex interaction of at least three factors, that is density difference between the phases, viscosities, and interfacial tension. Under laboratory conditions multiphase systems have been produced. Dextran, Ficoll® (a polysucrose) and polyethylene glycol give a three-phase system. The addition of hydroxypropyl dextran produces a fourth phase. An eighteen-phase system has been reported by Albertsson (1986).

Although a large number of phase-forming polymers are known most workers have used polyethylene glycol as the upper phase and dextran or an inorganic salt as the lower phase. For investigative purposes purified dextran fractions have normally been used because the crude material gives very viscous solutions. It will be noted that two-phase systems can be formed with fairly dilute polymer solutions, resulting in solutions of low osmotic pressure. This factor may be of importance when partitioning cellular material. Polyethylene glycol and dextran are well tolerated by most cells which can be separated without loss of viability, but some aggregation may occur due to polymer bridging. A phase diagram of a typical polymer–salt system is shown in Fig. 3.52.

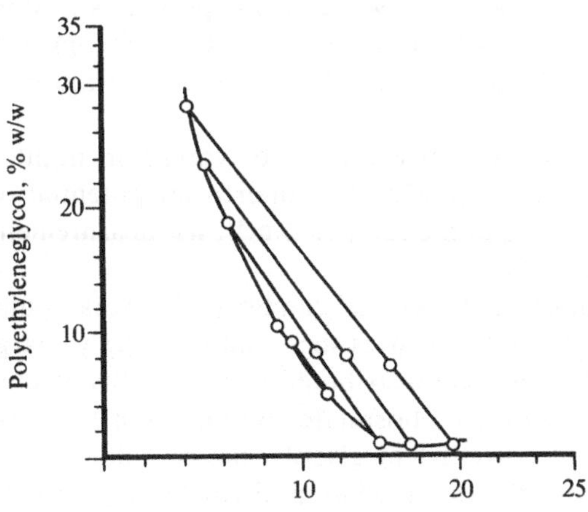

Fig. 3.52. Phase diagram of the system polyethylene glycol 6000 – potassium phosphate at 20°C. (From Albertsson 1986, p 319.)

Although the polymers are water soluble they display relative degrees of hydrophobicity. Polyethylene glycol is more hydrophobic than dextran. All the polymers which have been used occupy a very narrow band in the usual hydrophobicity ranking of solvents. Because the band is so narrow very precise fractionations are theoretically possible. A detailed discussion of the thermodynamics of partitioning is given by Brooks *et al.* (1985). The application of aqueous two-phase systems to biotechnology has been pioneered by Albertsson and coworkers who published the first comprehensive monograph on the subject, now in its third edition (Albertsson 1986).

Particulate material, for example bacterial cells, cell débris, subcellular particles, and viruses, usually partitions into the lower phase or forms a third, interfacial, phase. Mammalian cells partition between the interface and the top phase when using polyethylene glycol 8000/dextran 500, but in the lower phase when dextran 40 is used (N.B. polyethylene glycol 8000 was formerly designated polyethylene glycol 6000). Macromolecules of moderate size such as enzymes may partition into either phase depending on their degree of hydrophobicity and other factors. Small organic molecules and salts have a partition coefficient close to unity. This parameter is defined as $K = C_t/C_b$ where C_t and C_b are the equilibrium concentrations in the top and bottom phases respectively, that is a value of K less then 1 indicates preferential partition into the lower phase. Examples of particulate separations include viruses (Albertsson 1967) and cell debris (Kula *et al.* 1982).

Proteins are amphipathic molecules (that is they contain hydrophobic and hydrophilic domains) and will partition in an aqueous polymer two-phase system on the basis of their van der Waal's or hydrogen bonding interactions with the polymers thus effecting a separation from contaminants.

Partition of salts between two aqueous polymers is almost unity but in many cases just sufficient preferential partition of ions occurs to create an electrical potential between the phases which can aid certain separations (Johansson 1970).

An illustration of the consequences of variations in interfacial electrical potential is given in Fig. 3.53. The interfacial potential obtained in such systems appears to be in the range 1–5 mV; its measurement is discussed by Bamberger *et al.* (1985).

If required, interfacial potential can be usefully manipulated by the addition of salts or charged polymers such that negatively charged molecules or particles are preferentially distributed in the positively charged phase and *vice-versa*. The hydrogen phosphate (HPO_4^{-2}) ion has a very low K value (0.74 in 7 per cent polyethylene glycol-7 per cent dextran T500) (Johansson 1970) and can give the lower phase a distinctly negative charge.

Aqueous two-phase systems have a very low interfacial tension (1×10^{-4}N/m or lower) between the phases when using polyethylene glycol/dextran and a large interfacial area is readily formed when the phases

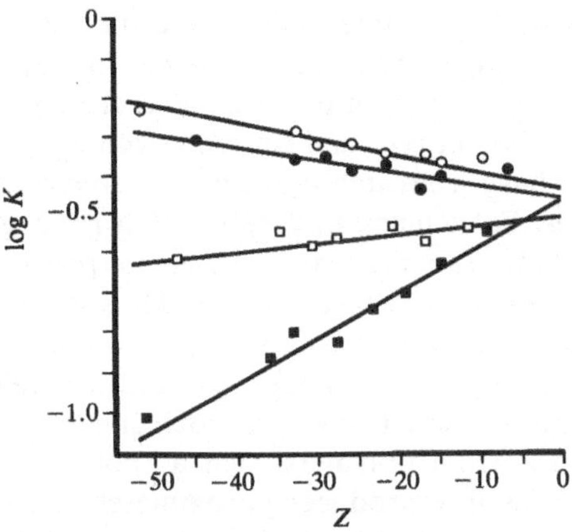

Fig. 3.53. Partition of bovine serum albumin as a function of its net charge in phase systems having different interfacial electrical potentials. ○ 0.05 M potassium sulphate; ● 0.1 M lithium chloride; □ 0.1 M potassium acetate; ■ 0.1 M potassium chloride. (From Albertsson 1977.)

are mixed. The value is much higher when salt forms the lower phase but is still low compared to aqueous/organic systems. This factor gives good mass transfer of molecules where with other separation methods diffusion is often limiting. The low interfacial tension is also useful in preventing potentially damaging conformational changes to the solutes. As mentioned above, all the systems fall in a very narrow band of the normal solvent polarity scale and are capable of further fine tuning by altering pH or ionic strength of the solutions. Selectivity can be enhanced by the attachment of various ligands to one of the polymers.

How does one design an aqueous two-phase system for a particular application? The essential features to be considered are the properties of the phase-forming polymers (average molecular weight, molecular weight distribution, and hydrophobicity), pH, ionic strength and nature of the co-occurring ions and also the temperature and the length of the tie line of the phase diagram. The solute properties to be considered are electrical charge, hydrophobicity, size, and conformation. Although many of these parameters can be measured their complex interaction has not been elucidated and there are currently no basic theories for prediction of partition behavior (Mattiasson and Ling 1987). The initial approach seems largely empirical and intuitive but with experience the interaction of certain factors can be envisaged.

The properties of the material to be partitioned may to some extent dictate the choice. The presence of certain ions may be mandatory in some instances (for example magnesium ions for preserving the integrity of ribosomes) and some delicate cellular material may not be compatible with high osmotic pressure (for example salt) solutions.

The requirements for soluble macromolecules are less restricting. Bamberger *et al.* (1985) suggest starting the search for a suitable system with 8 per cent dextran 40 or 5 per cent dextran 500 and 4 per cent polyethylene glycol 8000, then varying the type and concentration of salts used. The latter parameters usually have a greater influence on solute partition than the polymers themselves. For a macromolecule with a net negative charge in a polyethylene glycol/dextran system, increase in partition coefficient is related to charge density of the added ion. Thus the order of effect of monovalent cations is lithium > sodium > potassium > caesium and for anions fluoride > chloride > bromide > iodide. Divalent ions such as sulphate and hydrogen phosphate produce much greater effects because they are partially excluded from polymers such as polyethylene glycol. Thus divalent ions may be used to produce approximately the required partition and monovalent ions can be used for fine tuning. The most important aspect of the polymer system is that it should not be too close to the the critical point otherwise it is difficult to obtain reproducible results and the phase system will be more difficult to separate.

One way to devize a suitable system would be to set up a 'general purpose' system and optimize using a simplex procedure as reported by Backman and Shanbhag (1984). They investigated four of the possible variables (concentrations of dextran T70, polyethylene glycol, potassium chloride, and potassium phosphate) using forty vertices. The possible additional variables of pH and temperature were not investigated. The advantage of keeping to dextran and polyethylene glycol for pharmaceutical applications is that they are recognized as non-toxic substances and conform to many regulatory requirements.

Some practical points are worth mentioning here. Dextran can suffer from batch-to-batch variation with respect to average molecular weight, polydispersity, salt content, and polymer branching. For consistent results in a series of experiments the same batch should be used. Alternatively, for reproducible preparative results the polymer system should be optimized for each batch. Dextran is subject to microbial degradation and unless good housekeeping is practised there will inevitably be a buildup of dextranase-producing organisms in the immediate environment if the polymer is in routine use.

Polyethylene glycol is stable for most practical purposes but prolonged storage of aqueous solutions should be avoided as they will discolour due to oxidation of the antioxidants and acidic end groups may be formed.

Since the 'dry' polymers usually contain some water the solution concentration must be determined analytically. Optical rotation is suitable for dextran and refractive index determination or the colorimetric method of Hasko *et al.* (1982) for polyethylene glycol.

In many cases partition coefficients in unmodified two-phase aqueous systems are quite low (less than 10) and if after investigating a range of polymers together with manipulation of the pH and interfacial potential the required degree of separation cannot be attained the use of chemically modified polymers may be considered. Selectivity of extraction can be enhanced by chemically modifying one or both of the polymers. Some methods given by Albertsson (1977) are based on the use of polymers modified with covalently bound fatty acid residues to enhance partition of hydrophobic molecules. It was found, for example, that serum albumbin could be extracted into a palmitoyl polyethylene glycol (upper) phase while the majority of other serum proteins remained in the lower dextran phase. It appears that it is necessary for the polymer to be substituted with a fatty acid residue containing at least eight carbon atoms for this interaction to occur (Fig. 3.54) (Shanbhag and Axelsson 1975).

Certain salts used as phase modifiers can have ion-pair extraction properties. Thus the addition of sodium perchlorate will extract positively charged proteins into the upper phase and tetrabutylammonium phosphate extracts negatively charged proteins into the same phase (Johansson *et al.* 1973).

There is an increasing tendency for the polymer system to be investigated as a vehicle for specific ('affinity') extractants. This approach is discussed in the context of enzyme purification on p. 271.

The theory and biotechnological applications of aqueous two-phase systems are fully discussed in the book edited by Walter *et al.* (1985).

3.7.2 Applications

Large-scale preparative applications of two-phase aqueous systems have not been extensively developed and no industrial applications have been

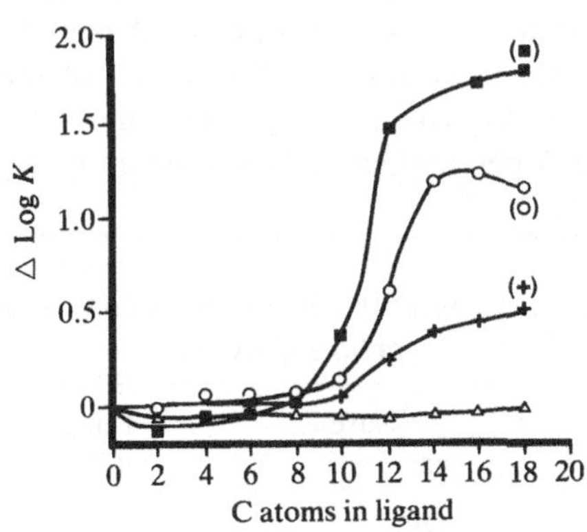

Fig. 3.54. Effect of polyethylene glycol bound fatty acids on the partitioning of proteins in the system 7 per cent dextran 500, 7 per cent polyethylene glycol 8000, 0.1 M potassium sulphate, 2 mM potassium phosphate, pH 7.0 ■ human serum albumin; ○ β-lactoglobulin; + carboxyhaemoglobulin; △ cytochrome C. (From Shanbag and Axelsson 1975.)

reported. A limited amount of work has been carried out on a pilot plant scale notably at the Gesellschaft fur Biotechnologische Forschung at Braunschweig. Most of this work has been concerned with the isolation and purification of enzymes.

3.7.2.1 *Enzymes*

(1) *Simple partition systems.* Enzymes of microbiological origin have traditionally been isolated from a clarified aqueous solution using precipitation or absorption techniques. The clarification stage is relatively straightforward if the enzyme is extracellular and the producing organism has favourable sedimentation characteristics. However, since many enzymes occur as intracellular products mechanical or biochemical disruption of the cell is required to release the enzyme prior to purification. This process generates an extra impurity load with respect to soluble and insoluble material. In addition to cell debris from the disruption process consisting of very fine particulate material in the sub-micrometre range, there will also be a variety of contaminating solutes, for example polysaccharides, nucleic acids, and possibly unwanted enzyme activities.

Two-phase aqueous systems can be used to effectively remove cell debris and simultaneously effect some purification of the required solutes. The procedure is particularly useful as proteins are generally stabilized by the phase-forming polymers and good recoveries can be obtained when operating at ambient temperatures. This approach has been used to purify a number of different enzymes produced by yeasts and bacteria by partitioning a cell homogenate in polyethylene glycol/dextran or, more conveniently, polyethylene glycol/salt (Kula *et al.* 1982). The twenty-two enzymes and phase systems listed in Table 3.7 give an indication of the scope of the method. The phase systems quoted are the nominal starting phases. After the addition of homogenized bacterial cells much additional polymer will have been introduced and will modify the phase system. The system 18 per cent polyethylene glycol 1550, 7 per cent sodium potassium phosphate was effective in a number of cases.

TABLE 3.6 *Solubility of some pharmaceutical compounds in supercritical carbon dioxide*

Compound	CO_2 pressure bar	Temp. (°C)	Solubility (%)
Naphthalene	228	45	10
Imipenem	340	40	0
Mevinolin	340	40	0.04
Efrotomycin	340	40	0.03

The composition of this mixture is below the binodal and polymers extracted from the cells are contributing to the formation of a two phase system. In such cases the concentration of biomass must be quoted to define the phase system (see below). The effect of cell polymers on phase composition is given in Table 3.8. The phase ratios change to a marked degree in the presence of cell polymer and density difference increases. As expected, the lower phase containing most of the cell polymers is considerably more viscous than the original phase. Since with most aqueous two-phase systems the heavy phase is the minor one, good recovery of enzyme can be achieved. This is often in excess of 95 per cent of the enzyme passed into the upper polyethylene-rich phase leaving insoluble matter and some soluble proteinaceous and other impurities in the lower phase. It is seen from Table 3.7 that in only two cases was the partition coefficient less than 1; it was usually in the range 3–10. Partition of the required enzyme into the upper phase is not inevitable and a certain amount of experimentation may be required. A change of pH can have a pronounced effect on partition especially in systems containing phosphate. In addition to altering the net charge on the protein the pH change affects the proportion of dihydrogen and monohydrogen phosphate ion present, with the latter giving a significant interfacial potential (see p. 258). An example of the effect of pH change is given in Fig. 3.55. This also represents an example of the cell polymers contributing to the phase system.

Using two-phase aqueous partition, the specific activity of the enzyme can be higher by a factor of 2.5 to 3 compared to a simple liquid/solid separation. Overall the most effective systems for use at this stage have been polyethylene glycol/salt. Dextran has been less frequently employed became cell debris does not always partition to the lower phase and in general less contaminating solutes are removed (Hustedt *et al.* 1985).

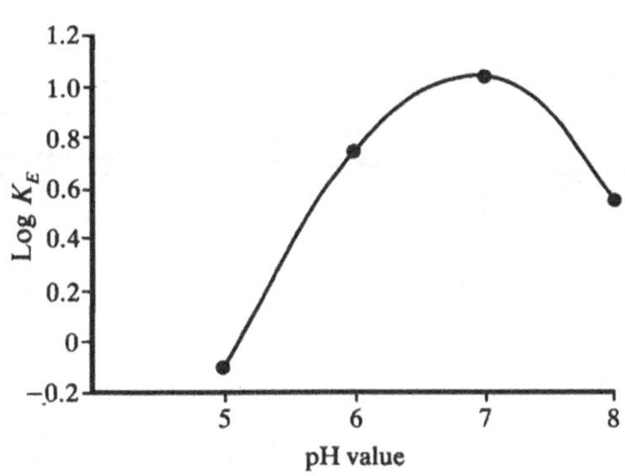

Fig. 3.55. Effect of pH on the separation of L-hydroxyisocaproate dehydrogenase from *Lactobacillus confusus* cell debris. Phase system: 18 per cent polyethylene glycol 8000, 7 per cent potassium phosphate, 20 per cent biomass. (From Hummel *et al.* 1984.)

TABLE 3.7 *Extraction of enzymes from cell homogenates*

Organism	Enzyme	Constituent of the phase system	K_{enzyme}	Yield (%)
Candida boidinii	Catalase	PEG 4000/crude dextran	2.95	81
	Formaldehyde dehydrogenase	PEG 4000/crude dextran	10.8	94
	Formate dehydrogenase	PEG 4000/crude dextran	7.0	91
	Formate dehydrogenase	PEG 1000/potassium phosphate	4.9	94
	Isopropanol dehydrogenase	PEG 1000/potassium phosphate	18.8	98
Saccharomyces carlsbergensis	α-Glucosidase	PEG 4000/dextran T-500	1.5	75
Saccharomyces cerevisiae	α-Glucosidase	PEG 4000/dextran T-500	2.5	86
	Glucose-6-phosphate dehydrogenase	PEG 1000/potassium phosphate	4.1	91
Streptomyces species	Glucose isomerase	PEG 1550/potassium, phosphate	3.0	86
Klebsiella pneumoniae	Pullulanase	PEG 4000/dextran T-500	2.96	91
	Phosphorylase	PEG 1550/dextran T-500	1.4	85
Escherichia coli	Isoleucyl tRNA synthetase	PEG 6000/potassium phosphate	3.6	93

	Leucyl tRNA synthetase	PEG 6000/potassium phosphate	0.8	75
	Phenylalanyl tRNA synthetase	PEG 6000/potassium phosphate	1.7	86
	Fumarase	PEG 1550/potassium phosphate	3.2	93
	Aspartase	PEG 1550 /potassium phosphate	5.7	96
	Penicillin acylase	PEG 4000/crude dextran	1.7	90
Bacillus sphaericus	Leucine dehydrogenase	PEG 4000/crude dextran	9.5	98
Bacillus species	Glucose dehydrogenase	PEG 4000/crude dextran	3.2	95
Brevibacterium ammoniagenes	Fumarase	PEG 1550/potassium phosphate	0.24	90
Lactobacillus cellobiosus	β-Glucosidase	PEG 1550/potassium phosphate	2.2	98
Lactobacillus species	Lactate dehydrogenase	PEG 4000/ dextran PL-500	6.3	95

Data from Kula *et al*. (1982).

TABLE 3.8 *The effect of cell polymers on phase composition*

Phase system	V_t/V_b	Upper phase density (g/cm^3)	Lower phase density (g/cm^3)	Upper phase viscosity (mPa s)	Lower phase viscosity (mPa s)
1. 9% PEG 4000, 2% Dextran T-500	8.80	1.010	1.057	3.0	94
2. as 1 + 20% *Klebsiella pneumoniae* cells	2.50	1.038	1.134	10	11,000
3. 7% PEG 4000, 1.25% crude dextran	7.45	1.046	1.138	2.7	~2,000
4. as 3 + 20% *Klebsiella pneumoniae* cells	2.70	1.063	1.159	9	~3,300

Both systems also contain 0.34M potassium phosphate (pH 7.5).
Data from Hustedt *et al.* (1985)

In the pilot plant scale studies the phases were separated in a small disc-stack centrifuge (Alfa-Laval 'Gyrotester') at 60 litres per hour or a nozzle separator (Alfa-Laval YEB 1334) at 130 litres per hour and in principle the procedure is capable of considerable scale-up. A further feature of this approach is that continuous operation is more readily achieved than is the case with many other methods and it could readily be linked to a continuous fermentation. After homogenization the suspension may be mixed in-line with the phase-forming reagents using a static mixer and, following a holding time of 5 minutes, it was found that equilibrium had been achieved (Hustedt *et al.* 1985). It is important to ensure that temperature is constant throughout partitioning and separation as the phase systems themselves are temperature dependant.

The costs of cell debris removal by centrifugation, ultrafiltration, or two-phase partition have been compared by Datar (1986). The aqueous partition method was four to five times more expensive but it must be remembered that the product obtained is partially purified, which offsets some of the additional cost. A true cost comparison can only be performed on products of similar purity.

Secondary purification stages can also be carried out using similar methods. Multistage aqueous two-phase partition was used by Kroner *et al.* (1982) to purify formate dehydrogenase to give a 70 per cent pure product with an overall yield of 75 per cent. Cells of the producing organism, *Candida boidinii* (50 kg), were passed through a glass bead mill to release the enzyme. The first partition stage was in polyethylene glycol 400 (18 per cent w/w), polyethylene glycol 1550 (7 per cent w/w) and potassium phosphate at pH 7.8 (8 per cent w/w). After stirring for about 2 hours to ensure equilibrium had been attained, the phases were separated in a YEB 1334 nozzle-type centrifugal separator (Alfa-Laval). The mixture could be separated at up to 180 litres per hour. The heavy phase containing cell debris and some other impurities was discarded. Potassium phosphate and potassium chloride was added to the light phase, giving another two-phase system which was stirred for 1 hour then separated under gravity. The lower phase containing a substantial part of the polysaccharide and nucleic acids was discarded. A fresh lower phase consisting of 4 per cent w/w polyethylene glycol 400, 2 per cent w/w polyethylene glycol 1550 and 18 per cent w/w potassium phosphate was added. The resulting phase system was such that the enzyme transferred to the lower phase, leaving most of the proteinaceous inpurity in the upper phase. The separated lower phase was treated with 10 per cent w/w potassium phosphate to transfer the required enzyme back to the polyethylene glycol phase in which it was stable for 6 to 8 months. For the intended application no further purification was required; in the final solution the enzyme represented more than 70 per cent of the total protein present. An indication of the efficacy and yield of the various stages is given in Table 3.9. Should the enzyme be required free of polyethylene glycol and salts the final solution may be subjected to ultrafiltration.

It was noted that the stage yields on this scale were almost the same as those initially obtained in the laboratory at 1/30 000 of the scale.

As mentioned before, high values of the partition coefficient are rarely obtained with simple two-phase aqueous systems but yields are high and, as shown above, useful separation from troublesome particulate material is obtained. Thus the systems have some promise for initial cleanup and preliminary purification of fermentation broths or cell homogenates. In any case the procedure would be less applicable for concentrated enzyme solutions because polyethylene glycol is a precipitating agent and enzyme solubility would necessarily be limited.

Where dextran has been used as the lower phase, laboratory scale work has invariably been performed using purified dextran fractions which would be uneconomic on a large scale: the problem with the cheaper alternative is that it forms a highly viscous solution when used at two-phase forming concentrations. An indication of the difference in viscosity is given in Table 3.8.

One way to alleviate the apparent problem is to perform a controlled acid hydrolysis of the crude dextran which will reduce the viscosity, but it also produces material of a wide molecular weight range and in theory less precise fractionations are possible. Kroner *et al.* (1982) compared the effect of substituting dextran T500 with either crude dextran or the acid hydrolysed material. The physical properties of the phases so produced are summarized in Table 3.10 and the effect on partition coefficient is shown in Table 3.11. It is seen that yields remained high and although partition coefficients were altered to some extent there was no consistent pattern.

In many cases high viscosity can cause handling problems but there are some instances where it can be used to advantage. Using a nozzle bowl separator, a viscous heavy phase is beneficial as it holds the separation line between the phases near the wall of the bowl. Thus there is relatively little viscous pure heavy phase in the region of the discs. For the optimization of any particular application careful geometrical design of the nozzles is required.

The general conclusion reached was that crude dextran can be used as an inexpensive phase-forming polymer for applications involving initial cleanup or cell debris removal provided due allowance is made for the high viscosity of the heavy phase.

A system containing crude dextran was 23 per cent of the cost of a system with similar partition characteristics containing dextran T500 although the crude dextran still accounted for approximately 50 per cent of the total phase system cost.

The frequently occurring potential problem of viscosity and cost of the lower phase, especially when dextran is used, can be ameliorated using

TABLE 3.9 *Stage yields: formate dehydrogenase purification*

Step	Method	Yield FDH (%)	Specific activity (unit/mg)	Yield of polysaccharide (%)	Yield of 'nucleic acid' (%)	Total activity of FDH (units)
1	Cell disruption	100	0.59	100	100	456 000
2	Heat denaturation	100	0.96	100	100	456 000
3	Upper phase A	94	1.08	89	90	429 000
4	Upper phase B	75	1.20	31	30	340 000
5	Lower phase C	70	2.15	11	19	320 000
6	Upper phase D	70	2.22	8	4	320 000

Data from Kroner *et al.* (1982)

TABLE 3.10 *Physical properties of crude dextran phases*

System	Volume ratio	Density ($\rho u/\rho 1$)	Density difference (g/cm^3)	Viscosity dispersion (mPa s)	Viscosity lower phase (mPa s)	Viscosity upper phase (mPa s)
I 7% w/w PEG 4000						
1.2% w/w crude dextran	7.45	0.92	0.09	4.0	1984.0	2.7
II 7% w/w PEG 4000						
1.5% w/w crude dextran hydrolysed	7.36	0.95	0.05	3.4	107.8	3.0
III 9% w/w PEG 4000						
2% w/w Dextran T-500	8.80	0.95	0.05	3.4	93.9	3.0

[a] All systems contain 0.3 M potassium phosphate (pH 7.5)

Data from Kroner *et al.* (1982)

hydroxypropyl starch (Tjerneld *et al.* 1986). The dynamic viscosity of the lower phase of the system 5 per cent polypropylene glycol, 14 per cent hydroxypropyl starch (Aquaphase PPT ®, Perstorp, Sweden) and 25 mM sodium phosphate, pH 7.0, was in the range 55.7–50.9 mPa s, approximately one half the value of a pure dextran-containing phase with similar partitioning properties. Systems containing hydroxypropyl starch can be manipulated in similar ways to the more familiar dextran-containing systems, for example by the addition of salts, and they have the advantage of lower cost; the polymer was 16 per cent of the cost of fractionated dextran (in 1986).

Despite the degree of sophistication which can be achieved in aqueous two-phase separations there are however a number of practical problems. The density difference between the phases is usually very low and this, combined with the high viscosity of one or both phases, may take phase separation rather difficult. Another important point is that the phase ratio is determined by the system itself. It may not be possible to control the degree of concentration as is the case with aqueous/organic solvent extraction. Additional problems concern recovery of the desired solute from the polymer solution. The technical problems may not be as serious as generally perceived as the group at Gesellschaft fur Biotechnologische Forschung have been able to overcome many of them. One important economic aspect which has received relatively little attention is that of polymer regeneration or recovery. It has been suggested by Michaels and Matson (1985) that membrane ultrafiltration procedures could be useful for this purpose. Thus it would be possible to separate residual impurities from the polymer using diafiltration, following which the polymer could be ultrafiltered to the required concentration. This procedure has been investigated by Datar and Edwards (1986). They found that ultrafiltration using cellulose acetate or polysulphone membranes with a nominal molecular weight cutoff 100 000 daltons recovered 75 per cent of the polyethylene glycol in the permeate using five exchange volumes of water while 77 per cent of the enzyme remained in the retentate.

Further development of aqueous two-phase systems has undoubtedly taken place as it is believed that several commercial scale processes are in operation. There have been no reports of the products concerned. Most processes use polyethylene glycol/salt phase systems.

(2) *Chemically assisted extraction*. The partition coefficient of enzymes in aqueous two-phase systems can be increased using chemical recognition ('affinity') techniques. For example, specificity can be achieved with affinity partitioning using polymers coupled to triazine dyes such as those widely use for protein chromatography (Lowe 1984). The ligand may be distributed in the free state or, more usually, covalently attached to one of the polymers. If a dye-polyethylene glycol complex is used the system can be designed such that the majority of the soluble protein passes into the lower phase, whereas

TABLE 3.11 *Comparison of enzyme partition coefficients in different dextrans*

Enzyme	System[a]	Volume ratio	A	ΔA	Yield (%)	M_r enzyme
α-Amylase	I	8.16	3.64	-	96.7	
	II	8.78	4.18	+0.54	97.8	97 000
	III	8.60	3.68	+0.04	97.2	
Glucoamylase	I	8.60	1.22	-	91.3	
	II	10.12	1.80	+0.58	95.0	93 000
	III	9.28	1.34	+0.12	92.6	
Isoleucyl-tRNA-	I	5.86	2.17	-	92.5	
synthetase	II	9.11	1.62	−0.55	93.6	105 000
	III	8.50	1.50	−0.67	92.6	
α-Glucosidase	I	6.31	2.84	-	94.7	
	II	10.88	3.37	+0.53	97.4	68 500
	III	9.67	2.67	−0.17	96.3	
Glucose-6-phosphate	I	8.20	0.62	-	84.9	
dehydrogenase	II	10.00	1.31	+0.69	92.9	128 000
	III	8.90	0.89	+0.17	89.9	
Formate	I[b]	8.30	0.94	-	88.6	
dehydrogenase	II[c]	8.00	1.36	+0.42	91.6	70 000
	III[c]	7.90	1.07	+0.13	89.4	

Formaldehyde dehydrogenase	I[b]	8.30	1.17	-	90.6	
	II[c]	8.00	1.29	+0.12	91.2	78 000
	III[c]	7.90	1.06	−0.11		
Catalase	I[b]	8.30	1.53	-	92.6	
	II[c]	8.00	1.29	−0.24	91.2	280 000
	III[c]	7.90	1.57	+0.04	92.5	
Pullulanase	I[b]	8.80	2.10	-	94.9	
	II[b]	7.90	1.78	−0.32	93.4	145 000
	III[b]	6.80	2.10	0.0	93.5	

(a) I = 9% w/w PEG 4000, 2% w/w Dextran T-500; II = 9% w/w PEG 4000, 1.25% w/w crude dextran; III = 9% w/w PEG 4000, 1.50% w/w crude dextran hydroly. All systems include 0.3 M potassium phosphate (pH 7.5).
(b) Including 0.05 M potassium phosphate (pH 7.5).
(c) Systems are composed of 7 per cent w/w PEG 4000 and include 0.05 M potassium phosphate (pH 7.5).
Data from Kroner *et al.* (1982).

the enzyme specifically interacting with the ligand will be extracted into the upper phase.

It was found by Johansson and Joelsson (1986) that the use of an affinity ligand (Cibacron blue F3-GA) for lactate dehydrogenase increased the partition coefficient more than seventy-fold compared with an unmodified phase system. Using an alternative ligand, Procion yellow HE-3G, the effect was even more striking; the partition coefficient was increased 760 times. Most other proteins present were not affected by the affinity reagent and retained their original partition behaviour. To release the enzyme from the ligand complex a 50 per cent aqueous solution of sodium dihydrogen phosphate/disodium hydrogen phosphate 1 : 1 was added. The enzyme transferred to the lower (salt) phase, leaving a small volume of viscous upper phase containing the affinity reagent which could be reused. It is not clear to what extent irreversible fouling of the reagent occurs, restricting multiple reuse.

Similar studies have been reported by Birkenmeier *et al.* (1986). In this case the protein of interest was thyroxine-binding globulin.

(3) *Enzymic reactions*. The concept of extractive enzymation has been investigated by several groups, notably those at the University of Lund. The approach is still in the investigative phase and no commercial applications are known.

Two-phases aqueous systems may be used as a vehicle for enzymic reaction. Systems can be devised where substrates and products reside in one phase and the enzyme in the second phase, thus rendering the enzyme reusable on a batch basis or giving scope for continuous operation. Compared to a packed bed immobilized enzyme reactor, a two-phase aqueous system is advantageous when the reaction requires addition of reagents to maintain the required conditions.

A reaction of this type, the hydrolysis of benzylpenicillin to 6-aminopenicillanic acid using aminohydrolase from *Escherichia coli*, was studied by Andersson *et al.* (1984) (Fig. 3.56). Although the reaction is the basis of well developed commercial processes using enzyme immobilized on solid matrices, it was used as a model for establishing the concept of a two-phase enzyme reactor. A two-phase system which contained virtually all the enzyme in one phase was found. It consisted of 8.9 per cent polyethylene glycol 20 000 and 7.6 per cent dipotassium hydrogen phosphate at pH 7.8. The partition coefficient for the enzyme was less than 0.01. A coefficient of greater than 0.01 was obtained with systems containing dextran as the lower

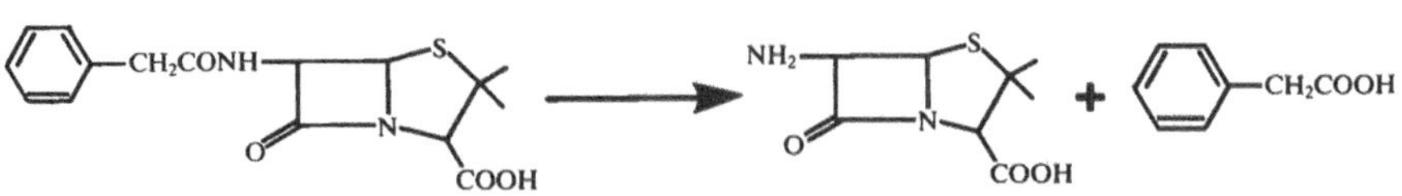

Fig. 3.56. Enzymic deacylation of penicillin G.

phase and with lower molecular weight polyethylene glycols as light phase; these were considered unsatisfactory. The reaction generates a carboxylic acid, and even with the high concentration of phosphate in the system titration with alkali was still required to maintain constant pH. This factor is important on two counts. Not only is it necessary to maintain the pH optimum for enzyme activity, it is also necessary for maintaining the phase system, became the monohydrogen phosphate ion is a stronger phase-forming component than dihydrogen phosphate.

During these studies a number of practical points emerged. Two advantageous aspects were that the enzyme could be reused a number of times, and, being in the lower phase, the enzyme was protected from the degradative effects of the titrant (7.5 M sodium hydroxide). Two drawbacks were that the enzyme activity was somewhat inhibited in the presence of high concentrations of phosphate, and the products of the reaction had partition coefficients near unity. This if the system is used in batch mode some product will remain in the enzyme-containing phase.

Another type of application could be the enzymic hydrolysis of a macromolecular substrate to low molecular weight products. A two-phase system would be particularly advantageous if the reaction is inhibited at high product concentration. High concentrations of enzyme and substrate could be allowed to react with continuous export of product into the second phase.

A development of this approach would be to use a macromolecular substrate as one of the phase-forming components.

3.7.2.2 *Other applications*

Two-phase aqueous partition has been used extensively in biochemical research to study interacting systems such as protein-ligand, protein-nucleic acid, antigen-antibody and receptor sites. The systems have also been employed to study the surface properties of cells and to isolate viruses. Some of these applications are briefly discussed below.

Before Albertsson's initial studies of aqueous two-phase systems, a related system was used by Kirby (1956) to separate polysaccharide from RNA. The aqueous sample solution was mixed with an equal volume of 2.5 M dipotassium phosphate, 0.05 volume 33 per cent phosphoric acid and one volume 2-methoxyethanol. The mixture separated into two phases (upper phase: lower phase 5 : 1). The lower phosphate phase contained the polysaccharide and RNA was in the upper phase from which it could be precipitated as the n-hexadecyltrimethylammonium salt (Ralph and Bellamy 1964). Dimethylsulphoxide could be substituted for 2-methoxy ethanol with similar results (Kirby 1962).

Two-phase aqueous systems have been investigated for the purification of β-interferon from human fibroblast cell cultures. This product has a molecular weight of 20 000, much lower than most enzymes, and normal phase systems are unlikely to give good separation. A large number of systems were investigated by Menge *et al.* (1983). Polyethylene glycol, substituted by some charged groups such as trimethylammonium or phosphate esters as well as the affinity ligands Cibracron blue or concanavalin A, gave enhanced partition coefficients although other components of the system had a marked effect. The crude interferon solution contained a substantial concentration of protein impurities and an objective of the investigation was to obtain high selectivity. pH was one important factor. In the system 8 per cent polyethylene glycol 1550 phosphate ester, 24 per cent sodium phosphate, 7.5 per cent sodium chloride at pH 3.7 partition coefficient for interferon was 23 with a specific activity of 18 000 u/mg. The values for pH 4.7 were 354 and 29 000 and for pH 5.7 they were 407 and 31 000 respectively.

Some of the effects seen are difficult to interpret. For example, the use of both positively and negatively charged polyethylene phases at the same pH value extracted interferon, whereas the uncharged polymer did not do so. Menge *et al.* (1983) suggested that the salt in the system was reducing the solubility of interferon in the lower phase and the charged substituents were exerting their effect largely by increasing the polarity of the polymer to make it a more effective solvent for the hydrophilic interferon molecule.

The most effective partition system found for β-interferon was 1 per cent polyethylene glycol phosphate ester, 19.5 per cent potassium phosphate, 3 per cent sodium chloride (pH 6.7) which gave a partition coefficient in excess of 150 for interferon but less than 1 for protein impurities. A number of fairly similar phase systems investigated gave very different partition coefficients, emphasizing that there is no substitute for extensive experimentation for optimizing a two-phase aqueous partition system.

The systems can be used to study molecular interaction. A pair of interacting molecules partitioned as their binary complex will have a different coefficient to the separate molecules. Thus the system can provide a convenient means of separating bound antigen from the free form in an immunological assay as reported by Mattiasson and Eriksson (1982). Similar studies were reported by Ling and Mattiasson (1984). The method can also be used for particulate samples such as bacteria. The identification of bacteria usually requires a time-consuming preculture stage, but this can be eliminated using various sensitive immunological methods. Ling *et al.* (1982) proposed the method of partition affinity ligand assay for the rapid identification of type B1 streptococci. Both streptococci and antibody partition into the lower phase in a wide variety of two phase systems. However, the bacterial cells can be modified without altering their immunological properties by coupling monomethoxypolyethylene glycol to their surface. They then partition into the upper phase of a polyethylene glycol/dextran system. Thus the sample (containing streptococci) was mixed with a fixed amount of monomethoxypolyethylene glycol-modified streptococcus B1 and the mixture was allowed to compete with a limited amount of ^{125}I-labelled antibody. The mixture was then partitioned in a two-phase system consisting of 5 per cent polyethylene glycol 6000 and 7.5 per cent dextran T250. The antibodies were bound to the chemically modified cells and the sample cells distributed into different phases and could be quantitated by conventional radiochemical methods and thus the cell count of the test sample determined.

Bioaffinity can also be used to design separating systems. DNA partitioned in one phase can be employed to concentrate proteins which are bound to it. It is reported that this procedure has been used for purifying DNA polymerase (Vehle and Ganesan 1973).

As with any single stage partition system, these aqueous systems can be used for separating a two-component mixture or for isolating a single component from a mixture. More sophisicated separations require multistage procedures. Countercurrent distribution using aqueous two-phase systems has been used for the separation of three isoenzymes (Blomquist and Wold 1974) and different types of cell population (Albertsson and Larsson 1976).

A number of studies were performed in the early days on the partition of virus particles (see, for example, Philipson *et al.* 1960; Albertsson and Frick 1960). Despite the encouraging results obtained, the studies have not been pursued, as can be judged from the relative paucity of recent literature.

Recently submicron-sized polymeric particles have become available and have potential as a bioseparation tool. These particles are typically 0.01–1 μm in diameter. They have a very slow sedimentation velocity and the surface area of 0.2 μm particles is 3×10^5 cm^2/g. The sedimentation velocity is of the same order as bacterial cells. The particles can be functionalized with various charged groups (such as quaternary amine or carboxylic acid) and the chemistry of the absorption cycle is similar to that of conventional ion exchangers. Once the particle has absorbed macromolecular material it flocculates and can be removed from solution quite easily. Elution with a suitable buffer releases the product and redisperses the particles which may be recovered by centrifugation.

It has been reported that these submicron-sized polymeric particles can be used in conjuction with aqueous two-phase systems to enhance partition. The outline of two procedures is given by Kim and Rha (1987). Using charged particles, certain proteins can be directed into a particular phase or, alternatively, accumulate at the interface from which they can be recovered in a small volume. The second method uses the particles initially in liquid/solid mode for adsorption then recovered using a polyethylene glycol/potassium phosphate system. Thus galactosidase was absorbed electrostatically from an *Escherichia coli* homogenate in 93 per cent yield and 1.5-fold purification. By placing the loaded particles in the two-phase system the surface charges were shielded and the enzyme was recovered in the upper, polyethylene-rich phase in 89 per cent yield and 4.7-fold purification. A standard liquid/liquid process is reported to give 77 per cent yield and 3-fold purification.

3.7.3 Summary

Aqueous two-phase systems have been known and used for about 30 years. Some work has shown that they can be employed on a pilot plant scale using scaled-down industrial equipment and there appears to be no major technical obstacle to production scale use. Mattiasson and Ling (1987) hint that there may be some industrial applications of the systems. The principal problems are economic rather than technical. The polymer systems are still quite expensive and the related aspect of phase regeneration or recycle has received very little attention.

Some advantages of two-phase aqueous extraction are:

(1) there is probably no limit to the scale of operation;

(2) separation processes can be operated continuously;

(3) the phase systems may stabilize some proteins;

(4) the polymers can be substituted with specific ligands;

(5) it is possible to combine cell debris removal with some purification;

(6) the systems are non-flammable and non-toxic.

Against the above advantages must be weighed some disadvantages:

1. It is difficult to predict the phase system required for a particular separation. Prediction becomes increasingly complicated with a decrease in molecular weight of the solute.
2. There is little information on polymer recycle or regeneration.
3. The materials are relatively expensive.

The main competing techniques are batch or chromatographic adsorption; currently these are favoured due to their greater familiarity. The disadvantages of aqueous two-phase systems listed above suggest that some of the factors impeding industrial application stem from a lack of knowledge rather than any inherent technical problem.

In a survey of 100 papers on protein purification, Bonnerjea *et al.* (1986) found that of the major methods of purification ion exchange chromatography, the most commonly used method, was employed in 75 per cent of the purification procedures. Affinity chromatography was second at 60 per cent. The papers surveyed, which appeared to be confined to laboratory scale preparations, contained no reference to aqueous two-phase systems. The authors commented that although these systems undoubtedly have potential for early stage fractionation, the ready availability of high efficiency centrifuges in the laboratory makes the technique less important. They did not extend their comments to larger scale separations.

3.8 Structure–activity relationships

A pharmaceutical substance can be administered in only a limited number of ways. The oral route is of direct relevance to treatment within the gastrointestinal tract. It is also a very convenient route for administering drugs which are absorbed and pass into the bloodstream to exert their action within that medium or at some other site in the body. Some drugs may be administered in a manner which ensures direct delivery to a particular tissue or organ. In most cases the drug will have to partition several times between a hydrophilic and a lipophilic phase in order to reach the site of action. Having reached the target site, it will be necessary for it to bind, probably by hydrophobic interaction to an enzyme or other biomolecule in order for its intended action to occur.

The initial research on a new drug substance may be performed using isolated enzymes but eventually it will be necessary to progress to tissue and finally to a whole animal in order to obtain the required data. Within any group or class of biologically active molecule a large number may show activity on isolated enzymes but relatively few are likely to show useful *in vivo* activity.

Over a period of time a set of data has accumulated which has given rise to an indication of structure–activity relationships which can often be correlated with partition data of the compound between two immiscible liquids. In other words, in any series of compounds relative lipophilicity may be a useful guide to the potential *in vivo* utility of an experimental drug substance.

The approach is not new and even last century it was noted that the activity of certain narcotic drugs was often related to their partition coefficient in an oil/water system (Meyer 1899). The natural oils (for example olive oil) which were used in some of the early studies may not be the most appropriate hydrophobic liquid as an equivalent to a biophase. Enzymes and biological membranes have, in addition to hydrophobic bonding properties, a substantial hydrogen-bonding characteristic and this led later workers to suggest that long chain alkanol/water systems should give more relevant data. The use of octanol was questioned by Rytting *et al.* (1972) who suggested that an inert hydrocarbon would be preferable to avoid the solvent interactions occurring in the highly associated octanol. However, it would appear that some polar drug substances are not amenable to this technique because of immeasurably low solubility in the hydrocarbon.

Dimyristoylphosphatidylcholine liposomes have recently been proposed as a more appropriate lipophilic phase for partition studies since they bear a closer resemblance to biological membranes. The system has been compared with the established octanol–water method using a range of β-blockers (Betageri and Rogers 1987).

Over the years a number of solvents and different procedures have been advocated for the determination of structural parameters which can be correlated with biological activity. Largely due to the work of Hansch, Leo and co-workers the parameter log P_{oct} (derived from the octanol–water partition coefficient) has become accepted and substantial amounts of data are now available in tabular form (Leo *et al.* 1971; Hansch and Leo 1979) and recently computer software has become available (Pomona College 1986).

In a series of compounds a group constant (F) can be defined as the ratio of partition coefficient of a derivative to its parent compound (assuming simple partition with no dimerization or ionization). Thus a library of F values for various substituents should assist in designing a drug molecule for a particular biological response.

For example, the contribution of the methylene group has been discussed by Davis *et al.* (1972). It was found that the mean difference in activity coefficient at infinite dilution in water ($\Delta\gamma\infty$) was 4 for a wide range of solvents and solutes. This correlates with the observation that aqueous solubility of a homologous series of simple alkyl substituted compounds often decreases by a factor of 4 between each member.

The primary method for the determination of reference partition data is still considered to be the shake-flask method where the two phases, together with the pure solute, are shaken in a stoppered flask until equilibrium is reached and one or both phases is assayed for concentration of drug. Using the same components, a more rapid determination of partition coefficient can be achieved using the AKUFVE apparatus (Davis and Elson 1974). The apparatus, which was developed by Reinhardt and Rydberg (1970) for nuclear chemistry studies, is described in section 3.9.2.3. It is reported that, using this apparatus, equilibrium is attained in a matter of minutes and with suitable solutes on-line analysis by UV absorbance is possible. A temperature or pH profile can be obtained in a matter of a few hours. Because of the rapidity which results can be obtained the apparatus would appear to be particularly appropriate for compounds which are of limited stability in aqueous solution. Alternatively, partition data may be obtained over a wider range of conditions, for example by extending the pH or temperature range into areas which would be unsuitable for studies requiring a long equilibration time.

Both the shake flask method and the AKUFVE method require moderate amounts of pure compound to obtain the required data. A number of workers have investigated the possibility of obtaining similar data using reverse phase high performance liquid chromatography (HPLC). In this method the compound is subjected to multistage partition between an alkane (in the form of an octyl or octadecyl group covalently attached to a silica matrix) and an aqueous solution of an organic solvent such as methanol, acetonitrile, or tetrahydrofuran. This secondary method is calibrated using a standard set of solutes for which log P_{oct} values have been determined. However, certain exceptions have been found as there is no theoretical relationship between log P and log k' (capacity factor in HPLC) or R_m (relative mobility in thin layer chromatography) (Terada 1986). The chromatographic method can be used at log k' values up to 2.5. Beyond this point peak retention and dispersion become excessive. Therefore different solvent compositions are employed to give practical retention times, but it has been questioned (Leo 1987) whether the polar phase where methanol concentration is in excess of 50 per cent is sufficiently water-like for the partition results to be meaningful.

Octanol itself has been used as a stationary phase in HPLC but the eluant which must be used in this system, octanol-saturated water (or buffer), is unsuitable for the elution of hydrophobic compounds.

It is possible to obtain a stationary phase consisting of controlled pore glass with glycerol covalently bonded to the glass (Electro-nucleonics, Fairfield, USA). The phase was developed for the chromatography of proteins but it has been found that it can be used to give data that correlate well with log P_{oct}. In a study using 44 compounds the relationship

$$\log P_{oct} = 2.58 + 2.26 \log k'$$

was established. The solvents included non-hydrogen bonding, hydrogen acceptors, and amphipathic molecules. The column could be used for compounds of a wide range of hydrophobicities as the slope of the line (2.26) was considerably less than that of other methods (Terada 1986). A diol-bonded stationary phase is also of value for these applications.

The log P_{oct} value is only one of several parameters used in structure–activity relationships. It can only indicate an overall property of the molecule and more detailed parameters such as electronic and steric effects are also relevant in predicting drug activity. Discussion of these factors is beyond the scope of this chapter.

In the future partition data as a separate entity may become less important with the increase in computer-assisted molecular design techniques.

3.9 Centrifugal extraction equipment

3.9.1 General considerations

When liquid–liquid extraction is used on any crude or partially purified natural product solution there is a marked tendency for the phases to emulsify and render gravity separation impractically slow. Moreover, many of the procedures discussed in the previous sections rely on rapid passage of the product through a stage where it is of very limited stability. For these reasons centrifugal extraction equipment is widely employed in the pharmaceutical industry. In some instances simple batch extraction or extraction columns are used but as these have few features of interest they will not be discussed.

Some centrifugal extraction or phase separation equipment used in the pharmaceutical industry has been adapted from that designed for other purposes. This is particularly the case with disc-stack and decanter centrifuges. However, the Podbielniak extractor evolved into its present form due to the requirements of penicillin extraction and it was subsequently used in other industries. The extractors are almost invariably constructed of stainless steel.

The machines can be divided into three groups:

(1) single stage liquid/liquid separators which may also incorporate a simultaneous liquid/solid separation function;

(2) countercurrent extractors which have discrete mixing and separating stages;

(3) countercurrent extractors with continuous countercurrent flow of both streams.

The choice of machine depends on the nature of the feedstock, distribution coefficient of the product, and the throughput required. As previously mentioned, the admixture of filtered fermentation broth or other natural product extract with a nominally immiscible organic solvent will often precipitate macromolecular impurities. Although much can be done to assist solvent extraction by way of 'broth conditioning', the machines should be designed to minimize deposition of solids. Alternatively, they should have simple solids-discharge or cleaning-in-place facilities. Cleaning is generally performed with hot 2 per cent sodium hydroxide containing, if necessary, additional detergent and chelating agents often formulated as a proprietary product. Some inorganic deposits may require acid treatment; 0.5 per cent nitric acid is normally recommended. This reagent has the additional property of repassivating the stainless steel of the plant. Occasionally an atypical batch of product may require that the extractor be dismantled for more thorough cleaning.

3.9.2 Single stage centrifugal separators

Single stage liquid–liquid separators may be preferred for a variety of reasons, for example:

(1) to hasten a separation which would otherwise be rather slow under natural gravity;

(2) to separate mixtures which have a tendency to emulsify;

(3) for process development or evaluation;

(4) if frequent cleaning is likely to be necessary;

(5) if the extraction train is required to deal with a variety of feedstocks;

(6) if the distribution coefficient is sufficiently high to give acceptable recovery in a single stage.

If the distribution coefficient is low two or more extractions may be performed, but product stability under extraction conditions must also be considered (cf. penicillin G). The disc-stack centrifuge is the most widely used centrifugal separator, but for some small-scale applications a tubular bowl centrifuge may be more suitable.

3.9.2.1 *Tubular bowl centrifuge*

A typical machine of this type is the Sharples centrifuge (Pennwalt Ltd). When used for the solvent extraction of natural products it has a typical throughput of 5 dm^3/min. The original version tended to emit an aerosol during operation and it was modified so that the separated phases discharged via weirs ('Vaportite' model). Further containment for hazardous materials is provided by the 'Pressurtite' version.

The general arrangement of the machine is depicted in Fig. 3.57 and consists of a hollow cylinder (C) (internal dimensions: 10 cm diameter, 72 cm long) suspended from a single bearing. It can be rotated up to 15 000 r.p.m, giving 13 200 g at the periphery. The lower end of the bowl is centred in a phosphor bronze bush (B).

The feed enters via a jet (J) inserted into the bottom of the bowl and passes up the wall, eventually reaching the ring dam (R) at the top which holds the required interface. The interface position can be altered by fitting a ring dam of a different diameter. Various feed rates are accommodated by fitting a jet of the appropriate bore (typically 3–7 mm). Should solid impurities collect in the bowl, it will eventually be necessary to perform a cleaning cycle. Some solid deposits may be amenable to cleaning-in-place procedures but they often consist of denatured macromolecular material which can be removed only by removing the bowl and cleaning manually.

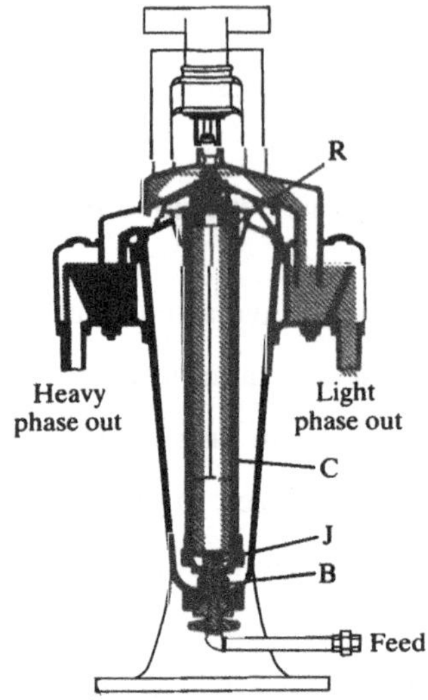

Fig. 3.57. Sharples super centrifuge. B = bottom bush; C = rotating cyliner; J = feed jet; R = ring dam.

3.9.2.2 *Disc-stack separators*

The disc-stack centrifuge was originally designed by Gustav de Laval for dairy use but it has since evolved into a wide variety of machines of differing sizes and applications produced by Alfa Laval Ltd. Disc-stack centrifuges are also produced by Westfalia Separator Ltd. The company has similar origins, being founded in Oelde, Westfalia, West Germany in 1893 to produce hand-operated cream separators. The nearby industrial area of the Ruhr provided the incentive to move into the chemical industry and over the years centrifugal separators have been developed for many diverse applications.

The simplest disc-stack separator is designed to separate two premixed liquid phases. As with the tubular bowl machine, the interface can be controlled by fitting ring dams of various diameters. The two-phase feed stream passes through a central inlet and is delivered to the base of the stack of conical discs. The discs contain holes at an intermediate radius and allow the mixture to travel up through the stack. The short path length between the discs and high centrifugal force ensure rapid separation. The flow pattern of the phases is shown in Fig. 3.58, which depicts one half of the rotor. The heavy phase collects at the periphery and is conveyed to the ring dam where it overflows and is discharged into a collecting cover. The light phase is displaced to the centre of the rotor and is discharged via the neck.

More sophisticated versions of the disc-stack centrifuge incorporate centripetal pump discharge of one or both phases. The interface in this case can

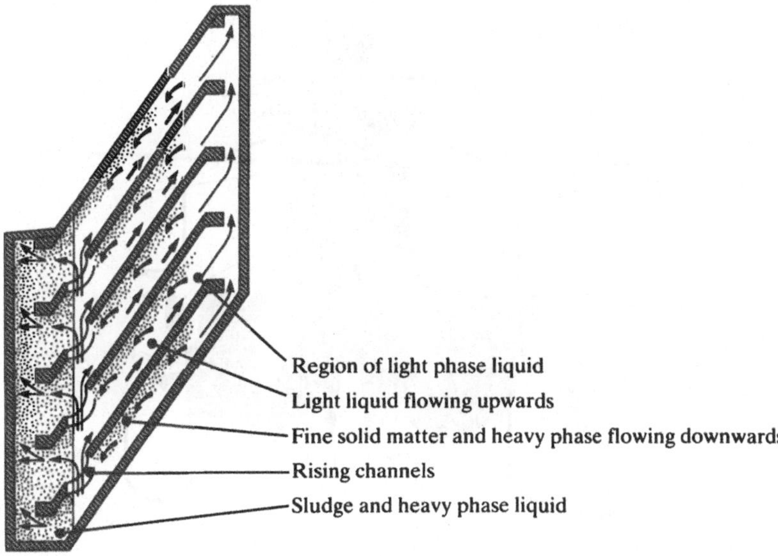

Fig. 3.58. Disc-stack centrifuge; flow pattern in rotor.

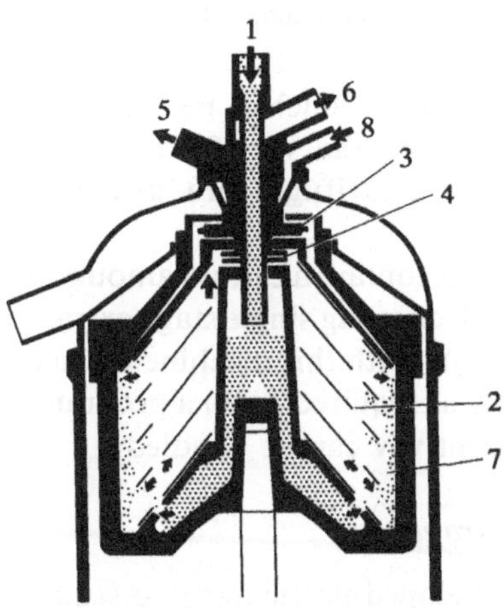

Fig. 3.59. Single stage extractor with inbuilt mixer. 1. Feed. 2. Discs. 3. Centripetal pump for heavy liquid and admixed light liquid. 4. Centripetal pump for light liquid. 5. Heavy liquid and admixed light liquid. 6. Light liquid. 7. Sediment holding space. 8. To mixing chamber.

be controlled by altering the back pressure on the light phase discharge line. For applications using flammable solvents the machine would be fitted with nitrogen purging and flameproof electrical equipment.

If insufficient extraction is obtained in a single equilibrium stage it is possible to mix the first stage raffinate within the extractor with further fresh

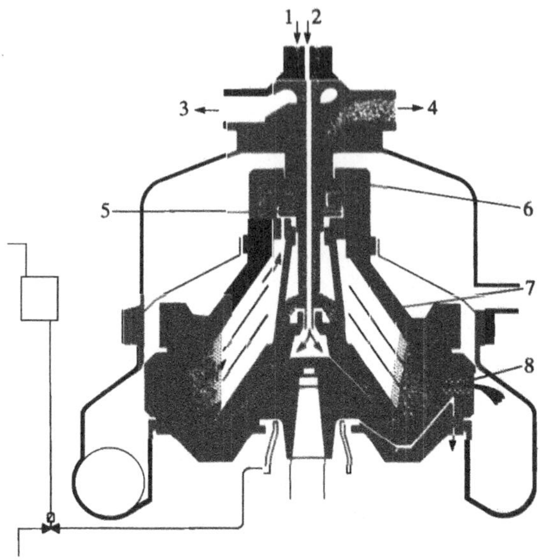

Fig. 3.60. Self-cleaning single stage extractor with piston valves. 1. Heavy liquid feed. 2. Light liquid feed. 3. Light liquid discharge. 4. Heavy liquid discharge. 5. Centripetal pump for light liquid. 6. Centripetal pump for heavy liquid. 7. Separating disc. 8. Solids discharge valves.

solvent to provide a mixture which is passed to a second separator. This arrangement, as typified by the Westfalia TA-40-04 separator, is shown in Fig. 3.59. Disc-stack centrifuges are available for a wide range of throughputs.

Should the feed stream contain a certain amount of suspended solid it may be desirable to use a self-cleaning single stage extractor such as the Westfalia VA 35-09-566 (Fig. 3.60). With this machine the accumulated solids may be ejected either manually as required or on a programmed basis. A solids content of up to 10 per cent v/v can be handled.

3.9.2.3 *Akufve apparatus*

The apparatus was developed particularly to study, on an analytical scale, short-lived radioactive species and other unstable intermediates, but, as mentioned in section 3.8, it has been found useful in pharmaceutical studies. The machine could also be used to economically determine extraction parameters prior to scale up

The essential feature of the AKUFVE apparatus is a small high efficiency centrifugal liquid–liquid separator fed from a mixing chamber in which the liquid mixture can be maintained under conditions which are predefined or changed as required. The arrangement of the apparatus is indicated in Fig. 3.61 (Reinhardt and Rydberg 1970). In operation a binary mixture

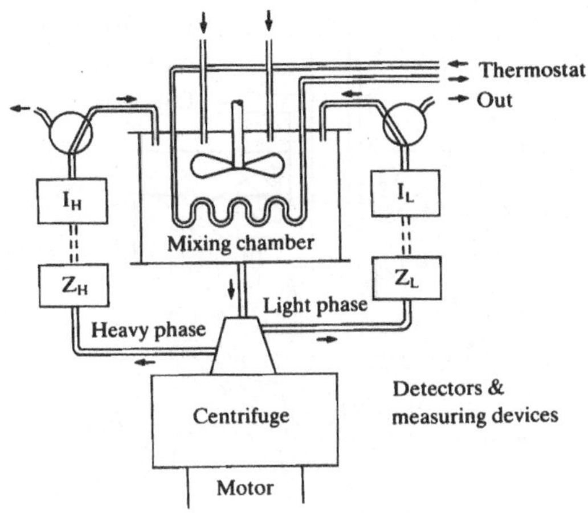

Fig. 3.61. Schematic Diagram of AKUFVE Apparatus.

under investigation is held in the thermostatted mixing chamber and vigorously stirred. The mixture is fed to the separator which can operate at up to 0.3 m^3/h (model H-33). The separated phases may be returned to the mixing chamber or directed elsewhere. Using suitable solutes, in-line spectrophotometric or other analyses of the phases may be performed. Thus when operating in the recycling mode the analysis reflects the composition of the individual phases within the mixture. If the conditions are changed, for example solute or reagent concentration, pH or temperature, the effect on the system can be immediately ascertained by means of the on-line analysis. Since the throughput of the separator is many hundred mixing vessel volumes per hour the apparatus can also be used to determine some kinetic parameters.

The success of the apparatus relies on the high efficiency of the centrifuge which achieves 'absolute' phase separation. The internal construction of the centrifuge is shown in Fig. 3.62, from which the flow pattern can be seen. The mixture enters the centre of the centrifuge and acquires the rotational speed in chamber A. The mixture then passes into the separation zone B containing eight chambers symmetrically placed around the axis and separated from each other. The droplets thus follow a zig-zag path dictated by the partition walls and the interspersed baffle ridges. The separated light phase is discharged from an upper collecting chamber C and the heavy phase from the lower chamber D by separate centripetal pumps.

The rotor may be driven by a variable speed pneumatic or electric motor. Two sizes are available; physical details and typical operating parameters are summarized in Table 3.12 (Rydberg *et al.* 1980).

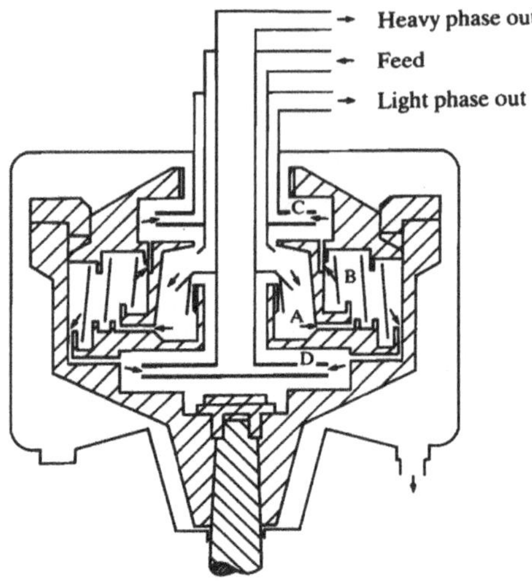

Fig. 3.62. Details of the AKUFVE H-Centrifuge. A. Inlet chamber. B. Separation zone. C and D. Collecting chambers with pump stators.

TABLE 3.12 *Details of AKUFVE H10 centrifuges*

	H-10	H-33
External diameter × length (mm)	90 × 250	165 × 310
Bowl internal diameter (mm)	54	110
Bowl volume (cm^3)	12	120
Hold-up time (s)	0.3	3
Rotational speed (rpm)	24 000	16 000
Gravitational force (*g*)	17 400	15 700
Flow capacity (dm^3/h)	150	300

3.9.3 Countercurrent extractors

Multistage countercurrent extraction may be performed using the required number of discrete mixer/separator units which may include some additional devices such as the inbuilt mixer for further extraction as shown in Fig. 3.59. It may be preferable to employ plant of this type if it is to handle different products or varying feedstocks. For the extraction of a single product using a reasonably consistent feedstock a multi-stage extractor offers the most economical use of floorspace and capital investment.

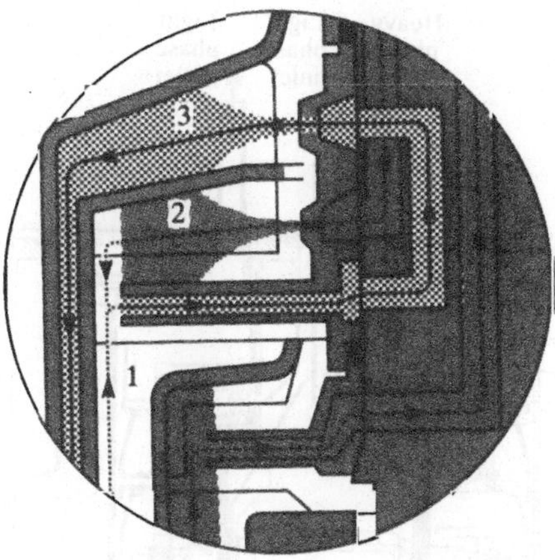

Fig. 3.63. Lurgi–Westfalia extractor; details of interstage transfer. 1. Heavy phase from inlet. 2. Light phase from stage 2. 3. Mixture to stage 1 separation.

The available machines can be divided into two types:

(1) those having a number of separate mixing and separating stages (usually two to four) on a common shaft;

(2) those with continuous countercurrent flow.

3.9.3.1 *Discrete stage countercurrent extractors*

The disc-stack configuration is used in the Lurgi–Westfalia extractor. Although no longer manufactured, there are doubtless many machines still in use. They were made in two or three-stage versions. Figure 3.63 shows details of the interstage fluid transfer. Phase separation in the two-stage machine is controlled by back pressure regulation and in the three-stage machine the separation in all three stages is controlled by means of a multi-piston valve incorporated in the central tube, which also carries the centripetal pumps.

A similar type of discrete stage arrangement is used in the Robatel extractor which was developed for use in the French nuclear power industry. A four-stage machine is depicted in Fig. 3.64. Phase separation in machines of two and more stages is controlled by back pressure adjustment. These machines are useful for the extraction of clean feed streams. However, they have very little solids-holding capacity and unless effective cleaning-in-place procedures are available they should not be used where solids deposition is likely to be a problem.

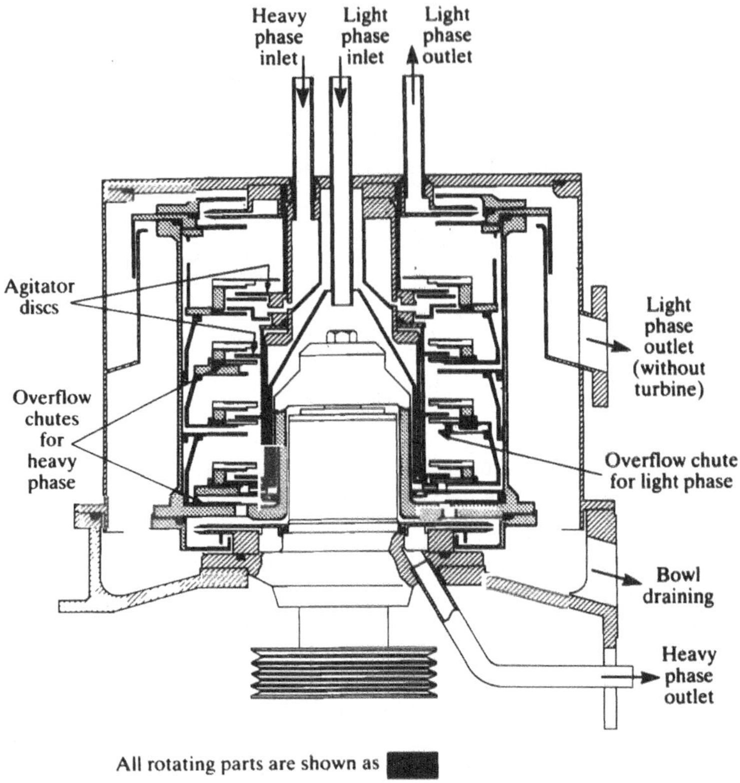

Fig. 3.64. Robatel four-stage countercurrent extractor.

3.9.3.2 *Continuous flow countercurrent extractors*

The most widely used machine of this type is the Podbielniak extractor, which bears the name of its inventor. It was first described in 1935 but few applications were found until the advent of penicillin production.

The original device, patented by Walter Podbielniak in 1935, consisted of a coil of tubing on a vertical rotating cone. It soon evolved into a horizontal axis machine with a hollow cylinder containing a 30 m spiral band with a 1.8 mm clearance. The first machine supplied for penicillin production in 1945 contained a 150 mm wide 50 m long spiral band and was rated for a combined flow capacity of 0.68 m^3/h. Within a year the extractor design had been modified to incorporate louvre type holes in the spiral band and this machine, despite its low throughput, was adopted as the standard for the pharmaceutical industry. A later design modification replaced the internal spiral with a series of concentric rings containing transfer ports together with integral clarification zones for the two phases and internal conduits for the

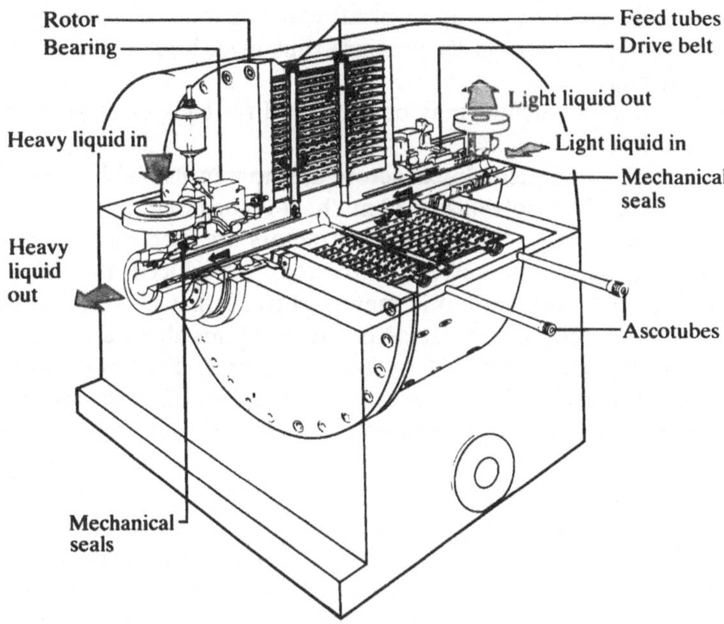

Fig. 3.65. Podbielniak extractor; general arrangement.

two inlet and outlet phases, as depicted in Fig. 3.65. The various modes of operation are described by Todd (1966).

As the cylinder rotates at several thousand revolutions a minute, heavy liquid is centrifuged outwards, displacing the light liquid inwards. As the phases pass through the slits in the concentric rings, countercurrent contact is made. The light liquid eventually reaches the inner zone of the rotor where a different pattern of perforations in the rings promotes phase clarification. A similar zone exists at the rotor periphery to produce clear heavy phase.

A machine of more open design has been used for whole broth extraction but, although most of the solids pass through the machine, it is reported that periodical cleaning is required (Todd and Davies 1973).

The efficiency of phase separation will depend on the depth of the clarification zone and the gravitational force experienced. For scale up and design purposes the concept of *g*-minutes is used. The Podbielniak extractor usually operates with a maximal *g*-minute value of 500–1000. As with the majority of extractors used for natural product isolation, it is necessary to operate with a compromise between liquid phase clarity and deposition of solids. Should the machine be required for experimental purposes or variable duties the drive unit may be fitted with speed control. Light phase discharge is controlled via a back pressure valve to give a suitable interface position within the rotor (typically, the back pressure is approximately 4 bar). The interface is adjusted to give an appropriate clarification zone for each phase; increasing the light phase outlet pressure thus gives a greater clarification zone for this phase.

A similar, smaller machine may be employed for back extraction into an aqueous solution.

For penicillins, Podbielniak extractors are now in use with a combined flow capacity (broth + solvent) of 30 m^3/h. It is suggested by the manufactures that in some plants Podbielniak extractors are used on whole broth.

The Alfa-Laval countercurrent extractor, introduced in 1962, was designed specifically for antibiotic manufacture. It has certain features in common with the Podbielniak extractor but it rotates about a vertical axis. The machine has been described by Larssen (1983). The principal model (ABE 216) with a typical throughput of 5–9 m^3/h has a 26-metre spiral contact flow path and operates with three to five theoretical stages depending on the properties of the fluids. The rotor consists of a series of concentric cylinders, each equipped with a spiral in the annular space which directs the flow into a fixed channel. Light liquid passes into the central channel of the spindle and initially travels towards the outermost extraction chamber. In the extraction channel the light phase then moves towards the centre of the bowl and the heavy phase follows the reverse path. At the end of each annular spiral the liquids pass through perforations into the next annulus. Thus regions of parallel countercurrent flow are alternated with zones of intensive mixing. The outer annulus serves to effect a final separation of traces of light phase from heavy phase.

The net extraction is influenced by the ratios of the two phases within the bowl. Some systems, for instance, require that the bowl be mainly filled with light phase. The relative amount of the two phases in the bowl is determined by the diameter of the gravity disc fitted and by adjustment of the light phase outlet pressure.

The extractor requires a feed pump generating 6 bar pressure for the heavy phase. The light phase feed pump is built in. For fermentation broth extraction a heavy : light phase ratio of about 5 : 1 would be used but the machine will accommodate phase ratios of 9 : 1 to 1 : 9.

It is recommended that the feed liquids should be free of suspended solids but light particles are carried through the machine due to the high linear velocity of the streams. Heavy solids may accumulate and if necessary the extraction elements can be removed as a complete unit to facilitate cleaning. A cleaning cycle can be performed in about 2 hours.

For feeds containing a higher concentration of solids a countercurrent extracting decanter may be used. This will handle a feedstock containing up to 60 per cent v/v solids. The machine is a relatively recent development of the horizontal bowl decanter centrifuge which has a scroll to transport the solids out of the machine. The configuration employed (Fig. 3.66), when

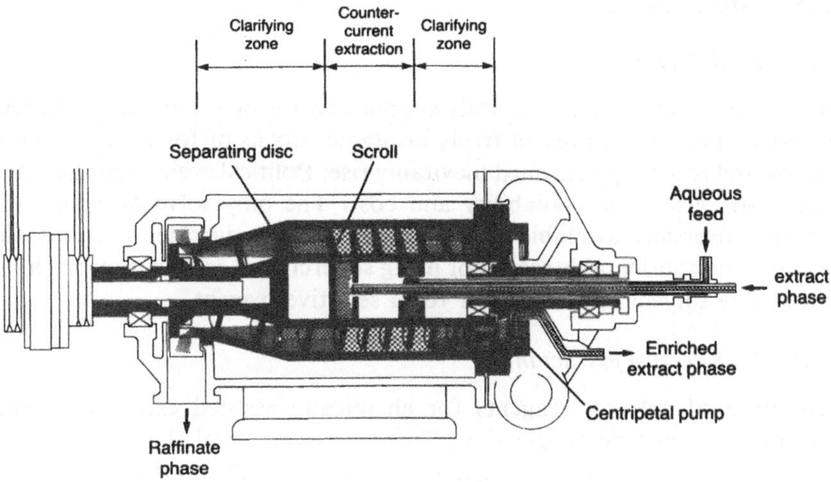

Fig. 3.66. Decanter for continuous countercurrent extraction.

used on whole broth, allows a practical countercurrent extraction corresponding to 1.7 theoretical stages which gives good yields of penicillin G and erythromycin, but for the highest recoveries two similar machines are operated in countercurrent mode. Operating details and advantages of the type of extractor are discussed in section 3.2.2.1.

3.9.4 Summary

A range of centrifugal liquid–liquid separators and extractors is available in the pharmaceutical industry. The above description outlines the main types of machine available. There are many variants of the basic machines for specific duties and full details can be obtained from the manufacturers' literature.

3.10 Future developments

As discussed in previous sections, solvent extraction procedures are well established for certain types of pharmaceutical product and there exists the potential for extension into other product areas. The processes of the twenty-first century may conveniently be discussed in the context of the shortcomings of those existing. Potential developments can be divided into three groups: reagents, equipment, and feedstocks.

3.10.1 Reagents

3.10.1.1 *Solvents*

Most organic solvents are currently supplied by the petrochemical industry. Whilst supplies are at present freely available, stocks of fossil fuels are not infinite and relative prices must inevitably rise. Political events can also have a dramatic effect on availability and cost. The only solvents which will remain universally available are water and carbon dioxide. There would appear to be considerable scope for using supercritical carbon dioxide either as a simple solvent or as a carrier for a selective reagent.

3.10.1.2 *Selective extractants*

Regardless of solvent, reagents for chemically assisted extraction could become more specific or selective. The current empirical choice of reagent could give way to extractants designed to interact with specific solutes. The development of various chiral complexing agents indicates the potential of such an approach which has been discussed by Cram and Cram (1974). Most work has been based on substituted crown ethers.

Resolution of racemic mixtures is not normally required when isolating natural products due to the chiral specificity of biosynthetic pathways. However, similar principles could be applied to the selective isolation of one component from a mixture of closely related compounds. It is not uncommon to find compounds with a common nucleus containing a variety of substituents where the balance of properties of the substituents makes members of the group very similar in terms of the normal parameters used for separation. It is not unreasonable to suggest that a specific complexing agent could be designed to selectively extract one component. This approach is a refinement of that long used in hydrometallurgy and is closely allied to biomimetic drug design studies. Wilcox *et al.* (1987) have designed relatively simple molecules with concave surfaces containing rigid chiral grooves in an attempt to give highly specific binding sites. They prepared a compound of the structure depicted in Fig. 3.67. The molecule has a concave surface of helical symmetry which can form a semi-clathrate with suitable guest molecules. The example reported could accommodate two ethanol molecules in the cleft. Recently a specific extractant, consisting of a crown ether attached to a polysiloxane, has been designed for the separation of codeine and norcodeine (Abed-Ali *et al.* 1987). Further work in this area is awaited with interest.

3.10.1.3 *Liquid membranes*
[See also Weatherley Chap. 5]

Liquid membrane emulsions can in principle offer the advantage of high surface area to promote rapid mass transfer. Applications were first described

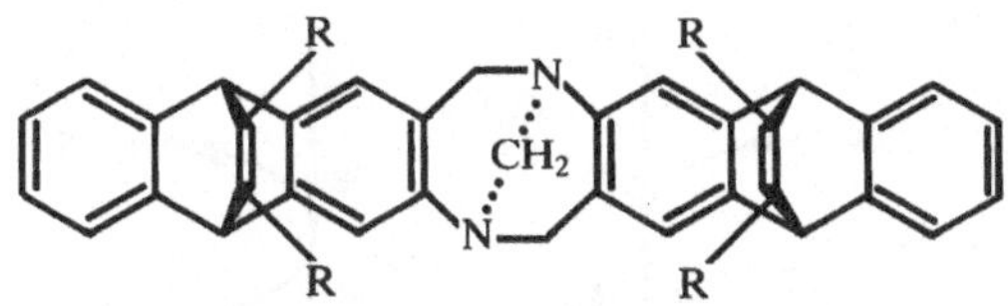

$R = -CO_2CH_3$

Fig. 3.67. Structure of ethanol receptor molecule (dibenzodiazocine derivative).

by Li (1971) and have since been investigated for several separations leading to at least one commercial application, the recovery of zinc from an effluent stream (Protsch and Marr 1984).

In general, the system consists of globules of water-immiscible liquid (the membrane) containing small droplets of aqueous phase, suspended in a continuous aqueous phase (Fig. 3.68). The membrane phase may be a simple solvent containing sufficient surfactant to stabilize the emulsion or it can also contain additional reagents to facilitate transport.

The extraction of acetic acid provides a simple example of liquid emulsion membrane technology. The continuous phase would consist of dilute aqueous acetic acid and the inner aqueous phase (which corresponds to the back extractant in conventional solvent extraction) could be sodium hydroxide. The driving force for extraction is provided by the pH gradient. For compounds which are already ionized chemically assisted extraction is used.

Reagents similar to those described in section 3.4 may be incorporated in the membrane phase. Reports on the application of this technique in biotechnology are now beginning to appear. Some significant studies on the extraction of phenylalanine have been reported (Thien *et al.* 1986). The basis of the method is outlined in Fig. 3.69 and a simple process flow sheet is given in Fig. 3.70. Liquid membrane studies on citric acid have also been reported (Boey *et al.* 1987). Apart from devising the initial emulsion, it appears that the main problem relates to swelling of the encapsulated water droplets and

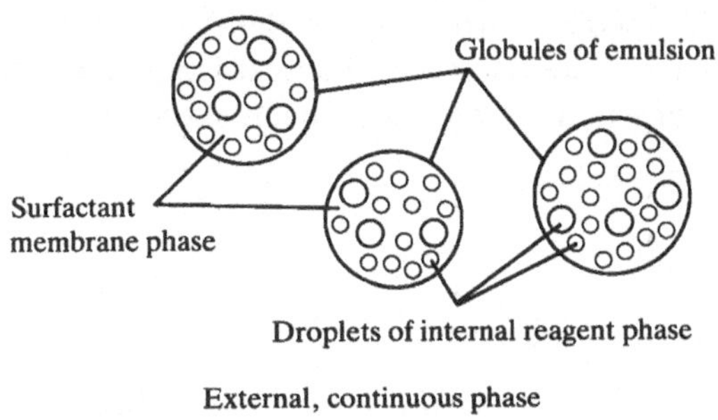

Fig. 3.68. Schematic diagram of liquid emulsion membrane.

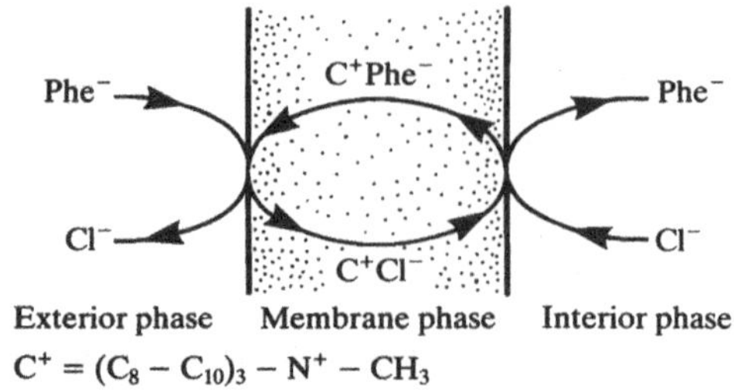

Fig. 3.69. Mechanism for facilitated transport in a phenylalanine/chloride system.

subsequent rupture of the membrane. This occurs because solutes are often transported across the membrane as a hydrate. Osmotic effects can also play a part in the process.

A high surface area extraction system is also provided by biliquid foams ('polyaphrons'; greek αψροσ, foam). The system has been studied by Sebba and coworkers (Wallis *et al.* 1985). They found that a predispersed solvent biliquid foam can be prepared to contain 95 per cent v/v solvent and surfactant as the dispersed phase. The remaining 5 per cent is water and the suspension can be diluted with water as required. Individual aphron particles can be of controlled size from submicron up to 10 μm in diameter, giving the system a high surface area for mass transfer. After extraction the loaded aphrons can be separated by flotation using colloidal gas aphrons and broken down using a suitable particulate material (Sebba 1984).

The perceived advantages of both liquid membranes and colloidal liquid aphrons are the low energy input and the possibility of producing concentrated extracts. Also, since the extracting solvent does not come into direct contact with the continuous phase, these extractants could, in principle, be used for the *in situ* extraction of fermentation products. Some preliminary results using an ethanol fermentation as an example are reported by Wallis *et al.* (1985). Possible disadvantages could include incompatibility of the surfactant with other components of the system and transfer resistance of the solvent-encapsulating membrane.

Both the above methods offer the possibility of reduced extractant usage and make selective reagents more practical. Liquid membrane emulsions have the advantage that forward and back extraction are performed as a single stage.

3.10.2 Equipment

3.10.2.1 *Centrifugal extractors*

It is generally found that an enhanced gravitational field is required for phase separation when solvent extracting natural products from aqueous

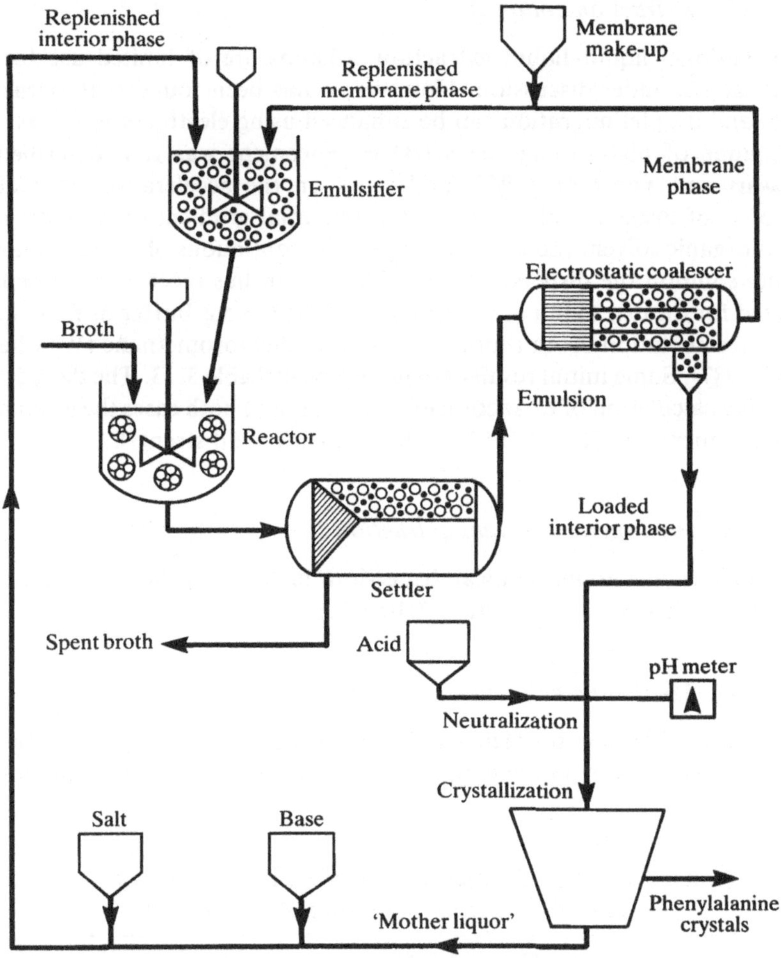

Fig. 3.70. Process diagram for liquid emulsion membrane-based recovery of phenylalanine.

solution. The machines which have been produced for the purpose are generally effective, but are they wasting energy? Most extractors have regions of intensive mixing and the resulting emulsion requires a high centrifugal field for separation. More control of mixing and separating stages could save some mechanical energy and additional economies could result from a reduced cooling requirement. Most fundamental studies on liquid–liquid mass transfer have been done using column systems; they should now be extended to centrifuges.

3.10.2.2 *Extraction columns*

Conventional liquid–liquid extraction columns are of limited use for the applications under discussion. However, it has been found that extraction rates and droplet migration can be enhanced using electrostatic forces. The tenhnique of high voltage electrical enhancement was first described by Stewart and Thornton (1967) and relies on the acceleration of charged droplets of dispersed phase through a potential gradient of 2–50 kV m^{-1}. Thus organic solvent (non-conducting) is the continuous phase and aqueous solute solution the dispersed phase. The system has recently been demonstrated for the extraction of penicillin G from aqueous buffer or from unfiltered fermentation broth (both at pH 4) into dichloromethane (Weatherley *et al.* 1987). Some initial results are presented in Table 3.13. The data do not permit a calculation of extraction efficiency, but in both cases the gross mass transfer increased considerably with increasing field strength.

3.10.2.3 *Predispersed solvent extraction*

As an alternative to mechanical dispersion, high interfacial area can be obtained in other ways (see section 3.10.1.3.).

3.10.2.4 *Membrane extractors and reactors*

The phase separation problem can be avoided using a membrane extractor. It has already been pointed out that conventional solvent extraction of

TABLE 3.13 *Continuous countercurrent extraction of penicillin G into dichloromethane at pH4*

Run no.	Nominal field strength (kV/cm × 10^2)	Gross mass transfer (kg/s × 10^{-9})
1	0	2.17
2	3.076	4.49
3	7.692	3.972
4	15.385	4.872
5	18.462	6.713
6	0	8.33
7	1.154	12.73
8	1.538	14.80

Runs 1–5 were performed on pure penicillin G solution (1 g/dm).
Runs 6–8 were performed on penicillin G whole broth (1 g/dm).

fermentation products suffers from a number of practical problems concerned with the presence of emulsion-stabilizing impurities. These give rise to entrainment of fermentation broth in the organic phase leading to low purity product. Additionally, some rich solvent may be entrained in the raffinate (spent broth), causing low yield and effluent treatment problems. Both problems can in theory be eliminated using a membrane solvent extractor. The principle is the same as liquid membranes but the membrane is in the solid state, used as a hollow fibre or a flat sheet which prevents the two phases from physically mixing. The solute partitions initially between the aqueous phase and the membrane and subsequently between the membrane and the extractant. A key aspect of this technology is obviously the choice of membrane.

It is reported that contact times can be as short as those attained in a Podbielniak extractor. The membranes currently available are generally unsuitable for low pH applications (below pH 3) and could not, for example, be used for direct penicillin G isolation. Coupled transport membranes have been proposed as a possible solution to this problem (Smith *et al.* 1977). An organic reagent is immobilized within the pores of a membrane, one side of which is in contact with the fermentation solution, while the other side is contacted with a suitable stripping solution. The chemical aspects are similar to those of liquid membranes and it would appear that most reagents used for ion-pair extraction could be used in this mode. Coupled transport membranes in theory use minimal amounts of reagent so more expensive and selective extractants could be employed. Currently these systems suffer from problems of limited stability caused by reagent leaching.

A more sophisticated version of the coupled transport membrane is the membrane bioreactor. Instead of acting as a device to keep the two phases apart, the membrane in this case incorporates an enzyme immobilized within its matrix. Using suitable solvents on each side of the membrane, selective reaction and product recovery can be achieved. This approach is being developed by Sepracor Inc. (Marlborough, USA) using hollow fibre membrane technology. The reactor consists of a bundle of thin (about 0.3 mm outside diameter) hollow fibres which are compatible with most organic solvents and are also hydrophilic with low protein binding characteristics. The enzyme is physically immobilized within the membrane matrix. Covalent linking agents are not used so the enzyme can retain its full activity and natural conformation and there is no danger of excess coupling reagent leaching from the membrane. The enzyme can be very stable in this system. A half life of 311 days is reported for lipase from *Candida cylindricea* when hydrolysing amyl acetate (Brandt 1987). It is claimed that the enzyme can be replaced by backflushing the reactor and recharging it with fresh enzyme.

Reactant solution flows through the fibres and product sweep solution flows concurrently along the outside surface of the fibres. Reactant and

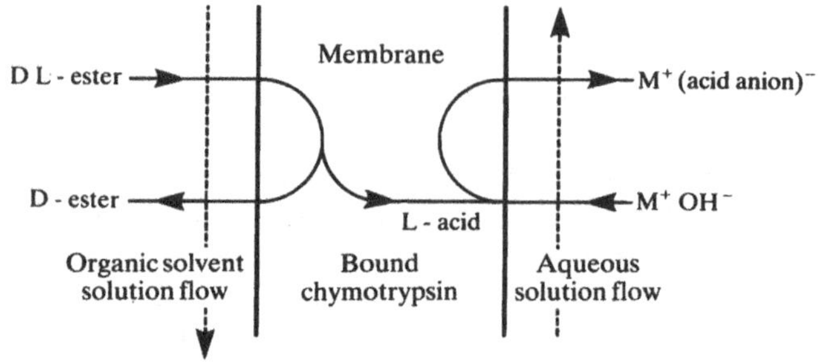

Fig. 3.71. Membrane enzyme reactor for enantioselective hydrolysis of an ester.

product transfer are mainly by diffusion and there is no substantial pressure gradient to give rise to membrane fouling such as can occur in ultrafiltration. The Sepracor workers report a model system consisting of the enantioselective hydrolysis of N-benzoyl-DL-tyrosine ethyl ester using chymotrypsin (Fig. 3.71).

Data have been presented to show that, using the same amount of enzyme, a continuous flow membrane reactor is more productive than a stirred reactor containing immobilized enzyme. The membrane system does not suffer from problems of emulsion formation which are said to be a drawback of the dispersed system. Additionally, product recovery is much simpler as one of the downsteam processing stages (separation of the required product) has already been performed in the reactor.

Among other successful two-phase applications of membrane reactor technology are the enantioselective hydrolysis of glycidol butyrate esters (Brandt 1987) and an indication that penicillin G can be extracted from fermentation broth and hydrolysed to 6-aminopenicillanic acid using a combination of coupled transport and bioreactor membrane systems (Sepracor 1987).

3.10.3 Feedstocks

The most dramatic biotechnological advance in recent years has been the development of recombinant DNA technology ('genetic engineering') and its subsequent application to the manufacture of products previously rather inaccessible, for example insulin, interferon. So far, liquid–liquid partition has played little part in these developments although it is possible that two-phase aqueous systems could be used.

Genetic engineering can be used to assist downstream processing. Thus if it is not possible to design the reagent to suit the product the converse may be investigated. It may be possible to the produce product in a form which renders it more amenable to a particular extraction procedure. After purifying, the pro-product can be converted into the required form. This approach has already been employed in the purification of urogastrone. Using genetic engineering techniques, the producing organism was modified so that it produced urogastrone containing a terminal polyarginyl group which made the compound more basic than most co-occurring impurities. The polyarginyl urogastrone was selectively absorbed on to a cation exchanger and, after elution, enzymically converted to the required compound using carboxypeptidase B and rechromatographed (Sassenfeld and Brewer 1984).

It is conceivable that similar techniques could be used to add a lipophilic tail to an otherwise hydrophilic compound to render it solvent extractable.

The 'new biotechnology' is developing pharmaceutical products which are active at very low dose levels and the volume of production is several orders of magnitude less than that of antibiotics. It follows that the equipment must be scaled down from that previously used. This aspect, combined with the fact that the products are potentially more hazardous, imposes severe design criteria. Thus engineering tolerances are smaller and the equipment is more complex due to the requirement for containment, sterilization, and aseptic operation (Aronsson 1987).

3.10.4 Integrated processes

Most fermentation and downstream processing operations are currently performed as discrete batch operations. There is however increasing interest in integrating a number of stages for continuous operation. This provides the opportunity to introduce improved process control and better utilization of facilities. Additionally, microbial production of secondary metabolites is often subject to feedback inhibition. The effect can be minimized either by selecting a producing strain which is more tolerant of the product or by continuous product removal. The former approach is normally favoured but with increasing interest in integrated processes the latter has recently been investigated. It is most readily achieved if the product is extracellular.

On-line fermentation/extraction processes based on solvent extraction must conform to the following criteria (Wang 1983):

(1) extraction must be selective;

(2) the extractant must have a high loading capacity for the product;

(3) the extraction system must be capable of aseptic operation;

(4) the extractant must be non-toxic to the growing culture;

(5) the extractant must be reusable if it is expensive.

These constraints are very demanding and eliminate many conventional extraction procedures. As far as the fermentation system is concerned, the least intrusive method would appear to be dialysis extraction. The fermenter contains a tube of dialysis membrane through which water or buffer circulates. Product may then be solvent extracted from the permeate using an extractor external to the fermenter. A typical arrangement is described by Kominck (1975). The possible use of polyaphrons or liquid membranes has already been mentioned. Fermentation in the presence of a non-toxic oil has also been proposed. Using an example from the food industry, it was found that capsaicin could be continuously removed from a plant cell culture of *Capsicum frutescens* by coupling the fermenter to an extraction column containing sunflower oil, thereby increasing product yield and stability (Mavituna *et al.* 1987).

3.11 Summary

Solvent extraction in the pharmaceutical industry has developed from the preparation of unpurified plant extracts into a complex, large-scale technological process. The scale may now have reached a maximum as the future is likely to see the development of more potent biologically active substances as medicinal products. The net result will be an increase in sophistication especially with respect to containment and process control but a reduction in scale.

The foregoing has indicated that solvent extraction processes are well developed for certain fermentation products and the potential exists for extending the technique to other types of compound.

Solvent extraction is particularly relevant to continuous processes and the requirements for carefully controlled and integrated processes may provide the incentive to increase the number of applications. Several potential uses are still at the laboratory scale or have been demonstrated only on model compounds but the concepts may eventually be incorporated into production scale processes. There could be a particularly bright future for membrane extractors and reactors. The use of aqueous two-phase partition can eliminate most of the mechanical problems connected with the recovery of product from cell homogenates and by using substantial purification can be obtained affinity ligands. Should petrochemical solvents become prohibitively expensive the untapped potential of supercritical carbon dioxide could be employed.

References

Abed-Ali, S. S., Brisdon, B. J. and England, R. (1987). *Chemical Communications*, 1565.

Abraham, E. P. (1968). In Perlman, D. (ed.) *Topics in Pharmaceutical Sciences*, Vol. 1, pp. 1–31. Wiley–Interscience, New York.

Abraham, E. P. (1974). *Dev. Ind. Microbiol.* **15**, 3–15.

Albertsson, P.-Å. (1967). *Biochim. Biophys. Acta* **27**, 378.

ALBERTSSON, P.-Å. (1977). *Endeavour* **1**, 69–74.
ALBERTSSON, P.-Å. (1986). *Partition of cell particles and macromolecules*, 3rd ed. Wiley, New York.
ALBERTSSON, P.-Å. and FRICK, G. (1960). *Biochim. Biophys. Acta* **37**, 230–7.
ALBERTSSON, P.-Å. and LARSSON, C. (1976). *Mol. Cell. Biochem.* **11**, 183.
ANDERSON, D. W. and LAU, E. F. (1955). *Chem. Eng. Prog.* **51**, 507–12.
ANDERSSON, E., MATTIASSON, B. and HAHN-HAGERDAL, B. (1984). *Enzyme Microb. Technol.* **6**, 301–6.
ANON (1986). *Scrip*, No. 1075 (10 Feb. 1986), p. 26.
ARONSSON, G. (1987). *Bio/Technol.* **5**, 394–5.
AUERBACH, M. E. (1943). *Ind. Eng. Chem., Anal. Edn.* **15**, 492.
BACKMAN, L. and SHANBHAG, V. P. (1984). *Anal. Biochem.* **138**, 372–9.
BAKER PERKINS (1987). *Technical Brochure*. Baker Perkins Inc., Saginaw, MI 48601, USA.
BAMBERGER, S., BROOKS, D. E., SHARP, K. A., VAN ALSTINE, J. M. and WEBBER, T. J. (1985). In *Partitioning in Two Phase Aqueous Systems* (ed. H. WALTER, D. E. BROOKS and D. FISHER), pp. 86–131. Academic Press, Orlando.
BEGNER, C. and WILDMAN, A. S. (1983). European Patent, 82674.
BEIJERINCK, M. W. (1896). *Zentbl. Bakteriol.* **2**, 627, 698.
BÉRDY, J. (1985). In *Discovery and Isolation of Microbial Products* (ed. M. S. VERRALL), p. 9. Ellis Horwood, Chichester.
BERKOVITCH, I. (1976). *Process Eng.*, Oct. p. 95.
BERLIN, J. (1984). *Endeavour, New Series* **8**, 5–8.
BETAGERI, G. V. and ROGERS, J. A. (1987). *Int. J. Pharmaceutics* **36**, 165–73.
BIRKENMEIER, G., EHRLICH, U. and KOPPERSCHLAGER, G. (1986). *J. Chromat.* **360**, 193–201.
BLOMQUIST, G. and WOLD, S. (1974). *Acta Chem. Scand. B* **28**, 56.
BLUNDEN, G., CULLING, C. and JEWERS, K. (1975). *Trop. Sci.* **17**, 139–54.
BOERIU, C. G., DORDICK, J. S. and KLIBANOV, A. M. (1986). *Bio/Technol.* **4**, 997–9.
BOEY, S. C., DEL CERRO, M. C. and PYLE, D. L. (1987). *Chem. Eng. Res. Des.* **65**, 218–23.
BONNER, D. P., MECHLINSKI, W. and SCHAFFNER, C. P. (1975). *J. Antibiot.* **28**, 132–5.
BONNERJEA, J., OH, S., HOARE, M. and DUNNILL, P. (1986). *Bio/Technol.* **4**, 954–8.
BRANDSTROM, A. (1977). *Adv. Phys. Org. Chem.* **15**, 267.
BRANDT, S. (1987). *Perf. Chemicals*, May, 4–8.
BREZHNIKOVA, M. G., KONSTANTINOVA, M. V., POMAZKOVA, V. A. and ZAKHAZOV, B. V. (1966). *Antibiotiki (Moscow)* **11**, 763–7.
BRINK, L. E. S. and TRAMPER, J. (1987). *J. Chem. Technol. Biotechnol.* **37**, 21–44.
BROCKMANN, H. and HENKEL, W. (1950). *Naturwissenschaften* **37**, 138.
BRODELIUS, P. and NILSSON, K. (1983). *Eur. J. Appl. Microbiol. Biotechnol.* **17**, 275–80.
BROOKS, D. E., SHARP, K. A. and FISHER, D. (1985). In *Partitioning in Two Phase Systems* (ed. H. WALTER, D. E. BROOKS and D. FISHER.), pp. 11–84. Academic Press, Orlando.
BROWN, A. G. (1982). *J. Antimicrob. Chemother.* **10**, 365–72.

BRUNNER, K.-H. (1984). *Technical and Scientific Documentation No. 4.* Westphalia Separator AG, Oelde.
BRUNNER, K.-H. (1985). In *Discovery and Isolation of Microbial Products* (ed. M. S. VERRALL), pp. 182–94. Ellis Horwood, Chichester.
BUCKLAND, B. C., DUNNILL, P. and LILLY, M. D. (1975). *Biotechnol. Bioeng.* **17**, 815–26.
BUITAR, C., MASCELTI, G. and GUERRA, G. (1973). German Patent Application 2311117.
BU'LOCK, J. and KRISTIANSEN, B. (1987). *Basic Biotechnology*, p. 527. Academic Press, London.
BUTT, S. and ROBERTS, S. M. (1987). *Chem. Britain* **23**, 127–34.
BUTTERWORTH, D., HOOD, J. D. and VERRALL, M. S. (1983). In *Advances in Biotechnological Processes* (ed. A. MIZRAHI and A. L. VAN WEZEL), Vol. 1, pp. 251–92. Alan R. Liss Inc., New York.
BYWOOD, R., ROBINSON, C., STABLES, H. C., WALKER, D. and WILSON, E. M. (1977). *Spec. Publ. Chem. Soc.* **28**, 139–44.
CHABALA, J. C., MROZIK, H., TOLMAN, R. L., ESKOLA, P., LUSI, A., PETERSON, L. H., WOODS, M. F., FISHER, M. H. and CAMPBELL, W. C. (1980). *J. Med. Chem.* **23**, 1134–6.
COLE, D. L. and GOEGELMAN, R. T. (1979). United States Patent 4160861.
COONEY, C. L. (1983). *Science* **219**, 728.
COONEY, C., PETRIDES, D., BARRERA, M. and EVANS, L. (1988). In *The Impact of Chemistry on Biotechnology*, ACS Symposium Series No. 362. (ed. M. PHILLIPS, S. P. SHOEMAKER, R. D. MIDDLEKAUFF and R. M. OTTENBRITE), pp. 39–61. American Chemical Society, Washington.
CRAM, D. J. and CRAM, J. M. (1974). *Science* **183**, 803–9.
CREMONESI, P., CARREA, G., FERRARA, L. and ANTONINI, E. (1975) *Biotechnol. Bioeng.* **17**, 1101–8.
DARVEAU, R. P. and HANCOCK, R. E. W. (1983). *J. Bacteriol.* **155**, 831–8.
DATAR, R. (1986) *Proc. Biochem.* **Feb.**, 19–26.
DATAR, R. and EDWARDS, K. (1986). Poster presented at Engineering Foundation Meeting, 'Recovery of Fermentation Products', Uppsala.
DATTA, R. and WILDMAN, G. T. (1978). German Patent Application 2808636.
DAVIS, S. S. and ELSON, G. (1974). *J. Pharm. Pharmacol.* **25 suppl.** 90P.
DAVIS, S. S. HIGUCHI, T. and RYTTING, J. H. (1972). *J. Pharm. Pharmacol.* **24 suppl**, 30P–46P.
DEKKER, J. and ARK, P. A. (1959). *Antibiot. Chemother.* **9**, 327–32.
DEKKER, M., VAN'T RIET, K., WEIJERS, S. R., BALTUSSEN, J. W. A., LAANE, C. and BIJSTERBOSCH, B. H. (1986). *Chem. Eng. J.* **33**, B27–33.
DUBOST, M., GANTER, P., MARAL, R., NINET, L., PINNERT, S., PREUD'HOMME, J. and WERNER, G. H. (1963). *C. Ŕ. Hebd. Séanc. Sci.* **257**, 1813–5.
EDMUNDOWICZ, J. M. and ALLANSON, E. E. (1982). United States Patent 4354971.
EGAN, B. Z., RHEAR, R. W. and KELMERS, A. D. (1970). *Anal. Biochem.* **38**, 139–47.
FINHOLT, P., JÜRGENSEN, G. and KRISTIANSEN, H. (1965) *J. Pharm. Sci.* **54**, 387–93.
FLICKINGER, M. C. (1985). In: *Comprehensive Biotechnology* (ed. M. MOO-YOUNG), Vol. 3, pp. 230–43. Pergamon, Oxford.
FLICKINGER, M. C. and SANSONE, E. B. (1984). *Biotechnol. Bioeng.* **26**, 860–70.

FRENCH, J. C., SHORES, ST. C., HOWELLS, J. D., WOODS, G. P. and ANDERSON, L. E. (1970). United States Patent 3551294.

FUJITA, Y., TABATA, M., NISHI, A. and YAMADA, Y. (1982). In: *Plant Tissue Culture 1982*, (ed. A. FUJIWARA), p. 399. Maruzen, Tokyo.

GALANOS, C., LÜDERITZ, O. and WESTPHAL, O. (1969). *Eur. J. Biochem.* **9**, 245–9.

GALLAHER, M. G., GUARCIO, N. and GUERCIO, P. A. (1964). United States Patent 3121714.

GREIN, A., SPALLA, C., DIMARCO, A. and CANEVAZZI, G. (1963). *G. Microb.* **11**, 109–18.

GRUNWALD, J., WIRZ, B., SCOLLAR, M. P. and KLIBANOV, A. M. (1986). *J. Am. Chem. Soc.* **108**, 6732–4.

HANSCH, C. and LEO, A. (1979). Substituent Constants for Correlation Analysis in Chemistry and Biology. Wiley, New York.

HARADA, S., TSUBOTANI, S., ASAI, M., OKONOGI, K. and KONDO, M. (1983). *J. Med. Chem.* **26**, 753–7.

HASKO, F., VAZILEVA, R. and HALASZ, L. (1982). *Biotechnol. Bioeng.* **24**, 1931–9.

HAYES, H. B. and HUFF, G. L. (1971). German Patent Application 2031754.

HAZEN, E. L. and BROWN, R. (1950). *Science* **112**, 423.

HEIGEL, W. and HUESCHENS, R. (1983). United States Patent 4387178.

HERSBACH, G. J. M., VAN DER BEEK, C. P. and VAN DILCK, P. W. M. (1984). In: *Biotechnology of Industrial Antibiotics* (ed. E. J. VANDAMME), p. 45. Marcel Dekker, New York.

HIGUCHI, T., MICHAELIS, A., TAN, T. and HURWITZ, A. (1967). *Anal. Chem.* **39**, 974–79.

HOOD, J. D., BOX, S. J. and VERRALL, M. S. (1979). *J. Antibiot.* **32**, 295.

HUMMEL, W, SCHÜTTE, H. and KULA, M.-R. (1984). *Ann. N.Y. Acad. Sci.* **434**, 195–205.

HUSTED, H., KRONER, K. H. and KULA, M.-R. (1985). In *Partitioning in Aqueous Two Phase Systems: Theory, Methods, Uses and Applications to Biotechnology* (ed. H. WALTER, D. E. BROOKS and D. FISHER), p. 529. Academic Press, Orlando.

JACKSON, A. T. and DE SILVA, R. L. (1986). *Proc. Biochem.*, December, 185–9.

JAHNKE, H. K. (1967). United States Patent 3318867.

JEFFERY, J. D'A., ABRAHAM, E. P., NEWTON, G. G. F., (1960). *Biochem. J.* **75**, 216–23.

JOHANSSO N, G. (1970). *Biochim. Biophys. Acta* **221**, 387.

JOHANSSON, G., HARTMAN, A. and ALBERTSSON, P.-A. (1973). *Eur. J. Biochem.* **33**, 379–86.

JOHANSSON, G. and JOELSSON, M. (1986). *Appl. Biochem. Biotechnol.* **13**, 15–27.

JOHNSON, D. A., SMITH, R. R., RICHARDSON, E. J., ROUBIE, J. M. and SILVESTRI, H. H. (1970). German Patent Application 1933187.

KADAM, K. L. (1986). *Enzyme Microb. Technol.* **8**, 266–73.

KAHAN, J. S., KAHAN, P. M., GOEGELMAN, R., CURRIE, S. A., JACKSON, M., STAPLEY, E. O., MILLER, T. W., MILLER, A. K., HENDLIN, D., MOCHALES, S., HERNANDEZ, S., WOODRUFF, H. B. and BIRUBAUM, J. (1979). *J. Antibiot.* **32**, 1.

KAMIHIRA, M., TANIGUCHI, M. and KOBAYASHI, T. (1987). *J. Ferment. Technol.* **65**, 71–5.

KERTEZ, A. S. and KING, C. J. (1986). *Biotechnol. Bioeng.* **28**, 269–82.

KIM, C. W. and RHA, C. K. (1987). *Enzyme Microb. Technol.* **9**, 57–9.

KING, M. L., FORMAN, A. L., ORELLA, C. and PINES, S. H. (1985). *Chem. Eng. Prog.* **81**, 36–9.
KIRBY, K. S. (1956). *Biochem. J.* **64**, 405–8.
KIRBY, K. S. (1962). *Biochem. Biophys. Acta* **55**, 545–6.
KIRBY, K. S. (1956). *Biochem. J.* **96**, 266–9.
KLIMOV, A. N., NIKIFOROVA, A. A. and ETINGOV, E. D. (1971). *Antibiotiki* **16**, 243.
KOMINCK, L. A. (1975). *Antimicrob. Agents Chemother.* **7**, 861–3.
KRONER, K. H., HUSTEDT, H. and KULA, M.-R. (1982). *Biotech. Bioeng.* **24**, 1015–45.
KRONER, K. H., SCHÜTTE, H., STACH, W. and KULA, M.-R. (1982). *J. Chem. Technol. Biotechnol.* **32**, 130–7.
KULA, M.-R., KRONER, K. H. and HUSTEDT, H. (1982). *Adv. Biochem. Eng.* **24**, 73–118.
KURATH, P., JONES, P. H., EGAN, R. S. and PERUN, T. J. (1971). *Experientia* **27**, 362.
LAANE, C., BOEREN, S. and VOS, K. (1985) *Trends Biotechnol.* **3**, 251–2.
LANGRAND, G., BARATTI, J., BUONO, G. and TRIANTAPHYLIDES, C. (1986). *Tetr. Lett.* **27**, 29–32.
LARSEN, K. A. and KING, M. L. (1986). *Biotechnol. Progr.* **2**, 73–82.
LARSSON, Å. (1983). *Alfa Laval Technical Brochure* RP 60908 E.
LEO, A., HANSCH, C. and ELKINS, D. (1971). *Chem. Rev.* **71**, 525–616.
LEO, A. J. (1987). *J. Pharm. Sci.* **76**, 166–8.
LI, N. N. (1971). *Ind. Eng. Chem., Proc. Des. Dev.* **10**, 215–21.
LIKIDIS, Z. and SCHÜGERL, K. (1987). *Biotech. Lett.* **9**, 229–32.
LING, T. G. I. and MATTIASSON, B. (1984). *Talanta* **31**, 917–21.
LING, T. G. I., RAMSTROP, M. and MATTIASSON, B. (1982). *Anal. Biochem.* **122**, 26–32.
LOWE, C. R. (1984). In *Topics in Enzyme and Fermentation Biotechnology* (ed. A. WISEMAN), Vol. 9, pp. 78–161. Ellis Horwood, Chichester.
LUISI, P. L. (1985). *Angewante Chemie International Edition* (English) **24**, 439–50.
MCALPINE, J. B., TUAN, J. S., BROWN, D. P., GREBNER, K. D., WHITTERN, D. N., BUKO, A. and KATZ, L. (1987). *J. Antibiot.* **40**, 1115.
MCHUGH, M. A. and KRUKONIS, V. J. (1986). *Supercritical Fluid Extraction, Principles and Practice*. Butterworths, Boston.
MATTIASSON, B. and LING, T. G. I. (1987). In *Separations for Biotechonology* (ed. M. S. VERRALL and M. J. HUDSON), pp. 270–92. Ellis Horwood, Chichester.
MATTIASSON, B. and ERIKSSON, H. (1982). *Clin. Chem.* **28**, 680–3.
MAVITUNA, F., WILKINSON, A. K. and WILLIAMS, P. D. (1987). In *Separations for Biotechnology* (ed. M. S. VERRALL and M. J. HUDSON), pp. 333–40. Ellis Horwood, Chichester.
MENGE, U., MORR, M., MAYR, U. and KULA, M.-R. (1983). *J. Appl. Biochem.* **5**, 75–90.
MEYER, H. (1899). *Arch. Exp. Path. Pharmak.* **42**, 110.
MICHAELS, A. S. and MATSON, S. L. (1985). *Desalination* **53**, 231–58.
MODIN, R. and SCHRÖDER-NIELSEN, M. (1971). *Acta Pharmaceut. Suecica* **8**, 573–84.
MORIMOTO, A., KARIYONE, K., KURITA, M., HASHIMOTO, M. and NISHIWAKI, O. (1972). German Patent Application 2164561.

NAKAYAMA, I. (1984). *Macrolide Antibotics* (ed. S. OMURA), p. 264. Academic Press, Orlando.
NORDGREN, T. and MODIN, R. (1975). *Acta Pharmaceut. Suecica.* **12**, 407–16.
OKABE, M. and AIBA, S. (1975). *J. Ferment. Technol.* **53**, 730–43.
OMATA, T., IIDA, T., TANAKA, A. and FUKUI, S. (1979). *Eur. J. Appl. Microbiol. Biotechnol.* **8**, 143–55.
OMATA, T., IWAMOTO, N., KIMURA, T., TANAKA, A. and FUKUI, S. (1981). *Eur. J. Appl. Microbiol. Biotechnol.* **11**, 199.
OMURA, S. and TANAKA, H. (1984). *Macrolide Antibiotics* (ed. S. OMURA). Academic Press, Orlando.
PANDEY, R. C., KALITA, C. C., WHITE, R. J. and TOUSSAINT, M. W. (1979). *Process Biochem.* **14**, (12), 6–13.
PHILIPSON, L., ALBERTSSON, P.-A. and FRICK, G. (1960). *Virology* **11**, 553–71.
PODBIELNIAK, W. J. (1935). United States Patent 2004011.
PODBIELNIAK, W. J., KAISER, H. R. and ZEIGENHORN, G. J. (1970). In '*History of Penicillin Production', Chemical Engineering Progress Symposium Series* 100, **66**, 44–50.
Pomona College (1986). Medchem Software Manual, version 3.4, March. Pomona College, Claremont, CA 91711, USA.
PROTSCH, M. and MARR, R. (1984). *Proceedings of the International Solvent Extraction Conference*, 288, Society of Chemical Industry, London.
QUEENER, S. and SCHWARTZ, R. (1979). In *Economic Microbiology*, Vol. 3: *Secondary Products of Metabolism* (ed. A. H. ROSE). Academic Press, London.
RALPH, R. K. and BELLAMY, A. R. (1964). *Biochim. Biophys. Acta* **87** 9–16.
RANDOLPH, T. W., BLANCH, H. W., PRAUSNITZ, J. M. and WILKE, C. R. (1985) *Biotechnol. Lett.* **7**, 325–8.
REINHARDT, H. and RYDBERG, J. (1970). *Chem. Ind.* 11 April, 488–91.
RESCHKE, M. and SCHÜGERL, K. (1984a). *Chem. Eng. J.* **28**, B1–B9.
RESCHKE, M. and SCÜGERL. K. (1984b). *Chem. Eng. J.* **28**, B11–B20.
RESCHKE, M. and SCHÜGERL, K. (1984c). *Chem. Eng. J.* **29**, B25–B29.
RESCHKE, M. and SCHÜGERL, K. (1986). *Chem. Eng. J.* **32**, B1–B5.
REUBEN, B. and SJÖBERG, K. (1981). *Chemtechnol.* **11**, 315–20.
RICKARDS, R. W., SMITH, R. M. and GOLDING, B. T. (1970). *J. Antibiot.* **23**, 603–12.
ROBINS, R. J. and RHODES, M. J. C. (1986). *Appl. Microbiol. Biotech.* **24**, 35–41.
RORSELIUS, W., VITZTHUM, O. and HUBERT, P. (1979) United States Patent 4153063.
RYDBERG. J.ERSSON, H., ARONSSON, P. O., SELME, A. and SKARNEMARK, G. (1980). *Hydrometallurgy* **5**, 273–81.
RYTTING, J. H., DAVIS, S. S. and HIGUCHI, T. (1972). *J. Pharmaceut. Sci.* **61**, 816–8.
SASSENFELD, H. M. and BREWER, S. J. (1984). *Bio/Technol.* **2**, 76.
SCHILL, G. (1972). *Acta Pharmaceut. Suecica* **9**, 623–4.
SCHILL, G. (1974). In *Ion Exchange and Solvent Extraction* (ed. J. A. MARINSKY and Y. MARCUS) **6**, 1–57.
SCHILL, G., BORG, K. O., MODIN, R. and PERSSON, B.-A. (1977). In *Essays in Analytical Chemistry* (ed. E. WAINNINEN), pp. 379–95. Pergamon Press, Oxford.
SCHRÖDER-NIELSEN, M. (1976). *Acta Pharmaceut. Suecica*, **13**, 145–56.
SCHWARTZ, R. W. (1985). In *Comprehensive Biotechnology* (ed. M. MOO-YOUNG), Vol. 3, pp. 7–47. Pergamon Press, Oxford.
SEBBA, F. (1984). *Chem. Ind.* May, **21**, 367.

SEPRACOR (1987). *Technical Brochure*. Sepracor Inc., 33 Locke Drive, Marlborough, MA 01752, USA.

SHANBHAG, V. P. and AXELSSON, C.-G. (1975). *Eur. J. Biochem.* **17**, 17–22.

SMITH, D. R., LANDER, R. J. and QUINN, J. A. (1977). In *Recent Advances in Separation Science* (ed. N. N. LI), Vol. 3B, pp. 225–41. CRC Press, Cleveland.

STAHL, E., SCHILZ, W., SCHÜTZ, E. and WILLING, E. (1978). *Angewante Chemie, Int. Edn.* **17**, 731–8.

STEYTLER D., RICHMOND, P., LOH, V. Y. and REYNOLDS, J. (1986). In *Bioactive Microbial Products 3: Downstream Processing* (ed. J. D. STOWELL, P. J. BAILEY and D. J. WINSTANLEY), pp. 209–30. Academic Press, London.

SUZUKI, N., SOWA, T. and MURAKAMI, M. (1976). German Patent Application 2602099.

TANIGUCHI M., SUZUKI, H., SATO, M. and KOBAYASHI, T. (1987). *Agric. Biol. Chem.* **51**, 3425–6.

TERADA, H. (1986). *Quantit. Struct.-Activ. Relation.* **5**, 81–8.

THERISOD, M. and KLIBANOV, A. M. (1986) *J. Am. Chem. Soc.* **108**, 5638–40.

THIEN, M. P. HATTON, T. A. and WANG, D. I. C. (1986). In *ACS Symposium Series 314, Separation, Recovery and Purification in Biotechnology* (ed. J. A. ASENJO and J. HONG). American Chemical Society, Washington, USA.

TJERNELD, F. BERNER, S., CAJARVILLE, A. and JOHANSSON, G. (1986). *Microb. Technol.* **8**, 417–23.

TODD, D. B. (1966). *Chem. Engng. Progrd.* **62**, 119–24.

TODD, D. B. and DAVIES, G. R. (1973). *Filtr. Separ.* **10**, 663–66.

TSUJI, A., MATSUDA, M., MIYAMOTO, E. and YAMANA, T. (1978). *Pharmac. Pharmacol.* **30**, 442–4.

VANDAMME, E. J. (1984). *Biotechnology of Industrial Antibiotics*. Marcel Dekker, New York.

VEHLE, C. O. and GANESAN, A. T. (1973). *J. Biol. Chem.* **248**, 7456.

VINING, L. C. (1983). *Biochemistry and Genetic Regulation of Commercially Important Antibiotics*. Addison-Wesley, Reading.

WALLIS, D. A., MICHELSEN, D. L., SEBBA, F., CARPENTER, J.K. and HOULE, D. (1985) *Biotechnol. Bioeng. Symp.* **5**, 399–408.

WALTER, H., BROOKS, D. E. and FISHER, D. (ed.) (1985). *Partitioning in Aqueous Two-Phase Systems: Theory, Methods, Uses and Applications to Biotechnology*, Academic Press, Orlando.

WANG, H. Y. (1983). *Ann. NY Acad. Sci.* **413**, 313–21.

WEATHERLEY, L. R., CAMPBELL, I., SLAUGHTER, J. C. and SUTHERLAND, K. M. (1987). In *Separations for Biotechnology*, (ed. M. S. VERRALL and M. J. HUDSON), pp. 341–50. Ellis Horwood, Chichester.

WILCOX, C. S., GREER, L. M. and LYNCH, V. (1987). *J. Am. Chem. Soc.* **109**, 1865–7.

WILDFEUER, M. E. (1985). In *Purification of Fermentation Products, Applications to Large Scale Processes*, ACS Symposium Series No. 271 (ed. D. LEROITH, J. SHILOACH and T. J. LEAHY), p. 155. American Chemical Society, Washington.

WILSON, D. (1976). *Penicillins in Perspective*. Faber and Faber, London.

WOODWARD, R. B. (1957). *Angewante Chemie* **69**, 50.

ZAKS, A. and KLIBANOV, A. M. (1984). *Science* **224**, 1249–51.

ZAKS, A. and KLIBANOV, A. M. (1985). *Proc. Nat. Acad. Sci. USA* **82**, 3192–96.

4

LIQUID–LIQUID EXTRACTION IN FOOD PROCESSING

W. HAMM

4.1 Introduction

The food industry is as much of a process industry as the mainstream chemical industry, using many of the unit operations exploited by the chemical engineer in that industry, but differing from it in a number of important respects. The raw materials handled by the food industry are highly complex and their components are seldom simple molecular species that can be conventionally processed. The complexity of the substrate usually makes separations, including liquid–liquid extraction, difficult, particularly where the structure of the raw material or other valuable attributes require conservation. This complexity affects separations technology on at least two levels—in the first instance it may be necessary to effect a liquid–solid separation in order to produce a relative crude fraction that can be subjected to a more targeted liquid–liquid extraction; and in this subsequent liquid–liquid extraction the separation is more likely to be between classes of compounds rather than individual components. Thus, in a typical case, the first stage may comprise the separation of lipids from carbohydrates and proteins, to be followed by the application of liquid–liquid extraction in order to recover specific types of lipids. The sensitivity of most raw materials of biological origin to high temperatures as well as to strong acid or alkaline conditions constitutes a further constraint on the applicability of conventional processing conditions. Liquid–liquid extraction is used in the food processing industry to concentrate valuable components of a natural raw material as well as to remove undesirable or potentially harmful components.

The dominant role of subjective assessment in food processing creates problems for the process engineer in process design and operation. The lack of quantitative data in areas such as product assessment leads to empirical design and also to intermittent, often batch operation in order to permit subjective product assessment to be carried out. The need for strict hygienic and toxicological control of food processing is a subject having many ramifications, both in design as well as in operation, and is dealt with in a later part of this text.

The importance of preserving characteristics such as taste, structure, and oxidative stability in the material being processed and to avoid damage to heat-sensitive components gives rise to the requirement for 'soft', or mild, processing conditions, applying to the extraction itself as well as prior processing of the raw material in preparation for the extraction. Increasingly, too, end-user *perception* of the application of such processing techniques is an important product attribute, even if the use of 'soft' processing techniques imparts no improvement in objectively measured quality (where this is feasible).

As the substrate from which extraction takes place, even when it is not the primary product, may have considerable value as a foodstuff after extraction it is important that mild processing conditions be applied throughout the process, extending from any preparative treatment to the final stages of solvent removal from the extracted material. Moderate processing temperatures and solvent inertness are the characteristics most often associated with the requirement for 'soft' processing. A corollary of this is that the storage conditions for the products of liquid–liquid extraction generally warrant careful consideration, exposure to light and oxygen being particularly undesirable. Control of environmental conditions may also be of considerable importance.

When compared with the mainstream chemical process industry the throughputs achieved in liquid–liquid extraction in the food industry are generally small. This has sometimes favoured batch processing, and the possibility that a raw material may be seasonal in availability (in some cases crops are not produced annually) can also influence the decision whether to process continuously or batchwise.

A special case of liquid–liquid extraction that is of particular interest to the food industry is extraction with compressed gases, either above or just below the critical point of the gas. This 'new' process—the underlying principle of the enhanced solvent power of gases above their critical point was in fact discovered more than 100 years ago—has been described as a hybrid between extractive distillation and liquid–liquid extraction, though its potential in the food industry appears to be greater in liquid–solid extraction than in liquid–liquid extraction. In view of the growing interest in this process in the food industry the more general aspects of its application are considered here separately from the specific applications in the fields of lipids and essential oils.

4.2 Hygiene and toxicology

Food products must be of a hygienically acceptable standard, that is, they must not contain foreign micro-organisms or other contaminants that can affect the health of the consumer. Micro-organisms, which require the presence of moisture for growth, may be present in the raw material, thus emphasizing the importance of controlling raw material storage conditions, or may be introduced as a result of non-sterile operation of plant. Heat treatment is used to render those present in the feedstream inactive. The growth of foreign micro-organisms in the plant, often due to environmental pollution, must be prevented by appropriate means. In the case of aqueous systems, regular cleaning of the plant with an industrial detergent is essential. For this purpose the process is interrupted at specified intervals in order to purge the system with the detergent solution. Where the plant-cleaning operation is carried out without dismantling individual plant items it is known as cleaning-in-place (CIP). The design of plant items should seek to minimise the potential for growth of micro-organisms by avoiding stagnant zones and specifying standards of surface finish that will facilitate cleaning.

Whilst inadequate attention to hygienic design and operation can lead to toxicological problems these may also arise in another form, such as the presence of potentially toxic residues as a result of incorrect choice of solvent. For this reason the use of solvents for processing is strictly controlled in most countries. A solvent may be potentially harmful because of its inherent toxicity or because of its ability to form harmful artefacts with constituents of the material being processed.

As in other process industries, emission of solvent into the working atmosphere and the wider environment is controlled by the appropriate regulatory authorities; in the United Kingdom by the Health and Safety Executive, whose publication *Occupational Exposure Limits* (1987), contains details of the long-term exposure limits (8-hour time-weighted average) and the short-term exposure limits (10-minute time-weighted average) for all solvents normally used. It should be consulted when designing or operating solvent plant. Of solvents commonly used in the food industry hexane requires the strictest standards with respect to emission, the respective figures for the long-term and short-term exposure limits being 100 and 125 p.p.m. In the USA the American Conference of Governmental Industrial Hygienists, Incorporated, recommends Threshold Limit Values (TLV) for all solvents and the responsibility for enforcement rests with the US Occupational Safety and Health Administration (OSHA).

4.3 Solvent selection

Concern for safety and public health has caused regulatory authorities to restrict the range of solvents acceptable for use in food processing, where they are mainly used for liquid–solid and liquid–liquid extraction. Carbon dioxide, the short-chain aliphatic hydrocarbons, some alcohols and ketones such as acetone having relatively low molecular weights and a very limited number of halogenated hydrocarbons now constitute the range of solvents used. Hexane is the most widely used of the hydrocarbons, while the solvent properties of propane have been studied in a number of investigations of extraction using supercritical fluids. Methanol, ethanol, and isopropanol have all found application in liquid–liquid extraction in the food industry, as have the esters ethyl and butyl acetate. Dichloromethane (methylene dichloride) is virtually the last representative of the class of halogenated solvents still in use in the industry—though in some countries use of trichlorethylene is permitted—but in view of its successful replacement by more acceptable solvents in a number of applications the eventual complete elimination of halogenated solvents from food industry applications seems probable.

In some cases it has been found expedient to use a solvent miscible with water in admixture in order to modify the polarity of the solvent, and this approach has also proved useful in reducing the risk of chemical reaction during solvent recovery. Carbon dioxide has so far only been used on a large scale in liquid–solid extraction, but the feasibility of using it for liquid–liquid extraction, either in its supercritical state or in sub-critical conditions, has been demonstrated (Schultz *et al.* 1974).

The selection of a solvent for liquid–liquid extraction in the food industry is made on the basis of a set of criteria, most of which are common to the process industries as a whole and for that reason do not require special discussion here. The criteria generally used are:

(1) solute capacity;

(2) selectivity;

(3) chemical inertness of solvent towards the extraction system;

(4) physical properties of the solvent:
   (a) density,
   (b) viscosity,
   (c) boiling point,
   (d) latent heat of evaporation,
   (e) interfacial tension;

(5) flammability;

(6) toxicology of solvent;

(7) cost.

The concepts of solute capacity and selectivity are just as applicable in the food industry as in any other process industry, though the complex nature of the system being processed will, in some cases, make it difficult to express the information available—particularly that concerning selectivity—in a quantitative manner. In the extraction of flavours, for example, the profile of components comprising the flavour is too complex to permit representation by a single component and the monitoring of extraction performance in such a case may therefore have to be carried out subjectively, for example by a taste panel.

The importance of removing virtually all solvent from the processed food at a temperature low enough to avoid damage to the nutritional value of the food makes it essential to use low-boiling, non-reactive solvents. Generally the amount of solvent permitted in the final product by the appropriate regulatory authorities requires the use of desolventizing techniques that are powerful enough to ensure substantially complete removal of the solvent but not harsh enough to damage the product; steam stripping has in many cases been found to satisfy these requirements. Density has to be taken into account in choosing a solvent in so far as phase separation after extraction depends on this parameter; and as most food systems have densities close to or a little below that of water it is advisable to seek a solvent having a density below that of the food system, the only heavier-than-water solvents available being the halogenated solvents. Whilst the other physical properties of solvents noted above are important in all applications of liquid-liquid extraction the significance of interfacial tension in food industry systems is of particular significance on two grounds: the interfacial tension between extract and raffinate phase may undergo significant change as the extraction progresses, and the presence of otherwise insignificant minor components in the feedstream—possibly as a result of seasonal variation in composition—can also have disproportionate effects on the interfacial tension.

The need to ensure consumer safety extends the framework within which the safety of solvents must be considered. Regulations governing operator and plant safety apply to all process plant, which means that when using flammable or otherwise hazardous solvents special precautions regarding storage and movement of solvents have to be taken. It is of particular interest to the food industry that the regulations governing hexane emissions have been progressively tightened in recent years, with the time-weighted average TLV now being 100 p.p.m. in the USA and the EEC.

The need to meet standards for solvent residues in food products laid down by various regulatory authorities constitutes an important consideration in solvent selection, although in a number of cases an extract may constitute a product without desolventization. Within the European Community a two-part approach to the use of solvents in food processing has been evolved. One group of solvents (see Table 4.1) may be used freely, the maximum solvent residue in the product being defined as consistent with good manufacturing practice, whereas a second, longer list comprises solvents for which the application as well as the residue in the processed foodstuff are defined (Commission of the European Communities 1983). It is generally accepted that the use of a steam stripper operating at low pressure after the main desolventization evaporators will easily reduce solvent content of the fractions to 10 p.p.m. or lower.

TABLE 4.1 *Solvents proposed for unrestricted use by Council of the European Communities (1983)*

Propane	Ethanol
Butane	Carbon dioxide
Ethyl acetate	Acetone
Methanol	Nitrous oxide

No liquid–liquid extraction processes appear in the second list.

In recent years the food processing industry has come under pressure to reduce its use of halogenated solvents. This pressure accounts for much of the interest in alternative solvents, in particular in the use of supercritical and liquid carbon dioxide. In the USA permission to use trichlorethylene was withdrawn in 1976 and more recently the suitability of dichloromethane as a food industry solvent has been questioned. In 1985, however, the US Food and Drug Administration determined that the risk from dichloromethane is so low 'as to be essentially non-existent' (Federal Register 1985). A toxicology study found that dichloromethane caused no adverse effects on rats and mice (Serota *et al.* 1986), but further studies are in progress.

The short-chain alcohols are suitable solvents for many of the more polar compounds to be extracted, though their miscibility with water means that they cannot be used for extraction from an aqueous system. Admixture of alcohols with water in order to limit solvent reactivity has already been noted, and this approach can also be used to adjust the density difference between extract and raffinate phase. Ethyl and *n*-butyl acetate are both used in liquid–liquid extraction, being far less miscible with water than the corresponding alcohols. Acetone is a suitable solvent where the second solvent is sufficiently apolar to give a two-phase system for the required separation, though obviously the effect of the partitioning component on the phase relationships has to be taken into consideration. The lower aliphatic hydrocarbons are used for the extraction of apolar constituents, propane and hexane being of particular interest. The most important physical properties of these solvents are given in Table 4.2.

The extensive work on supercritical fluid extraction in the last decade has meant that carbon dioxide is now a recognized solvent in the field of food processing, although the actual applications are as yet very limited, largely because the high cost of plant associated with supercritical gas extraction has acted as a deterrent for many interested parties.

In certain applications the use of the feedstream from which a minor component is to be removed as the second solvent can be advantageous, since it simplifies solvent recovery and therefore process costs. The cost advantages

TABLE 4.2 *Physical properties of solvents*

Solvent	Molecular weight	Density kg/m^3 at 20°C	Boiling point, °C at 1 bar	Mass (%)		Solubility parameter $MPa^{0.5}$
				Solubility in water	Water uptake	
Methanol	32	792	65	Miscible		29.6
Ethanol	46	789	78	Miscible		26.5
2-Propanol	60	785	82	Miscible		23.5
ethyl acetate	88	901	77	8.08	2.94	18.1
n-Butyl acetate	116	871	126	0.43	1.86	17.4
Acetone	58	790	56	Miscible		20.0
Methyl ethyl ketone	70	805	80	27.5	12.5	19.0
Dichloromethane (methylene chloride)	87	1324	40	1.3	0.2	20.3
Trichlorethylene	131	1468	87	0.11	0.32	19.0
Pentane	72	625	36	0.004	0.012	14.5
Hexane	84	659	69	0.001	0.011	14.9

thus obtained must be balanced against a possible reduction in sharpness of separation, as well as the possibility of increased difficulty in phase separation during and after extraction.

Regular solution theory may be used to guide solvent selection in liquid-liquid extraction (Ashton *et al.* 1983), but the limitations on solvents available for use in food processing may make the procedures advocated less useful than in an unconstrained situation. The disparities in molar volume between the solvents normally used and the component(s) to be extracted are shown to have a bearing on the application of the theory.

4.4 High-pressure gas extraction

[See also Verrall Chap. 3, Weatherley Chap. 5]

4.4.1 Background

The quantitative determinations of the solubilities of different solutes in compressed gases have shown that the solubility profiles of solids and liquids in supercritical gases change significantly with pressure and temperature. Dalton's Law predicts that the mole fraction of a dissolved solute in a gas is given by P_s/P, where P_s is the vapour pressure of the solute at a given temperature, and large deviations from that law occur when high pressures are used. The concentration of a component extracted at high pressure is often 10^5 and up to 10^{12} higher than if it were calculated from the vapour pressure of a dissolved component.

In the food industry extraction with compressed gases has mainly taken the form of extraction from solid substrates, thus belonging to the category of liquid–solid extraction; but studies on extraction of flavours from liquids such as fruit juices using compressed carbon dioxide (Schultz *et al.* 1974) were undertaken at a relatively early stage in recent investigations of the potential applications of this technique. The interest in the use of compressed gases in the liquid, sub-critical state has persisted (Bunzenberger *et al.* 1984) but operation at above the critical point is the more attractive process, largely because the important properties of the solvent are enhanced by working in the supercritical region.

4.4.2 Process analysis

A knowledge of phase equilibria at high pressures is essential for an understanding of extraction with compressed gases. For this purpose equations of state such as the Redlich–Kwong and Peng–Robinson equations, expressing pressure as a function of temperature, molar volume, and parameters characteristic of the system components, have been used (Stephan and Schaber 1979; Coppella and Barton 1987) and have been shown to represent experimentally determined data satisfactorily. Another approach to data prediction depends on relating the solubilities of the solutes in the compressed gas directly to the density of the gas used (Chrastil 1982). By application of the law of mass action Chrastil derived the equation

$$C = d^k \exp(a/T + b)$$

to represent the solubility of a solute in a gas, where C = concentration of the solute in the gas, d = density of the gas, k = an association number indicative of the complex formed between solute and solvent, a, b = constants based on thermodynamic considerations. In the above equation both the concentration of the solute in the gas, C, and the density of the gas, d, are in g/l.

Excellent agreement of experimentally determined solubility data with the equation was obtained, thus emphasizing the importance of gas density in solubility estimations, though the presence of the exponent in the equation indicates that solubility is not dependent on gas density alone. The solubility–density log–log relationship was shown to be linear up to relatively high solute concentrations (100–200 g/l) and generally the relationship was found to remain linear up to very high pressures.

Although most of the work recorded has dealt with binary systems, multicomponent systems have also been studied and selectivities have been determined, thus allowing estimation of the separating power required for a given degree of fractionation. For this purpose a volatility relative to a key component was calculated from retention times found in chromatographic analysis of the fractions produced (Coppella and Barton 1987).

4.4.3 Solvent selection

A number of gases have been studied for their solvent properties in the supercritical region, but from the point of view of application of the technique in the food industry carbon dioxide is obviously the most interesting, being not only considerably cheaper than conventional solvents but completely inert (towards materials as well as people) and therefore unreservedly acceptable in the industry. Its critical temperature (T_c: 31°C) is low enough to permit supercritical operation to be carried out without going to high temperature.

In practice the only other supercritical solvents of potential interest are the low molecular weight hydrocarbons, and in this group propane has evoked most interest, at times in admixture with carbon dioxide. The critical properties of these solvents is shown in Table 4.3.

Supercritical carbon dioxide has been shown to be a relatively poor solvent

TABLE 4.3 *Critical data of solvents used in supercritical extraction suitable for use in the food industry*

Solvent	Critical temperature [°C]	Critical pressure [bar]	Critical density [g/cm^3]
Carbon dioxide	31.1	73.8	0.468
Ethane	32.3	48.8	0.203
Propane	96.9	42.6	0.220
Ethylene	9.9	51.2	0.227

for polar molecules, and the presence of polar groups such as carboxyl, hydroxyl, or nitro groups usually reduces the solubility of a compound. Thus sugars, amino acids, fatty acids, glycerides, and most alkaloids exhibit low solubility in carbon dioxide, whereas oxygenated organic compounds of low to medium molecular weight and most non-polar low-molecular-weight organics such as alkanes, terpenes, and other hydrocarbons are readily soluble in supercritical carbon dioxide.

In the case of carbon dioxide its solute capacity can be modified in a number of ways besides the use of temperature and pressure variation. The addition of propane produces a solvent having enhanced solute capacity for triglycerides and fatty acids and as it also raises the critical temperature the resulting mixed solvent is more versatile than pure carbon dioxide. Addition of small amounts of water can be used to modify the properties of carbon dioxide and other supercritical gases, as water affects both the density and the polarity of the pure solvent.

A conventional solvent can also be used as an entrainer to boost the solute capacity of the supercritical solvent, and for this purpose acetone was used when fractionating partial glycerides with supercritical propane (Peter *et al.* 1976). In this context the 'entrainer' is in fact a second solvent, fully miscible with the primary solvent, which serves to increase solute–solvent miscibility.

4.5 Applications

4.5.1 Lipids

4.5.1.1 *Deacidification of edible oils*

The field of lipids, which is the class of compounds using the glycerine skeleton as the basic building block with fatty acid radicals as the principal groups attached to the three hydroxyl groups, covers a wide range of molecular configurations having a considerable diversity of physical and physicochemical properties. Separations technology is extensively applied in their processing, although the current contribution of liquid–liquid extraction is limited. The most relevant application in the present context is the removal of the free, that is unesterified, fatty acids present in crude vegetable oils. The recovery of some of the potentially valuable minor components found in vegetable oils by liquid–liquid extraction has at times been suggested. The production of fractions enriched in the more unsaturated triglycerides has been of interest when such fractions have been perceived to have had high value or when their removal has enhanced the value of the remaining oil. The former is the case when enrichment in polyunsaturated fatty acids has been sought in vegetable oils (Parsons 1976) and in fish oils (Wagener and Hearshaw 1986), whereas the removal of the most unsaturated fraction of tallow (Moore 1950) may be regarded as an example of the latter, though in that case the removal of colour bodies was at least as important as the removal of the highly unsaturated triglycerides.

The separation of fatty acids from crude oils of vegetable or animal origin is of fundamental importance in the refining of these oils for edible use. World production of oils and fats exceeds 75 million tons and is expected to approach 100 million tons by the end of the present decade, with more than 80 per cent of this being used for edible purposes. The free fatty acid content of crude oils is mostly in the range 1–5 per cent, though negligent treatment of the material from which the oil is recovered may lead to a substantially higher free fatty acid content, this being particularly the case where lipolytic enzymes are present in the oil-rich raw material. Rice bran and the palm fruit are examples of materials where delay in inactivating the enzyme present will give rise to high levels of free fatty acid in the recovered oil. In view of the fact that the separated fatty acids have a value substantially below that of refined oil, a major objective in refining is the minimization of the quantity of triglycerides entrained with the separated fatty acids or, in the case of deacidification with alkali, saponified by the excess of alkali used.

For the purpose of deacidification several process options are available. Neutralization of the free fatty acids with aqueous alkali (approx. 4 N) has historically been the most widely used technique, but in the last decade steam stripping of the fatty acids at elevated temperature has replaced alkali neutralization in a growing number of cases because of its superior yield of refined oil and the simplification of the process made possible by combining removal of fatty acids and of the volatile components in one operation. Steam-stripping, now generally described as physical refining, cannot however be used if oil quality is likely to be damaged by exposure to high temperature. Whilst in most cases this means that alkali neutralization is used for such oils, in certain circumstances liquid–liquid extraction of fatty acids into a solvent phase becomes a potentially interesting alternative.

The neutralization of the free fatty acids present in edible oils using aqueous alkali has been analysed as a case of liquid–liquid extraction coupled with chemical reaction at the interface (Smits 1977). The significant enhancement of the mass transfer rate due to the chemical reaction taking place was shown to be related to the amount and type of fatty acid present. Since the reaction of fatty acid with aqueous alkali has a very high equilibrium constant, giving virtually complete neutralization of the fatty acids present in the oil, only a single-stage contactor for the oil and aqueous phases is required. In the range 30–100°C increase in system temperature has been shown to reduce the effect of spontaneous emulsification, a major cause of transfer and therefore downgrading of triglycerides ('neutral oil') to the aqueous phase. The process was earlier carried out batchwise in most edible oil refineries, but modern technology predominantly uses in-line centrifugal oil–alkali mixers, a low-volume holding zone, and continuous centrifugal separators to give refined oil as the light phase and the aqueous soapstock as the heavy phase. A second centrifugal separator is used to wash the oil free of residual dissolved soap.

In the 1950s the Podbielniak extractor was used as a contactor in the refining of edible oils with aqueous alkali but this development was not sustained in the face of the extensive adoption of bowl and tubular centrifugal separator systems.

In deacidification with aqueous alkali the principal concern of the oil refiner is to minimize the quantity of triglycerides transferred into the aqueous phase by

(1) entrainment;

(2) saponification by excess alkali present;

(3) spontaneous emulsification.

The emphasis in process design has been on efficient phase separation and short contact time to limit the saponification of triglycerides and the small amounts of partial glycerides sometimes present in oils. For most oils the reaction is carried out at between 80 and 90°C (Brunner 1986).

For the refining of crude cottonseed oil deacidification with alkali is sometimes carried out in the presence of solvent, since it facilitates the removal of colour bodies from the oil (Hendrix 1984). Hexane is generally used for this purpose, the miscella recovered from the seed extraction plant being concentrated to 50–60 per cent oil content prior to contacting with alkali. As in the case of conventional refining, the chemical reaction between fatty acids and aqueous alkali virtually goes to completion. Operation of the process in hexane also leads to reduced downgrading of triglycerides by transfer to the aqueous phase, though whether this is due to reduced spontaneous emulsification or to reduced saponification of triglycerides has not been elucidated. Recent work on this form of deacidification (Zaher and El-Nomany 1987) has demonstrated the effects of the choice of solvent in 'miscella refining' on the various parameters involved in judging refining performance. Obviously the fact that seed extraction is carried out using hexane affects the economics of solvent refining.

Deacidification by extraction of the fatty acids into a solvent without chemical reaction becomes a useful process option when it is important to reduce downgrading of 'neutral oil' to a minimum and at the same time avoid exposing the oil to the temperature conditions of physical refining. This is the case when the value of the refined oil is significantly greater than is normally the case in edible oil refining, an example of this situation being the production of certain specialized confectionery fats which are subsequently used to produce high-value cocoabutter equivalents. It has also been claimed that liquid–liquid extraction of the fatty acids from crude palm oil or its fractions using aqueous ethanol as solvent leads to improved bleachability of the raffinate (Swoboda 1984), and this may provide further scope for the application of liquid–liquid extraction in edible oil processing.

The short-chain alcohols are the preferred solvents for extraction of fatty acids and it is possible to operate such a process without using a second solvent, that is, the 'neutral' oil (the triglycerides) acts as a solvent for free fatty acids and contact with the added solvent leads to the partitioning of the acids between the triglycerides and the solvent. Limited miscibility of triglycerides and solvent is obviously required, and this miscibility can be adjusted by addition of water to the solvent. Other solvents have, however, been studied from time to time (Thomopoulos 1971). The data for the systems oleic acid–triolein–methanol and oleic acid–triolein–aqueous ethanol (Rigamonti *et al.* 1951) permit comparison of the effectiveness of the two solvents, with and without the addition of water (Figs. 4.1a–c).

Methanol shows greater immiscibility with trioleine, leading to a lower trioleine content in the extract, but, surprisingly, the addition of water does not appear to alter the binodal curve significantly. The slope of the tie lines is however changed appreciably by the addition of water, making the separation more difficult. Methanol is also preferable from the point of view of solvent recovery and solvent replacement costs. The Hand plot (Ashton *et al.* 1983) provides a good correlation of extract and raffinate compositions for the above system with solvent water content as parameter. It is known that the selectivity is affected adversely by the presence of significant amounts of partial glycerides in the oil, which may be the case when an oil has a high content ($>$ 5 per cent) of free fatty acids.

Deacidification to the level required in a refined oil (free fatty acid content $<$ 0.05 per cent) would require considerable separating power and a high solvent ratio, whereas the reduction of the free fatty acid content to 1 per cent can readily be accomplished in a contactor having a separating power

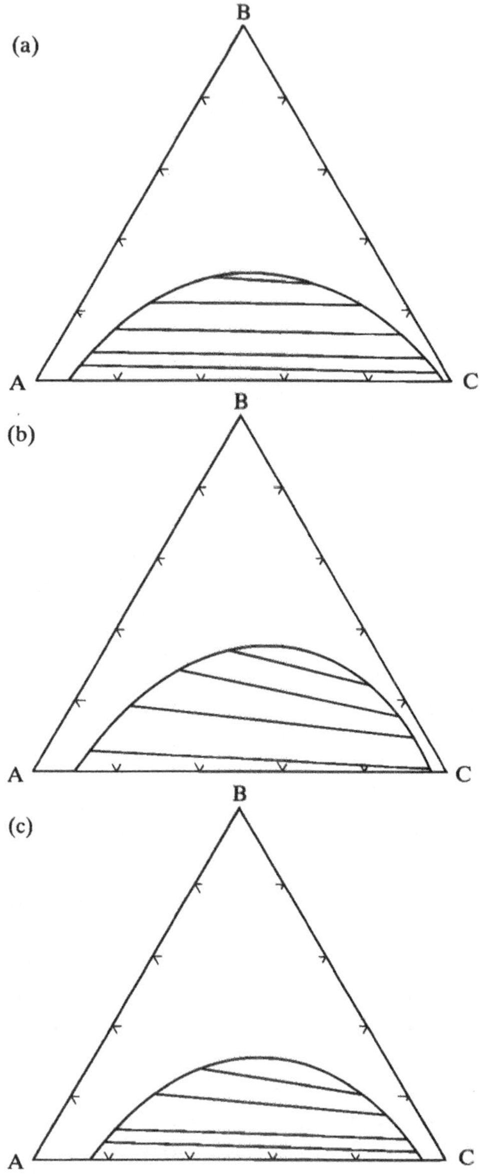

Fig. 4.1. Fatty acid distribution between trioleine and primary alcohols at 20°C. (a) Trioleine (A), oleic acid (B), methanol (C). (b) Trioleine (A), oleic acid (B), aqueous methanol (5% w/w water) (C). (c) Trioleine (A), oleic acid (B), aqueous ethanol (5% w/w water) (C). (Based on Rigamonti *et al.* (1951).)

equivalent to ten to fifteen theoretical stages provided a solvent : oil ratio of at least 2 is used. The remaining free fatty acids are then removed by neutralization with alkali after desolventization.

The process is shown in outline, with a typical mass balance, in Fig. 4.2. Desludging of the oil is necessary to avoid fouling the extraction contactor internals, which can affect the mass transfer performance. The extractor operating temperature should also be above the melting point of the components of the oil, which in practice means extracting at approximately 40°C. The slightly elevated operating temperature reduces the region of immiscibility but it is nevertheless possible to process oils containing up to 15% free fatty acid.

It is of course important to maintain a density difference between the phases (rather than between solvent and oil!) that is sufficient to ensure rapid phase separation. When using methanol as a solvent the water content should be kept below 10 per cent, but a very low water content increases 'neutral oil' solubility in the extract phase. It has been reported (Hamm 1983) that the interfacial tension for the methanol–fatty acid–neutral oil system falls sharply as the fatty acid content of the extract phase rises, and this must be taken into consideration when assessing the power needed to control the dispersed phase droplet size. When using a mechanically agitated contactor for this separation the recognized design procedures can be used, but it is advisable to provide a variable speed agitator so as to allow for fine-tuning of the process conditions.

The design of the solvent recovery units follows conventional practice, and the oil-rich raffinate stream can be desolventized in a single-effect falling film evaporator. The solvent-rich extract stream is best desolventized in a multiple-effect evaporator. Both streams require final desolventization in a steam stripper in order to produce material that is safe for further processing. The solvent recovered from the evaporators is rectified in order to eliminate excess water from the solvent and the vapour streams from the strippers, though very lean in solvent, are sometimes fed to a low point in the rectification column for reasons of safety and also in order to take advantage of their energy content.

The semi-refined oil (stream 10′) undergoes final refining in an alkali refining plant, the refining 'loss', i.e. downgrading, then being relatively modest for an oil of low free fatty acid content. The extract (stream 9′), which has a free fatty acid content of 85–90 per cent on a solvent-free basis, is used in soap production or as a feedstock in a fatty acid splitting plant.

The relatively low number of theoretical stages required for this extraction facilitates the choice of contactor. Contactors requiring re-dispersion of the disperse phase in each stage may, however, be considered unsuitable because of the risk of formation of fine droplets that will resist coalescence at the

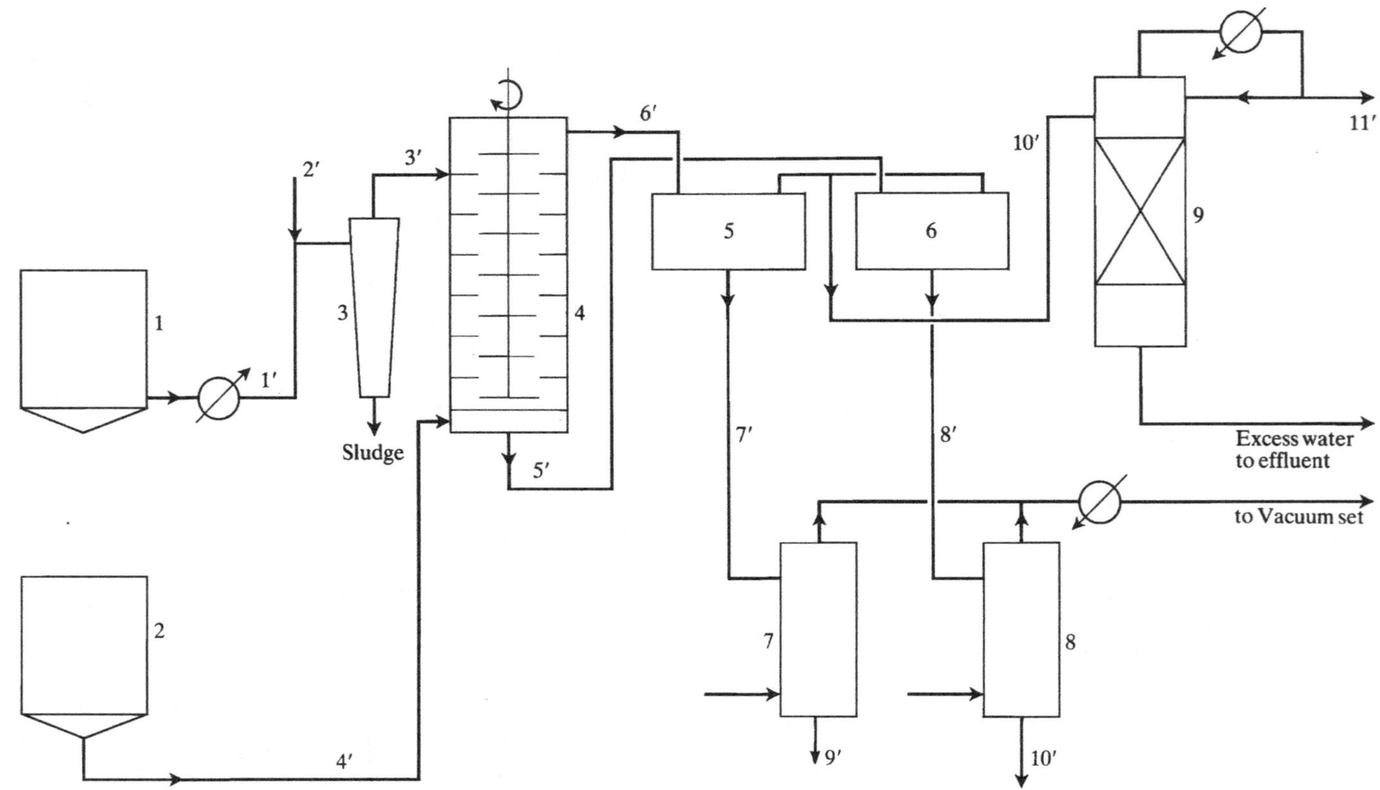

Fig. 4.2. Extractive deacidification of edible oils. 1 Crude oil storage; 2 solvent storage; 3 desludging centrifuge; 4 mechanically agitated contactor; 5 desolventizing evaporator system – extract; 6 desolventizing evaporator system – raffinate; 7 steam stripper, extract; 8 steam stripper, raffinate; 9 solvent rectification column.

Stream No.	1′	2′	3′	4′	5′	6′	7′	8′	9′	10′	11′	12′
Function	Crude oil to desludger	Water to desludger	Desludged oil to Extractor	Solvent to Extractor	Raffinate from Extractor	Extract from Extractor	Desolventized extract to stripper	Desolventized raffinate to stripper	Fully de-solventized extract	Fully desolventized semi-refined oil	Wet solvent to rectification column	Rectified solvent to solvent storage
Flow rate kg/h	1000	10	1000	2000	1030	1970	102	898	102	898	*2002	2000
Composition												
Water		10	2	100	7	95	< 1000 ppm	< 1000 ppm	< 10 ppm	< 1 ppm	*102	100
Methanol				1900	130	1770	< 1000 ppm	< 1000 ppm	< 10 ppm	< 1 ppm	1900	1900
Fatty acids	100		100		10	90	90	10	90	10		
Neutral oil	900		900		888	12	12	888	12	888		

* Stream may contain additional moisture from desolventization plant

contactor outlet. It is also likely that a contactor of low fouling potential will be preferred for this duty.

The economics of edible oil refining by liquid–liquid extraction makes it unlikely that the process will find widespread application in the industry unless special advantages, possibly associated with the partitioning of minor components, are found. The operating cost for the process will be higher than that for the competitive processes already referred to for the following reasons.

1. The capital charges for the process will be higher due to the required investment in solvent recovery plant.
2. The energy costs will be higher than those incurred in steam-stripping the free fatty acids since the energy required to desolventize extract and raffinate streams is, even when using highly efficient evaporator trains, substantially greater than that used to steam-strip fatty acids from the crude oil.
3. The difficulty of extracting fatty acids to a final level of less than 0.1 per cent means that an additional process step (removing the remaining fatty acids by alkali neutralization) is required.

The problem of neutral oil transfer into the extract phase will in certain cases have a greater effect on the economics of the process than the three factors listed above, especially when processing the more expensive fats. Information on the composition of the extract phase is therefore of considerable importance in any assessment of the economics of the process and monitoring of this composition is a key activity in controlling an extraction plant used for edible oil processing.

The use of a double-solvent system (hexane-aqueous ethanol) for the deacidification of high-acidity oils has also been studied and, as expected, an improved selectivity for fatty acids is obtained, as shown by the lower loss of neutral oil in the extract phase (Thomopoulos 1971). The greater cost of the more complex solvent recovery system must be set against the benefits of improved selectivity.

4.5.1.2 *Fractional separation of triglycerides*

Liquid–liquid extraction as a means of partitioning crude oils on the basis of the degree of unsaturation of the fatty acid chains in the molecule was used in the 1950s to refine tallow, using propane near its critical point (Moore 1950). The use of liquid–liquid extraction for the production of oil fractions of enhanced unsaturation was later examined again by Parsons

(1976), using solvents such as dimethyl formamide and aqueous *n*-methyl pyrrolidone. At high solvent ratios some enrichment of the more unsaturated acids was found in the more polar solvent phase, as indicated by the partition of linoleic acid between the light and heavy phases. This subject has recently generated renewed interest as a result of the observation that certain types of the highly unsaturated fatty acids present in fish oils are potentially beneficial to health. A recent patent (Nippon Oils and Fats 1984) suggests that it is possible to produce fractions enriched in eicosopentaenoic acid (EPA) by extraction with aqueous acetone. An increase of 85 per cent in eicosopentaenoic acid content was achieved by extracting the original oil, containing 12.88 per cent of this fatty acid, with ten times its weight of a 9 : 1 acetone-water mixture. No commercial application of this finding has, as yet, been reported. An alternative approach to enrichment in these acids has been reported from South Africa (Wagener and Hearshaw 1986), where it was shown that extraction of the methyl esters of the fatty acids of anchovy oil with a solution of silver nitrate in aqueous ethanol produced a fraction enriched in eicosopentanoic acid. In this case the EPA concentration was increased from 25.9 per cent in the original oil to over 60 per cent in the fraction richest in poly-unsaturated fatty acids.

A single-stage fractional separation of butterfat using supercritical carbon dioxide has also been reported (Kaufmann *et al.* 1982). The extracted fraction (operating conditions : 200 bar, 80°C) is, as expected, enriched in the short-chain fatty acids but, surprisingly, also in saturated fatty acids.

4.5.1.3 *Removal of minor components by liquid–liquid extraction*

Apart from its application to the extraction of free fatty acids and the partitioning of neutral glycerides into fractions of different degrees of unsaturation, liquid–liquid extraction has also been advocated for the extraction of some of the minor components found in oils. The separation of lecithin from crude soyabean oil gives a product containing approximately 60 per cent lecithin, most of the balance being the triglycerides of soyabean oil. An oil-free lecithin fraction can be produced by liquid–liquid extraction of the crude lecithin with aqueous (75–99 per cent) ethanol (Liebing 1972), the alcohol strength and alcohol : crude lecithin ratio determining the yield of oil-free lecithin and the phosphatide composition of the product. Using a solvent-to-lecithin ratio of between 1 : 3 and 4 : 1, a single-stage batch extraction was used to produce oil-free lecithin in yields of up to 20 per cent.

The separation can also be carried out in a continuous process and in practice both batch and continuous processes are in use. This type of separation is relatively simple due to the fact that the polar lipids are insoluble in acetone and precipitate once the solvent has been added.

Fractionation of the various phosphatide components of lecithin is carried out using ethanol as solvent. This produces a soluble fraction enriched in phosphatidylcholine, while the insoluble fraction is enriched in phosphatidylinositol and phosphatidic acid. Both fractions have emulsifying properties, but the ethanol-soluble fraction promotes oil-in-water emulsions, whereas the insoluble fraction may be used to produce water-in-oil emulsions. A full discussion of lecithins has been provided by Szuhaj (1989).

The use of carbon dioxide–propane mixtures near the critical point of the mixture for extracting oil from the crude lecithin has recently been advocated (Peter *et al.* 1987). In this case the propane component of the extractant serves to liquefy the soyabean oil–lecithin mixture while the carbon dioxide, which alone has very little solvent power for lecithin, enhances the selectivity of the mixed solvent. The conditions recommended for extraction of the oil are 80 bar pressure, 50°C and a propane content of 80 per cent in the extractant mixture. An alternative approach to the problem of lecithin de-oiling (Stahl and Quirin 1985) makes use of acetone both as diluent for the viscous crude lecithin and subsequently as entrainer to facilitate removal of oil from the acetone–oil–lecithin mixture employing supercritical carbon dioxide. This process requires far less acetone than the conventional de-oiling process based on acetone alone, but the two streams leaving the extractor still require conventional desolventization in order to reduce their acetone content to an acceptable level. Oil extraction from crude lecithin using only carbon dioxide was shown to require jet injection of the crude lecithin into the extraction chamber at very high pressure (900 bar) in order to achieve acceptable extraction times and yields.

The separation of alpha-tocopherol from oils has also interested investigators of the minor components of vegetable oils and here too the use of carbon dioxide as a solvent has been advocated (Ohgaki and Ketayama 1984); in this case in its liquid state at below the critical temperature of the gas.

4.5.1.4 *Other glyceride separations*

The concentration of monoesters of glycerol, which are widely used as emulsifiers, from the crude reaction product level of approximately 40 per cent to above 90 per cent has interested research workers at various times, and it has been shown (Monick and Treybal 1956; Arida *et al.* 1979) that a dual solvent system of ethanol plus a straight-chain hydrocarbon will separate the monoglycerides from the diglycerides which constitute the main reaction by-product. In practice, however, molecular distillation has proved to be a more economical method of producing material having a monoglyceride content of over 90 per cent.

The removal of lipids from fish protein concentrate has attracted attention in the past in view of their negative effect on the flavour stability of the protein. Liquid–liquid extraction using aqueous alcoholic solvents was found to be effective in reducing the lipid content to an acceptably low level (Mc Phee *et al.* 1972), the water content of the alcohol used proving to be an important parameter in process design.

4.5.2 Flavours and essences

4.5.2.1 *Citrus oils*

The citrus oils — principally orange, lemon, and lime — are amongst the most important of the essential oils, being highly valued as flavouring agents in a variety of foodstuffs. The oils are located in the peel of the fruit as well as being found in very small quantities in the juice, and in most cases are recovered simultaneously but separately with juice extraction. Orange oil is produced on a larger scale than lemon oil, world production being estimated as being in excess of 15 000 tons per annum, whereas that of lemon oil is estimated at 2000–2500 tons (International Trade Centre 1986). The amount of lime oil produced is considerably lower than this latter figure. Brazil produces more than 50 per cent of world orange oil production, whereas the USA is the major producer of lemon oil. The export markets for the citrus oils are mainly found in the more affluent industrialized countries.

Citrus oils consist principally of hydrocarbons which, in the form of terpenes (C_{10} hydrocarbons) and sesquiterpenes (C_{15} hydrocarbons) comprise more than 90 per cent of the oil. The remainder consists of an extensive list of oxygenated compounds of which the aldehydes are the most important, as these are largely responsible for the flavour of the oil. The acyclic monoterpene aldehydes neral and geranial, which are cis- and trans-isomers and are collectively referred to as citral, comprise approximately 0.2 per cent of orange oil, but in the case of lemon oil account for 2.5–5.5 per cent of the oil, depending on the area of cultivation of the fruit. They are regarded as the key flavour components. On the other hand, the hydrocarbons, of which limonene, a terpene, is the principal representative, contribute little to flavour and, in addition, oxidize readily to produce undesirable flavours. Their insolubility in water is a further factor favouring their removal from the oil, particularly where this is used to enhance the flavour of a juice concentrate. Deterpenation of citrus oils is therefore used to enhance their value.

Deterpenation is carried out in a number of ways. Fractional vacuum distillation produces an orange oil which has the terpene (limonene) content reduced from > 95 to 70–80 per cent, while at the same time the citral content may be increased to 1–1.5 per cent (Owusu-Yaw *et al.* 1986). Complete removal of terpenes can be obtained by using batch extractive distillation or liquid–liquid extraction, the solvent in both cases being aqueous ethanol, in which the terpenes and sesquiterpenes are insoluble. In the case of liquid-liquid extraction the oil to be deterpenated is mixed with 95 per cent ethanol, and water is then added to the mixture in sufficient quantity to reduce the ethanol concentration to 35 per cent. The terpenes, being insoluble in this solvent mixture, separate as an upper layer. The lower layer, essentially a solution of the deterpenated oil in aqueous ethanol may be used directly as a soluble essence, for example in the soft drinks industry, or it may be processed further to give a solvent-free deterpenated citrus oil. Other alcohols permitted as food additives may be used to process citrus oils in the same way. The extraction solvent–essential oil system has two pairs of partially miscible components, that is it belongs to the Type 2 category according to Treybal's classification (Blatterman 1957).

A two-solvent system for the enrichment of citrus oils in citral and removal of terpenes uses the solvent pair 70 per cent aqueous ethanol and a paraffinic hydrocarbon (Ruys 1957). Reman (1951) studied the use of the rotating disc contactor for such systems and obtained a citral concentration of 51 per cent in a contactor estimated as having a separating power equivalent to ten theoretical stages. The removal of the waxes present in the original oil with the terpenes is an additional benefit of liquid–liquid extraction. The terpene-rich fraction is marketed as a 'washed citrus oil' and is used in cheaper citrus oil formulations.

The use of propylene glycol together with either hexane or heptane as a solvent pair has also been advocated for the production of a fraction enriched in citral (Kiley and Kiley 1974). By varying the ratio of propylene glycol to hexane over a wide range, it was found possible to obtain good citral enrichment when using either orange oil or its concentrates as the starting material. A more recent approach to the problem of producing terpeneless citrus oils makes use of a combination of liquid–liquid extraction and chromatographic methods (Fleisher *et al.* 1987), but details of the process have not been disclosed.

The application of supercritical gas extraction, using carbon dioxide as solvent, to the deterpenation of citrus oils has attracted considerable interest. At a gas density above 0.6 g/cm^3 the components of lemon oil are fully miscible with carbon dioxide, thus making separation impossible, but at lower densities two phases are formed. The greater volatility of the monoterpene hydrocarbons in the oil leads to their preferential extraction in a two-phase fluid–liquid system (Coppella and Barton 1987). The solubility of the sesquiterpenes in carbon dioxide, on the other hand, is too close to that of the oxygenated compounds to permit separation by extraction, but saturating the carbon dioxide with water (and thus increasing the polarity of the solvent gas) enhances the separability of these two classes of compounds, particularly in the pressure range 80–90 bar (Gerard 1984). Where reduction of the sesquiterpene content of a citrus oil is necessary a two-stage separation

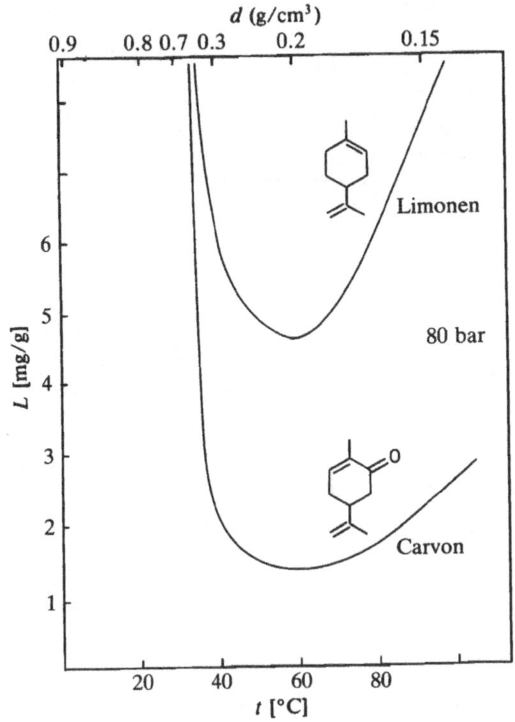

Fig. 4.3. Effect of temperature on solubilities of terpenes and oxy-terpenes in carbon dioxide. (Reprinted by permission from Gerard 1984.)

would be required. It has been shown that the solubilities of the major components of citrus oils pass through a minimum as the temperature is raised at constant pressure (Fig. 4.3). Depending on whether the temperature of extraction is chosen to be above or below this minimum, reflux can be obtained by respectively cooling or heating the overhead product. Phase equilibria reported for the pseudo-ternary system limonene–citral–carbon dioxide (Kalra *et al.* 1987) show the effect of temperature on gas-phase composition with respect to the liquid phase dissolved in it but also demonstrate the increased selectivity for the hydrocarbon limonene as the extraction temperature is raised.

On the basis of a process simulation it has been estimated (Robey and Sunder 1984) that by operating at 100 bar, 60°C, giving a gas density of 0.2–0.3 g/cm^3, a twelve-stage column fed with an oil containing 2.5 per cent citral and 95.0 per cent limonene would produce a raffinate containing 24.8 per cent citral and 49.9 per cent limonene (Fig. 4.4). Reflux is provided by cooling and decompressing the limonene-rich overhead fraction to 55 bar, 20°C. The solubility characteristics of the hydrocarbons and oxygenated

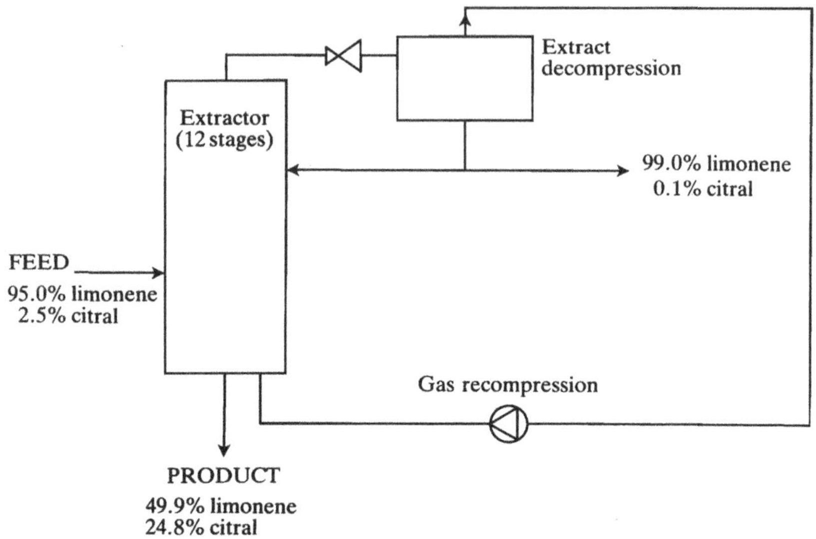

Fig. 4.4. Simulation of citral extraction process. Extractor: 100 bar, 60°C. Extract decompression: 55 bar, 20°C. (From Robey and Sunder 1984.)

compounds present in lemon oil also make separation possible by raising the temperature of the fluid phase above that at which extraction was carried out.

Using slightly milder conditions (65 bar, 40°C) than the above, on pilot-plant scale, extraction with carbon dioxide was shown to give significant enrichment in the volatile fraction of an orange oil (principally ethanol, ethyl butyrate, hexanal) but not affecting the limonene content appreciably (Japikse *et al.* 1987). The use of reflux improved the concentration of the volatile components significantly.

4.5.2.2 *Extraction of other flavours and food additives*
[See also Weatherley Chap. 5]

The extraction of flavours in most cases involves the application of liquid-solid extraction, as these are recovered as spices or essences by subjecting the natural raw material—seeds, fruits, etc.—to extraction. The extraction of onion flavour, which has recently been applied on commercial scale (Heath 1984), is a particular illustration of the fine dividing line between liquid-liquid and liquid-solid extraction in the food industry, as the flavour is present in the aqueous phase of the onion and can be extracted with a suitable solvent once the cell walls in the onion have been broken down. The minced onions are allowed to stand for some hours before extraction is begun in order to allow development of the flavour. After extraction in specially designed contactors both the extract phase containing the flavour and the sugar-rich residue are concentrated by low-pressure evaporation and are then recombined to give an onion flavour that is 35–100 times stronger than the original, depending on the ratio of the fractions used in the final blending. The use of such flavour concentrates in the food industry allows the providers of flavours to offer a more consistent quality of product, both organoleptically and hygienically, which at the same time is easier to handle.

The potential for further application of supercritical gas extraction is shown by the recent application of this process to the production of enriched celery and ginger extracts (Anon 1987). As yet little quantitative information on the separations carried out has been released, but in the case of celery extract it has been reported that the limonene content was reduced from 73.1 per cent in the seed extract to 6.9 per cent in the final extract after removal of carbon dioxide.

Hop extracts, used in the production of beer, are produced by solvent extraction of the dried hops. In addition to the flavour-giving soft resins and volatile oils the extract obtained in the conventional solvent extraction process contained the undesirable 'hard' resins, and their removal required the introduction of a liquid–liquid extraction stage, which has been described in the patent literature (Mitchell *et al.* 1972). The introduction of extraction using carbon dioxide, particularly when this is used in sub-critical conditions, has made the liquid–liquid extraction stage in the process redundant, as the first (liquid–solid) extraction stage is more selective and the extract is therefore relatively free of hard resins.

The extraction of propionic acid from fermented broth has been studied (Hindustan Lever Research Centre, personal communication 1988) in connection with the development of flavoured foods, ethyl acetate being used as the extracting solvent. At 30°C the mean distribution coefficients of acetic, propionic, and butyric acids between ethyl acetate and water were respectively 0.78, 2.35, and 4.46, with the distribution of butyric acid between aqueous phase and solvent being significantly affected by its concentration in the aqueous phase (see Table 4.4). In the presence of 3.5 per cent propionic acid and 10 per cent acetic acid the aqueous phase dissolved 4.5 per cent ethyl acetate at 30°C, whereas the solvent phase contained 3.3 per cent water.

Further useful information on systems obtained as a result of clostridium fermentation is given by Dadgar and Foulch (1985). The partitioning of aliphatic acids is discussed by Rublo *et al.* (1986).

The Raining Bucket contactor was used for the separation of propionic acid from the fermentation broth on the grounds of its ability to handle

TABLE 4.4 *Distribution coefficients of volatile acids between ethyl acetate and water at 30°C*

Acetic acid		Propionic acid		Butyric acid	
Concentration in aqueous phase (%w/v)	K_{AA}	Concentration in aqueous phase (%w/v)	K_{PA}	Concentration in aqueous phase (%w/v)	K_{BA}
0.46	0.83	0.32	2.5	0.18	2.72
0.88	0.74	0.65	2.18	0.27	3.67
1.60	0.76	0.89	2.32	0.41	4.20
2.32	0.77	1.17	2.38	0.59	6.07
3.12	0.78	1.48	2.36	1.04	6.8

$$K = \frac{\text{Concentration (\%w/v) of volatile fatty acid in solvent phase}}{\text{Concentration (\%w/v) of volatile fatty acid in aqueous phase}}$$

Geometric average K = 0.78	2.35	4.46

Reduction in distribution coefficient in mixtures of acetic and butyric acids:
For butyric acid concentration of 0.1–0.8 per cent K_{AA} decreases by 27 per cent.
For acetic acid concentration of 1–1.3 per cent K_{BA} decreases by 23 per cent.
(Courtesy of Hindustan Lever Limited.)

liquids having a tendency to emulsify and also those containing small amounts of solids. It was found, however, that the contactor gave only 30 per cent of the design mass transfer performance and this was ascribed to short-circuiting of the two streams through the clearance between the wall and the rotor as well as to the dead volume present in the settling zones, resulting in loss of effective contact of the two phases.

The availability of the enzyme glucose isomerase has for some time now been responsible for a reduction in the cost of sweeteners used in industry, as a concentrated fructose syrup can be produced from corn starch using this enzyme. Adsorption and reverse osmosis are processes currently available for the separation of glucose and fructose, and studies on the five-component system glucose–fructose–water–ethanol–carbon dioxide have shown (D'Souza and Teja 1988) that it is possible to separate glucose and fructose using supercritical carbon dioxide and ethanol to extract an aqueous solution of the two sugars.

4.5.3 Caffeine extraction

4.5.3.1 *Background*

The extraction of caffeine (1,3,7 trimethyl xanthine) from coffee or coffee extract has, in the eighty years since the invention of the first process for doing so, become an important process in the coffee processing industry. In the USA the amount of non-decaffeinated coffee consumed has shown a steady decline since 1970, whereas the volume of decaffeinated coffee drunk has increased considerably in that time (Katz 1987). In Europe the interest in decaffeination is less marked than in the USA, even in those countries that are major consumer of the caffeine-rich robusta variety of the coffee bean. Caffeine content of the coffee bean is approximately 2 per cent (dry basis) in the case of the Robusta bean but only 1 per cent in the Arabica species. In the European Community the caffeine content of decaffeinated instant coffee must be below 0.3 per cent (dry basis), and in practice it is often below 0.2 per cent (dry basis) (Macrae 1985). The caffeine recovered is extensively used in soft drinks formulations.

Caffeine is also recoverable by liquid–liquid extraction from the aqueous extract from Kola nuts, though the scale on which this extraction is carried out makes it an insignificant contributor to overall caffeine production.

4.5.3.2 *The properties of caffeine*

Caffeine (MW 194.19, melting point 236°C) is only sparingly soluble in water at ambient temperature but the solubility rises rapidly with temperature (Fig. 4.5a). In the presence of coffee extract, however, the solubility of caffeine in water is enhanced considerably, as shown in Fig. 4.5b, thus affecting to a considerable extent the partitioning of caffeine between the aqueous solution and an organic solvent. The formation of a caffeine–chlorogenic acid complex, more soluble in water than caffeine alone, is believed to account for the enhanced solubility.

Macrae (1985) lists the solubility of caffeine in a number of organic solvents, including various halogenated solvents. Those which are of interest in caffeine extraction are listed in Table 4.5. Treybal (1963) suggests that for a system comprising two largely immiscible solvents and a solid solute the partition coefficient of the solute between the two solvents will be close to the ratio of the solubilities of the solute in the individual solvents, provided the miscibility of the two solvents is low. In the case of caffeine partitioning between two solvents this approach must be used with considerable caution as the solubility of caffeine in water is, as has been shown, influenced by the presence of other soluble material. A second factor likely to affect the partition coefficient between solvents and therefore the separating power required is the extent of solvent miscibility, which has been documented by Hampe (1986). The partition coefficients of caffeine between water and a number of organic solvents have also been reported (Ulusoy *et al.* 1984). Data on partition coefficients for caffeine between chloroform and water as well as for the system caffeine–dichloromethane–water and are given in Table 4.6 (a) and (b) (Bichsel 1979).

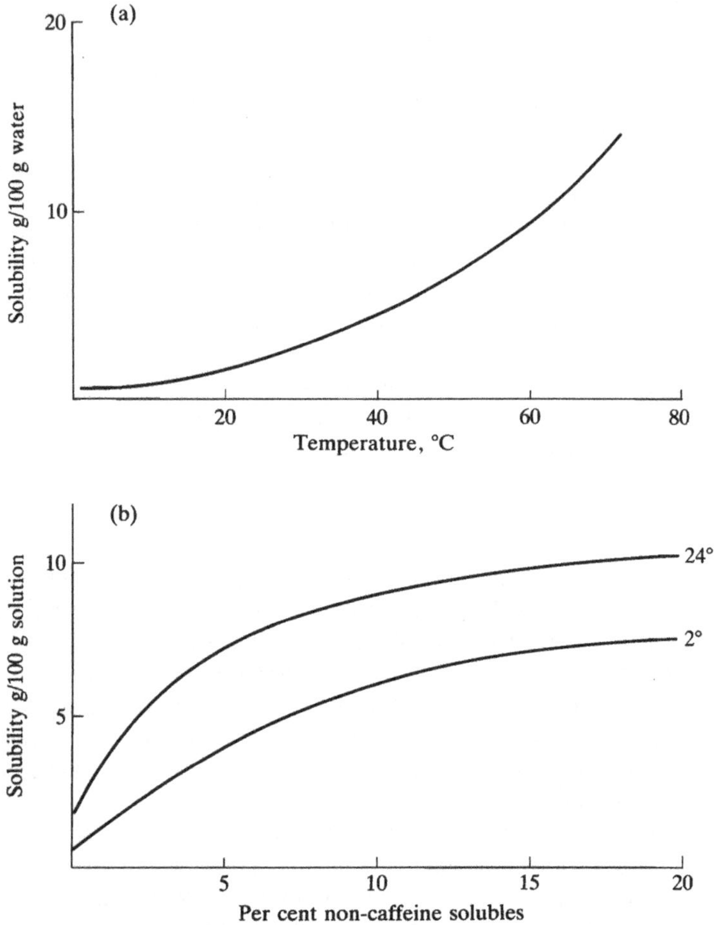

Fig. 4.5. Caffeine solubility in aqueous systems: (a) water, (b) aqueous coffee extract.

TABLE 4.5 *Solubility of caffeine in organic solvents (Katz 1987)*

Solvent	Temperature (°C)	Solubility (g/100 g solvent)
Chloroform	17	12.9
Dichloromethane	33	9
Trichlorethylene	15	0.76
Ethyl acetate	18	0.73

TABLE 4.6 *Partition coefficients for caffeine in the systems*

(a) Water-chloroform-caffeine

Temperature (°C)	Total caffeine content (mmol/50 ml)*				
	4.8	1.2	0.31	0.077	0.0048
20	18.7	21.4	22.4	22.6	22.8
30	15.5	17.3	19.5	21.0	19.4
40	13.9	17.1	16.8	17.4	17.4
50	12.1	13.9	14.4	14.4	14.6

(b) Water-dichloromethane-caffeine

Temperature (°C)	Total caffeine content (mmol/50 ml)				
	10.3	1.29	0.33	0.0824	0.0206
20	4.85	9.44	9.54	9.69	8.40
25	5.00	8.28	9.18	9.43	8.91
30	5.18	8.39	9.13	9.09	8.50
35	5.48	7.89	8.86	9.00	8.68

*Equal volumes of water and solvent (B Bichsel, personal communication, 1988)
(Courtesy of Elsevier Applied Science Publishers Limited)

4.5.3.3 *Decaffeination technology*

Decaffeination may be carried out on the coffee bean prior to or after roasting, but processing of the green (unroasted) coffee is preferred, since the true coffee flavour is undeveloped at that stage and will therefore not be markedly affected by the decaffeination process. Caffeine extraction from the roasted bean, on the other hand, reduces the contact time required to achieve substantial decaffeination and has therefore attracted considerable attention. Dichloromethane is now the most widely used organic solvent used in direct extraction of caffeine from green coffee, which needs, however, to be presoaked in water before extraction. This indicates that caffeine extraction from the bean really occurs via transfer from the aqueous phase. In the light of modern thinking about the toxicology of solvents used in the food industry extraction with organic solvents is being increasingly challenged by extraction of the pre-wetted green beans with supercritical carbon dioxide and other processes.

Direct decaffeination with water, introduced in 1941, was seen as a means of increasing the rate of extraction but it requires the addition of a liquid–liquid extraction step to recover the caffeine from the extract leaving the liquid–solid extractor. Since the extraction of caffeine by water is accompanied by the extraction of other water-soluble components of the coffee bean caffeine is recovered from an aqueous phase containing 15–40 per cent dry matter, of which only 0.6–1.0 per cent is caffeine. The decaffeinated extract is recycled. In selecting a solvent for the extraction of caffeine from the caffeine-rich extract the solubility of coffee solids in the organic solvent must also be taken into account in view of the effect thereof on overall soluble solids recovery and the purity of the recovered caffeine. In recent years adsorbents (for example activated carbon) have been used to recover caffeine from such aqueous extracts, thus avoiding the need for organic solvents altogether.

4.5.3.4 *Liquid–liquid extraction in decaffeination*

Many of the solvents investigated for suitability in the process for decaffeination by liquid–solid extraction have also been considered for use in caffeine removal by liquid–liquid extraction, though in practice dichloromethane and ethyl acetate almost certainly account for the bulk of liquid–liquid extraction capacity for caffeine removal from coffee extract at the present time. Toxicological pressures may contribute to a change in this situation in the near future (see also section on solvent selection). The value of the recovered caffeine and the importance of minimizing the amount of coffee solids leaving the system with the caffeine stream point to a double advantage in producing a relatively high-purity caffeine, and Katz (1987) points out that the caffeine content of the desolventized extract should be in excess of 70 per cent for refining of the caffeine to be economically justified. The first extraction is normally followed by a back extraction for caffeine recovery and recycle of decaffeinated solvent, or the caffeine-laden solvent may be evaporated to recover crude caffeine. A factor relevant to process design is the need to conserve the volatile flavour components in the primary extract, and this may be accomplished by stripping these from this extract before extraction of caffeine.

The field is covered by numerous patents dealing with solvent choice, the operating conditions, and the possible process combinations (Table 4.7). When extracting caffeine with dichloromethane the first liquid–liquid extraction stage may be operated at about 80°C with a solvent-to-coffee extract ratio of between 4 and 8 : 1 (Katz 1987). This gives a caffeine removal from the aqueous extract of above 99 per cent and at the same time causes less than 2 per cent of the soluble coffee solids to transfer to the solvent phase. The caffeine-rich extract is generally dispersed in the organic solvent for maximum contactor efficiency.

Where back extraction from dichloromethane into water is used to recover caffeine and non-caffeine solids an operating temperature of 20–30°C for this extraction and a solvent-to-water ratio of 1.6 : 1 should result in recovery of

TABLE 4.7 *Solvents and extraction conditions for caffeine extraction from coffee beans*

Bean condition	Primary solvent	Secondary solvent	Recommended extraction temperature (°C)	Recommended secondary to primary solvent ratio	Reference
Roasted	Water	Dichloromethane	21	6 : 1	USP 2 933 395
Roasted	Water	Ethyl acetate	not disclosed	1.4–2.1 : 1	USP 4 352 829
Green/roasted	Water	Partially hydrogenated soyabean oil	80	15 : 1	USP 4 446 162
Green	Water	Corn oil	82	40 : 1	USP 4 545 998
		Corn oil + 2% water	82	25 : 1	
Roasted	Water	Dichloromethane	Ambient	4.5 : 1	USP 4 547 378
Green	Water	n-Butyl Acetate	80	5.2 : 1	USP 4 562 083
Roasted	Water	Dichloromethane	38–43	3.5–5 : 1	EP 0159829 A1
Roasted	Water	Supercritical carbon dioxide	40	100 : 1	USP 4 341 804

caffeine of greater than 90 per cent purity and a soluble coffee solids loss of less than 1 per cent. As a result of the changed phase ratio it is advantageous to make the aqueous phase the continuous phase in the extractor. Solvent removal from the primary extract stream is an important part of the process in order to meet the requirements of the regulatory authorities.

The use of dichloromethane (density 1324 kg/m^3 at 20°C) ensures that an adequate density difference is maintained between the phases throughout the contactor used for the extraction. No data on interfacial tension between the organic and aqueous phases for this system have been published, but it is known that there is some risk of emulsification in using a continuous contactor for this separation. A number of conventional continuous counter-current contactors, including the rotating disc contactor (RDC), the Karr column, the Oldshue–Rushton and the York–Scheibel contactors have been tested and found satisfactory. The RDC is known to have found widespread application in decaffeination plants. A typical mass balance for decaffeination of the extract from roast, ground beans followed by back extraction to recover a relatively pure caffeine as well as to minimize loss of non-caffeine solids is given in Fig. 4.6 (Jones *et al.* 1985).

Ethyl acetate may be used to remove caffeine from the extract produced by water decaffeination of the coffee bean, and has the advantage of being more acceptable toxicologically than dichloromethane. This solvent is, however, less selective than dichloromethane in its separation of caffeine from other soluble coffee solids and its caffeine capacity is also considerably lower than that of dichloromethane. The solvent : extract ratio for operating a contactor is therefore somewhat higher than when using dichloromethane. The use of *n*-butyl acetate has been advocated as an improvement on ethyl acetate by virtue of the fact that its miscibility with water is considerably lower and that it has a higher selectivity for caffeine (Gottesman 1985). In the first stage of extraction (at 80°C) a solvent-to-coffee extract of 4.33 : 1 is used to transfer more than 95 per cent of the extracted caffeine to the solvent. The caffeine-enriched solvent is then concentrated by evaporation to give a five-fold increase in concentration, whereupon reverse extraction with water is used to strip out caffeine and the soluble coffee solids. This solvent is suitable for the extraction of caffeine from green coffee extract rather than from the extract from R&G coffee.

Trichlorethylene is not now of industrial significance in decaffeination in view of its toxicology. However, its efficacy as a decaffeination solvent has been demonstrated in a study (Escardino *et al.* 1977; Escardino *et al.* 1978) which also provided operating data for a packed column operating on the system caffeine–water–trichlorethylene (other solids normally extracted from coffee were not present). When operating the pilot-scale column packed with Raschig rings at 20°C the authors were able to achieve 97 per cent decaffeination at a flow rate of 9.6 m^3/m^2.h, giving an HTU of 0.14 m.

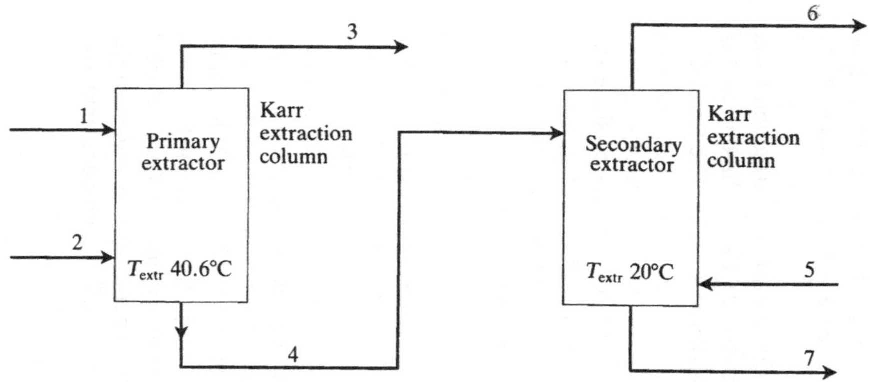

Stream No.	1	2	3	4	5	6	7
Content/Function	Caffeine solvent	Coffee solids extract	Decaffeinated coffee solids extract	Caffeine-rich solvent	Water for stripping ncs* from solvent	ncs-enriched aq. phase for recycle	ncs-depleted solvent for caffeine recovery
Flow (kg)	580	100	98	582	350	355	576
Composition (kg)							
Water	–	75	74	1.2	350	350	1.2
Dichloromethane	580	–	1.3				
				579	–	4.5	574
Non-caffeine solids	–	23.9	23.4	0.5	–	0.4	0.1
Caffeine	–	1.07	0.03	1.04	–	0.07	0.97

* ncs / non-caffeine solids

Fig. 4.6. Mass balance for decaffeination of aqueous coffee extract. (Based on Jones *et al.* 1985.)

The use of various triglyceride oils, including coffee oil, for the extraction of caffeine from the caffeine-rich aqueous extract has also been investigated. The poor partition coefficient for the system caffeine–vegetable oil–water would require a very high solvent ratio, and a novel means of overcoming this disadvantage – at least to some extent – consists of dispersing a more powerful solvent as a disperse phase in the principal solvent, in this case the vegetable oil, and thereby using the principal solvent as a barrier to the transfer of solutes other than caffeine (Katz and Skiff 1985). In this work it was shown that the dispersion of 2 per cent water in corn oil made it possible to reduce the solvent-to-extract ratio for 97 per cent decaffeination from 40 : 1 to 25 : 1 when operating at 82°C.

The use of supercritical gas extraction for direct caffeine extraction from green coffee was one of the earliest applications of this technique and is particularly attractive because of the inertness of the solvent. The feasibility of using supercritical carbon dioxide to extract caffeine from concentrated extract has been established (Prasad *et al.* 1980), and the resulting process may be regarded as distillative extraction or liquid–liquid extraction, since the supercritical fluid is neither gas nor liquid. At present direct supercritical gas extraction of caffeine from the bean is preferred to aqueous extraction of caffeine from the bean followed by supercritical fluid extraction.

The oil present in roasted coffee or spent coffee grounds may be used as a caffeine solvent in the same way as other triglyceride oils but it is also employed as a carrier for headspace aroma components in soluble coffee. In this latter application the oxidative stability of the oil is particularly important (though in practice instant coffee 'plated' with coffee oils are packed under inert gases to give a low oxygen content in the headspace) and the removal of diterpene esters from the oil is seen as essential in imparting maximum stability to the oil. The principal diterpene ester present in coffee oil is cafestol, and it has been shown (Roychoudhury 1985) that this can be removed by extraction of the oil with aqueous iso-propanol. A solvent-to-oil ratio of 6 : 1 is suggested and the extraction was carried out in a York-Scheibel pilot-plant column. It is claimed, however, that a prior heat treatment of the oil is essential if the extraction of cafestol is to confer significant oxidative stability on the oil.

4.6 The prospects for liquid–liquid extraction in the food industry

In considering the future of liquid–liquid extraction in the food industry it seems obvious that the shift towards 'soft' processing noted here is almost certainly irreversible, and there will therefore be increasing emphasis on finding ways of applying this concept in practice. Essentially that means that the number of acceptable solvents will diminish, possibly excluding some classes of solvents altogether, in future. Other solvents may also come under more rigorous scrutiny, and emission standards as well as limits on solvent residues in foodstuffs will continue to be tightened from time to time, thus making processing with solvents more demanding. This suggests that the food industry will concentrate its attention on finding solvents and conditions of processing that will maximize acceptability of processing on the part of consumers and regulatory authorities. Concurrently the industry will at all times be highly conscious of processing costs, since most of the industry operates on very low profit margins and the economics of solvent recovery, though unlikely to dominate an assessment of alternatives, will not be overlooked.

Extraction with supercritical gases, in particular with inert gases such as carbon dioxide, seems to meet such requirements, but the versatility of the short-chain alcohols, in particular ethanol and isopropanol, will also continue to prove valuable to the industry. Despite the promising early work on the use of liquid carbon dioxide (Schultz *et al.* 1974) little has been reported on further developments in this field more recently. The density of liquid carbon dioxide is close to that of many non-aqueous liquids of interest to the food industry, and phase separation could therefore become a problem in using this solvent in liquid–liquid as distinct from liquid–solid extraction.

The process is likely to be of interest primarily where the value added in processing is high, though regulations governing the use of conventional solvents will in some other cases encourage the use of supercritical and subcritical gas extraction. In some cases the products obtained by gas extraction have proved superior in quality to the established products and the perceived improvement will at times be sufficient to justify a higher processing cost. The view that processing with supercritical gases is more costly than conventional processing has been disputed (Peter *et al.* 1987; King *et al.* 1987) and with careful design the energy requirements of extraction with a supercritical gas may be lower than those obtained in conventional processing (King *et al.* 1987). The main drawback to the greater application of extraction with supercritical gases or high-pressure liquefied gases has almost certainly been the greater investment required and the lack of familiarity with high-pressure processing. Further development of the technology should see these factors diminish in importance.

The close links between liquid–liquid and liquid–solid extraction in the food industry have already been commented upon. The two processes are likely to become increasingly integrated in processing its raw materials, thus enabling a fractionated product to be recovered directly from the raw material. At the same time there appears to be increasing interest in combining liquid–liquid extraction with other separation processes, particularly those involving a variety of sorbents, and developments in this direction will undoubtedly be helped by the creation of more specific sorbents.

The applications covered in this review do not reflect the growing interest of the food industry in opportunities related to biotechnology, but the developments in this field may in due course create an opportunity for separation processes for macromolecules of interest to the food industry. For some time the development of separation processes suitable for the purification of proteins, in particular enzymes, by partitioning them between two aqueous phases has been actively pursued. Two immiscible aqueous phases may be generated by mixing aqueous solutions of various polymers or of polymers and certain salts (Albertsson 1986). Systems that have been extensively investigated and exploited comprise the polyethylene glycol (PEG)–dextran–water system and that comprising polyethylene glycol solution mixed with solutions of electrolytes such as potassium phosphate or ammonium sulphate. An example of phase compositions at equilibrium for a specific dextran–PEG mixture together with important physical properties of the system is given in Fig. 4.7; Fig. 4.8 shows the binodal curve for the system PEG–dextran–water. As in the case of more common liquid–liquid equilibria, the tie lines connect points representing the compositions of complementary phases. The area below this curve is homogeneous and therefore of no interest for liquid–liquid extraction. In studies of the thermodynamic relationships of PEG–dextran–water systems attempts have been made to use both the Flory–Huggins model and the UNIQUAC model for prediction of phase equilibria (Kang and Sandler 1987). The work led to the conclusion that either model could be used for this purpose, but that the Flory–Huggins model would be easier to apply. Interaction parameters were obtained from binary osmotic pressure and plait point data.

These systems may be used to separate enzymes from cell debris or to frac-

Light phase 4.9% PEG 1.8% Dextran 93.3% water	Light phase Density : 1.0114 g/cm^3 at 20°C Viscosity : 4.9 relative to water
	Interfacial tension :0.46$\times 10^{-3}$ mN/m
Heavy phase 2.6% PEG 7.3% Dextran 90.1% water	Heavy phase Density : 1.0326 g/cm^3 at 20°C Viscosity : 15.7 relative to water

Composition of the two phases formed by mixing 5 per cent W/W dextran (Dextran 500) solution with 3.5 per cent W/W polyethylene glycol (PEG 6000) solution.

Fig. 4.7. Composition and physical properties of complementary phases in aqueous phase partitioning. (Based on Albertsson 1986.)

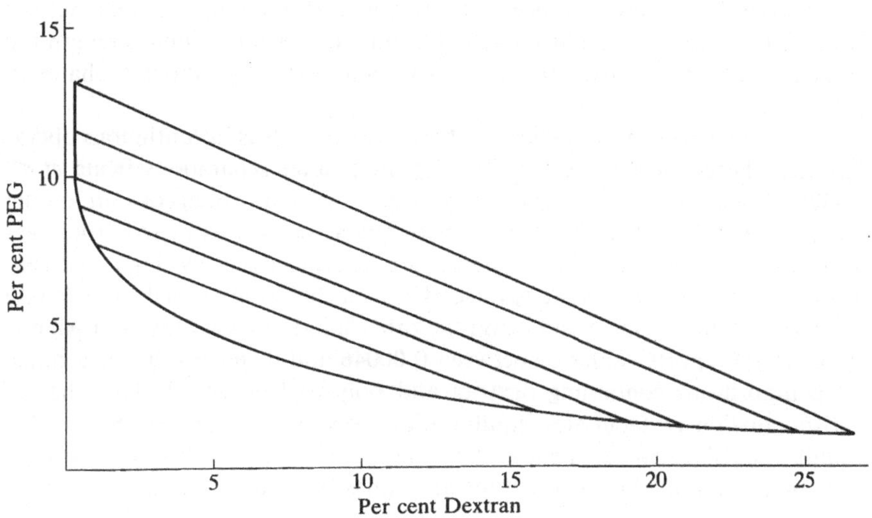

Fig. 4.8. Equilibrium diagram and binodal curve for ternary system polyethylene glycol (PEG 4000)-Dextran (D48)-Water. (Based on Albertsson 1986.)

tionate enzymes, and it follows that the two operations can also be carried out successively. The enzymes or proteins normally go into the light phase, whereas the other components of the cell or of the feed go into the heavy phase.

The partition of proteins between the phases depends on several factors, including the molecular weight of the dextran and of the PEG, pH, and the presence of electrolytes. It has been demonstrated that the partition coefficient of a protein in the dextran–polyethylene glycol–water system will increase if the molecular weight of the polyethylene glycol component is decreased or the molecular weight of the dextran used is increased, suggesting that the partitioned protein behaves as if it is preferentially attracted by smaller polymer molecules and repelled by larger polymer molecules.

Because of their acceptability from a toxicological point of view, these two polymers have been extensively used in experimental and development work, but the high cost of dextran has led to assessments of cheaper (less refined) types of dextran for use in these separation processes (Kula *et al.* 1982), and in several enzyme separations the partition coefficient and yield of product were not significantly affected by use of a cruder dextran. The PEG/electrolyte system has been employed for many of the enzyme separation studies reported (Kroner *et al.* 1984). The use of ligands bound covalently to or esterified with the PEG sharpens the separation obtained, and triazine dyes may be bound to PEG to enable 'affinity partitioning' to be carried out. In other cases fatty acids have been esterified with PEG in order to enhance the separating power. As the ligand–PEG bond requires a low ionic strength the system in this case must consist of PEG and dextran solutions (Johansson 1985).

Separations have been carried out batchwise as well as in continuous operation, the latter having been studied for multi-stage separations (Kula *et al.* 1982). Cross-current extraction has been found advantageous in model separations (Hustedt *et al.* 1987) where fumarase was extracted from saccharomyces cerevisiae. The systems are characterized by high viscosity (5–10 mPa/s), which has an adverse effect on diffusion as well as on phase separation. Interfacial tensions, on the other hand, are very low and Albertsson (1986) reports values of between 0.00046 and 0.066 mN/m for a range of compositions containing dextran and polyethylene glycol. This characteristic obviously facilitates equilibration between the phases but it also entails the risk of entrainment at the end of extraction and therefore calls for the application of additional settling after phase separation or of centrifugally aided separation.

The scale of extractive enzyme purification has been varied from gram quantities to extraction from 50 kg wet cells. The biomass concentration in the system is best held in the range 20–35 per cent and should be as high as possible in order to minimize the cost of chemicals, but the product yield can be expected to fall as biomass concentration increases, leading to a trade-off between yield of product and the cost of chemicals used (Hustedt *et al.* 1986).

It has been found that rotary columns are most effective for enzyme purification, for example the Scheibel and Kuhni contactors have given satisfactory results provided only low–medium rotor speeds are employed. The Podbielniak centrifugal extractor has also been used for separations using the PEG–salt system, but in using this extractor the high temperature rise due to the pressure drop in the machine must be taken into account in choosing processing conditions. It has been reported (Kula *et al.* 1984) that the RTL (formerly the Graesser 'raining bucket') contactor required reduction in agitation and additional phase separation capacity at discharge for the system polyethylene glycol–salt solution.

A system fundamentally analogous to the mixer-settler concept has also been used, disc centrifuges taking the place of the settler in order to avoid the slow phase separation already referred to, and this system has been employed for numerous enzyme purifications on pilot-plant scale as well as some larger-scale applications (Kroner *et al.* 1984). In this form the separation process should present few scale-up problems. In comparing the first and second stage of this separation it is noteworthy that increase of the density difference between the phases from over 50 to 90 kg/m^3 was considered to have influenced separation performance significantly.

A detailed analysis of processing costs is provided by these authors, and their calculations demonstrate the advantages of this method of purifying enzymes over the more traditional methods involving cell disruption followed by physical separation, such as centrifugation or filtration. Furthermore, these calculations also demonstrate the important contribution of the cost of chemicals used, particularly in the first stage when separating the enzymes from cell debris, and the results reported elsewhere on chemicals re-use (Hustedt *et al.* 1986) indicate that considerable savings on replacement of chemicals can be made. The proposed separation and solvent recycle scheme is shown in Fig. 4.9. For the extraction of the enzyme fumarase from a batch of 100 kg cell mass the overall process cost was calculated as being US $16.35/kg (1983 data). Of this 40 per cent is due to raw material costs and 29 per cent accounted for by the two extraction stages. The chemicals consumption in the first stage (cell debris separation) was an order of magnitude higher than that in the second and accounted for 11.5 per cent of total costs.

Since the solvent characteristics in a two-aqueous-phase system can be varied to give the process considerable flexibility it seems very probable that new applications will continue to be developed, but their interest to the food industry is likely to depend on the demand for macromolecules that require a considerable degree of purification before use or re-use.

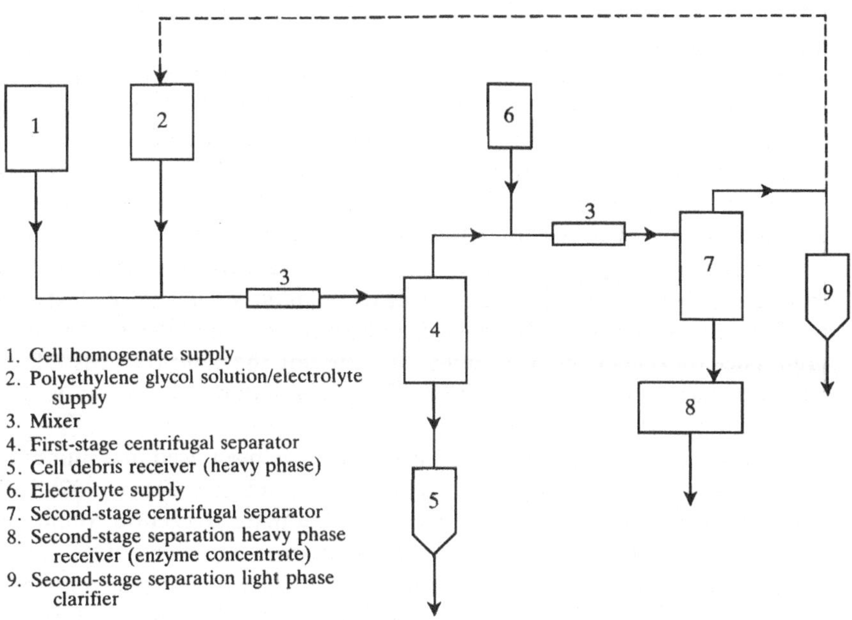

Fig. 4.9. Aqueous phase extraction of enzymes using polymer solution recycle. (Based on Hustedt *et al.* 1986.)

4.7 Equipment used and its design

The applications described in earlier sections demonstrate that in the majority of cases conventional liquid–liquid extraction contactors can be used. It is generally prudent to avoid contactors using high-intensity shear for dispersion or re-dispersion (except where this is essential for purposes of providing effective mixing of the phases, as in the processing of lecithin) because of the tendency to emulsification where naturally occurring surface-active agents are present. The Rotating Disc Contactor is almost certainly the most extensively used of the current generation of liquid–liquid extractors, but in employing this extractor low rotor speeds are essential where the system shows any tendency to emulsify. It is therefore advisable to provide for variable shaft speed in the design of these units. In the design of continuous contactors for food industry use special attention must also be paid to the density difference between the phases, particularly at that end of the contactor where the solute concentration in the extract phase is greatest, as the relatively small density differences that follow from the feed and solvent characteristics can be responsible for significant entrainment problems. In such cases centrifugal extractors may offer considerable advantages. As already indicated, interfacial tension problems may arise due to the presence of traces of natural surface-active agents in material being processed. Batch extraction is now rarely installed, although the more traditional industries may retain some batch extraction equipment.

Design of the contactor usually requires, as has already been discussed, pilot-plant studies on a system as close as possible to the system to be processed on full scale. The lack of data on the complex systems normally encountered in the food processing industry makes design from first principles virtually impossible, and the designer must take great care to avoid unquestioningly equating analytical replication with subjectively assessed acceptability. This generalization also applies in reverse—when changing from one solvent to another the profile of secondary components extracted may be quite different but the overall assessment may indicate no significant preference for one or other product. In designing the plant the possibility of seasonal changes in the composition of the raw material must also be taken into account.

The Rotating Disc Contactor is known to have been used in a number of liquid–liquid extraction applications in the food industry (Ruys 1957; Sivetz and Desrosier 1979) and the Karr reciprocating plate extraction column has been sucessfully tested for caffeine removal from aqueous coffee and tea extracts (AE Karr, personal communication 1987), with caffeine removal of up to 99.9 per cent being reported. It is known that the pulsed column can be used for separations such as the deterpenation of citrus oils, but attempts to use the RTL (Graesser 'raining bucket') contactor for food industry separations appear to have met with very limited success. The very low interfacial tensions and relatively low density differences experienced in the application of aqueous phase extraction systems restrict the choice of contactor for such applications since the energy input per unit of throughput must be kept to a level at which subsequent phase separation can only be achieved by use of long settling periods. One solution to this problem is provided by the use of centrifugal force, either in continuous two-phase contacting or for single-stage separation (Kula *et al.* 1982).

Operation in the high pressure range obviously calls for vessels that are inherently expensive to construct. Although Rotating Disc Contactors may be designed to operate at pressures up to 50 bar it seems unlikely at this stage that attempts would be made to use contactors requiring rotary shaft seals for supercritical gas extraction. A number of column packings have been assessed for their effectiveness in countercurrent fluid–liquid extraction (Peter and Tiegs 1984) and Sulzer packings (CY) were the most favourable from the point of view of capacity, though wire spirals showed greater separating power per metre of contactor height. In the case of the Sulzer packing, a separating power of 3.5 theoretical stages per metre height was reached at a gas flow rate of approximately 240 kg/m^2h, equivalent to a gas velocity of 12 mm/s, though at this loading the packing had not yet reached flood point. In the case of the wire spirals flooding was observed at a loading of 220 kg/m^2h, the separating power at this gas flow rate being approximately twice that of the Sulzer packing. The performance of Raschig rings was considerably poorer than the wire spirals and the Sulzer packing. The capital cost of plant for extraction at high pressure is considerably greater than that for conventional extraction and even at relatively modest pressures for example 100 bar the major items are likely to cost twice as much as similar items designed for operation at near-atmospheric pressure (Hederer and Heidemeyer 1985).

Even a relatively inexpensive solvent such as carbon dioxide must be recovered and recompressed to the operating pressure, pointing to the need for costly compressors and considerable energy costs. Technical aspects of plant design as well as the operating costs for a plant to extract hops using supercritical carbon dioxide are discussed by King *et al.* (1987). Their calculations show that carbon dioxide make-up accounts for 14% of the operating cost but that 46% of the operating cost is related to capital investment in the form of maintenance, depreciation and return on capital.

Acknowledgements

The author is indebted to Mr R. J. Clarke, of Chichester, Sussex, UK, for guidance in the preparation of the section on caffeine removal. He has also benefited from discussions and correspondence with various experts in the processing of essential oils.

References

ALBERTSSON, P. A. (1986). *Partition of Cell Particles and Macromolecules* (3rd edn.). Wiley-Interscience, New York.

ANON. (1987). *Parfums, Cosmetiques, Aromes*, No 75, 71.

ARIDA, V. P., CUADRA, F. B. and LEGASPI, S. L. (1979). *Philipp. J. Cocon. Stud.* **4**, 25.

ASHTON, N. F., MCDERMOTT, C. and BRENCH, A. (1983). In *Handbook of Solvent Extraction* (eds. T. C. LOH, M. H. I. BAIRD and C. HANSON). Wiley-Interscience, New York.

BICHSEL, B. (1979). *Food Chem.* **4**, 53.

BLATTERMAN, J. M. (1957). *Perfum. Essent. Oil Rec.* **48** (10), 514.

BRUNNER, K.-H. (1986). *Fette, Seifen, Anstr-mittel* **86**, 523.

BUNZENBERGER, G., LACK, E. and MARR, R. (1984). *Ger. Chem. Eng.* **7**, 25.

CHRASTIL, J. (1982). *J. Phys. Chem.* **86**, 3016.

Commission of the European Communities: Proposal for a Council Directive on the approximation of the laws of member states on extraction solvents used in the production of foodstuffs and food ingredients, COM (83).626 final, Brussels, October 1983 and amendment, March 1985.

COPPELLA, S. J. and BARTON, P. (1987). *Am. Chem. Soc. Symp.* Series 329, 202.

DADGAR, A. M. and FOULCH, G. L. (1985). *Biotechnol. Bioeng. Symp.* Section 15, 611.

D'SOUZA, R. and TEJA, A. S. (1988). *Fluid Phase Equilibria* **39**, 211.

ESCARDINO, A., MONZO, M. and COSTA, J. M. (1977). *Rev. Agroquim. Tecnol. Aliment.* **17**, 372.

ESCARDINO, A, AUCEJO, A. and COSTA, J. M. (1978). *Rev. Agroquim. Tecnol. Aliment.* **18**, 224.

Federal Register (1985). **50**, 51551.

FLEISHER, A, BIZA, G., SECORD, N. and DONO, J. (1987). *Perfumer and Flavorist* **12**(2), 57.

GERARD, D. (1984). *Chemie Ingr-Tech.*, **56**, 795.

GOTTESMAN, M. (1985). US Patent 4 562 083 to General Foods Corporation.

HAMM, W. (1983). In *Handbook of Solvent Extraction* (eds. LOH, BAIRD and HANSON). Wiley-Interscience, New York.

HAMPE, M. J. (1986). *Ger. Chem. Eng.* **9**, 251.

Health and Safety Executive (1987).*Guidance Note EH 40/87*, from the *Occupational Exposure Limits 1987*. Her Majesty's Stationery Office, London.

HEATH, H. B. (1984). *Food, Flavourings, Ingredients, Packaging and Processing* **6** (11), 35.

HENDRIX, W. BLAKE (1984). *J. Am. Oil Chem. Soc.* **61**, 1369.

HEDERER, H. and HEIDEMEYER, H. (1985). *Ger. Chem. Eng.* **8**, 112.

HUSTEDT, H. (1986). *Biotechnol. Lett.* **8** (11) 791.

HUSTEDT, H., KRONER, K. H., SCHUETTE, H., MENGE, U. and PAPAMICHAEL, N. (1986). *BioEngineering* (*Graefelfing, FRG*) **1**, 12.

HUSTEDT, H., BOERNER, B., KRONER, K. H. and PAPAMICHAEL, N. (1987). *Biotechnol. Techniq.* **1**, 49.

International Trade Centre (1986). *Essential Oils and Oleoresins*, Geneva.

JAPIKSE, C. H., VAN BROCKLIN, L. P., HEMBREE, J. A., KITTS, R. R. and MEECE, D. R. (1987). US Patent 4 647 466 to Procter and Gamble Co.

JOHANSSON, G. (1985). *J. Biotechnol.* **3**, 11.

JONES, G. V., MUSTO, J. A. and MEINHOLD, J. F. (1985). European Patent 0 159 829 A1 to General Foods Co.

KALRA, H., CHUNG, S. Y.-K. and CHEN, C.-J. (1987). *Fluid Phase Equil.* **36**, 263.

KANG, C. H. and SANDLER, S. I. (1987). *Fluid Phase Equil.* **38**, 245.

KATZ, S. N. (1987). *Decaffeination of Coffee*. In *Coffee*, Vol. 2: *Technology* (ed. R. J. CLARKE and R. MACRAE). Elsevier, Amsterdam.

KATZ, S. N. and SKIFF, H. (1985). US Patent 4 545 998 to General Foods Corporation.

KAUFMANN, W., BIERNOTH, G., FREDE, E., MERK, W., PRECHT, D. and TIMMEN, H., (1982). *Milchwissenschaft* **37**, 92.

KILEY, J. M. and KILEY R. T. (1974). US Patent 3 809 757 to Nestle Co.

KING, M. B., BOYES, A. P., BOTT, T. R. and MUBARAK, A. D. (1987). *Inst. Chem. Eng. Symp.* Series No. 103, 351.

KRONER, K. H., HUSTEDT, H. and KULA, M.-R. (1984). *Process Biochem.* **19**, 170.

KULA, M.-R., KRONER, K. H. and HUSTEDT, H. (1982). In *Advances in Biochemical Engineering* (A. FIECHTER ed.), Vol. 24, p. 73.

LIEBING, H. (1972). *Seifen-Oele-Fette-Wachse* **98** (12), 359.

MACRAE, R. (1985). *Nitrogenous Compounds*. In *Coffee* (ed. R. J. CLARKE and R. MACRAE), Vol. 1. Elsevier, Amsterdam.

MCPHEE, A. D., DUBROW, D. L. and HENDERSON, L. O. (1972). *J. Chem. Eng. Data* **17** (2), 244.

MITCHELL, W., HUMPHREY, A. M. and FINCHER, J. M. (1972). UK Patent 1 274 678 to Bush, Boake, Allen Ltd.

MONICK, J. A. and TREYBAL, R. (1956). *J. Am. Oil Chem. Soc.* **33**, 193.

MOORE, E. B. (1950). *J. Am. Oil Chem. Soc.* **27**, 75.

Nippon Oils & Fats Company (1984). Japan Patent 84 : 71 396 (*Chem. Abs.* 101 : P56848).

OHGAKI, K. and KATAYAMA, T. (1984). GB Patent 2 152 837 to Kuraray Co., Japan.

OWUSU-YAU, J., MATTHEWS, R. F. and WEST, P. F. (1986). *J. Food Sci.* **51**, 1180.

PARSONS, A. M. (1976). UK Patent 1 444 551 to Unilever Limited.

PETER, S., BRUNNER, G. and RIHA, R. (1976). *Fette, Seifen, Anstr-mittel* **78**, 45.

PETER, S. and TIEGS, C. (1984). Separation Efficiency of Different Packings in the Extraction with Dense Gases in Countercurrent Columns. Paper presented at International Symposium, '*High Pressure Chemical Engineering*'. Erlangen.

PETER, S., SCHNEIDER, M., WEIDNER, E. and ZIEGELITZ, R.(1987). *Chem. Eng. Technol.* **10**, 37.

PRASAD, R., GOTTESMAN, M. and SCARELLA, R. A. (1980). US Patent 4 341 804 to General Foods Corporation.

REMAN, G. H. (1951). *Proceedings of the 3rd World Petroleum Congress*, Section III, The Hague, 1951.

RIGAMONTI, R., VACCARINO, G. and DUZZI, A. (1951). *Chimica Ind., Milano* **33** (10), 619.

ROBEY, R. J. and SUNDER, S. (1984). Paper presented at Annual Meeting of American Institute of Chemical Engineers.

ROYCHOUDHURY, R. N. (1985). US Patent 4 517 120 to Nestec S. A.

RUBLO, F. C., RODRIGUEZ, V. B. and ALAMEDA, E. J. (1986) *Ind. Eng. Chem. Fund.* **25**, 142.

RUYS, A. H. (1957). *Perfum. Essent. Oil. Rec.* **48** (12), 584.

SCHULTZ, W. G., SCHULTZ, T. H., COULSON, R. A. and HUDSON, J. S. (1974) *Food Technol.* **28** (6), 32.

SEROTA, D. G., THAKUR, A. K., ULLAND, B. M., KIRSCHMAN, J. C. BROWN, N. M., and COOTS, R. H. (1986). *Food Chem. Toxicol.* **24**(9), 951.

SIVETZ, M. and DESROSIER, N. W. (1979). *Coffee Technol.* AVI, Westport, Conn.

SMITS, G. (1977). Losses in Alkali Neutralization of Edible Oils. Dissertation, University of Delft.

STAHL, E. and QUIRIN, K.-W. (1985). *Fette, Seifen, Anstr-mittel* **87**, 219.

STEPHAN, K. and SCHABER, K. (1979). *Ger. Chem. Eng.* **2**, 38.

SWOBODA, P. A. T. (1984). UK Patent Application No. 8 416 478.

SZUHAJ, B. F. (ed). (1989). *Lecithins: Sources, Manufacture & Uses*. American Oil Chemists Society.

THOMOPOULOS, C. (1971). *Rev. Cps gras* **18**, 143.

TREYBAL, R. (1963). *Liquid Extraction*. McGraw-Hill New York.

ULUSOY, E., AYDIN, A. and ERGIN, M. (1984). *Chim. Acta Turc.* **12**(1), 60.

WAGENER, W. W. D. and HEARSHAW, K. D. (1986). South African Fish Industry Research Institute, *Technical Report No. 1*.

ZAHER, F. A. and EL-NOMANY, H. M. (1987). *Seifen-Oele-Fette-Wachse* **113** (14), 465–6.

5

SOME CURRENT DEVELOPMENTS AND EXTRACTION TECHNIQUES

L. R. WEATHERLEY

5.1 Introduction

The evolution and development of new liquid–liquid extraction processes has been strongly dependent in the recent past upon the structural changes which have occurred within the chemical industry as a whole. The early 1980s have seen significant changes in distribution of chemicals manufacture and product mix which have resulted from important fluctuations in feedstock patterns and market demand for various products. These changes have given rise to much new process development including major re-examination of separation processes and have also stimulated a new demand for more elegant and complex separation techniques in areas like biotechnology and in the hydrometallurgical industry.

Liquid–liquid extraction currently plays a very important role in four areas of chemical processing:

(1) nuclear fuel extraction and irradiated fuel processing;

(2) hydrometallurgical extractions, including both precious and non-precious metal extractions;

(3) bulk petrochemicals;

(4) speciality products.

In recent years the development of new solvents and improvement in equipment design and flowsheet technology have dominated developments in the first two of this group. However, even more scope for radical changes in separation processes appears to be in prospect in the area of bulk petrochemicals and specialities where there is an expanding role for new product development and replacement of conventional feedstocks with novel renewable resources. A recent review by Malpas (1982) outlines the economic context of changes in these two areas which seem likely to occur in a 20-year period up to the year 2000.

Developments in liquid–liquid extraction in the nuclear industry have been dominated by two criteria: economics and concern for environmental factors. In a recent report, Naden *et al.* (1985) record that uranium prices on the world market have slumped to US $20 per lb and this has been paralleled by a slump in production from an all-time high of 44 000 te in 1981 to 38 000 te in 1983. In view of falling world energy prices, it appears unlikely that this situation will change for some time and the simultaneous fall in price and demand means that only the most economic plant and processes will be competitive. Development of new extraction processes will be directed towards reduction of process costs and increase of profit margins.

Continuing pressure on the nuclear industry worldwide to reduce environmental discharges and demonstrate more efficient waste management techniques has had an important influence on extraction flowsheets and solvent chemistry. Wilkinson (1986) indicates that although the Purex process (and its variants) are likely to continue to dominate irradiated fuel reprocessing flowsheets, the production of tractable wastes, for example salt-free waste liquors, and the removal of long-lived actinides will be development priorities in the liquid extraction process area. The environmental impact of uranium mine tailings and other waste streams has also spurred development activity in leach liquor extraction processes.

The uranium recovery industry has also exploited developments in recovery processes from wet-process phosphoric acid (Flett 1981), utilizing newly developed solvents. This is reflected in a number of new plants coming on-stream utilizing this technology both in the United States and Belgium.

Liquid extraction is also finding increasing applications in other more conventional hydrometallurgical product recoveries. Development of new solvents and modifiers have opened up treatment routes for nickel/cobalt solutions and the commercialization of selective cobalt separation from mixtures where nickel is now established. The precious metals industry has also reaped benefit by the application of solvent extraction to the separation of Platinum group metals. The major benefit here (Barnes and Edwards 1982) is one of process simplification in which complex chemical separations including many precipitation and solid–liquid separations have been replaced by a liquid extraction scheme involving few operations and lower material inventories.

The rapid growth of the biotechnology industries from the late 1970s onwards has had a continuing impact on the role of solvent extraction as a separation technique. Manufacture of specialist products by biological reaction pathways has been well established in the pharmaceutical industry. Here the objective of separation techniques should be seen against a background of relatively small volume but high value products being recovered from very dilute reactor product solutions accompanied by very high purity specifications.

There has been a rapid growth in the application of biological reactions to produce high value products such as antibiotics, enzymes, vitamins, polymer modifiers, specialist agrochemicals, fungicides and herbicides, and flavours. Many of the processes used require a highly selective separation technique and solvent extraction has present and future potential alongside other separation processes. Surprisingly, one of the spurs for the development of new, high-value products has been changes in the chemicals market place worldwide. The recent wisdom (BP 1984) indicates that there will be a shift away from bulk chemicals production in the developed world with greater concentration of production capacity in the resource-rich countries of the Middle East, Malaysia, and Canada. Such changes are already taking place and in response to these changes in production patterns a long-term growth in speciality chemicals production and development of new products and processes will ensure a continuation in the overall growth rate of chemical production in financial terms in the developed world. This shift will be accompanied by significant changes in world trade with large increases expected.

It is against such a background that a vigorous search for new and more sophisticated products and manufacturing processes is taking place in the developed world and the use of biotechnology exploiting developments in genetic engineering (OTA Report 1982), undoubtedly has a key role to play. The effective application of separation processes technology to this area is vital for success, and solvent extraction alongside other techniques has an important role to play.

A further major factor which is likely to affect the development of new solvent extraction processes is the longer term potential application of biological processes for the production of bulk organic chemicals. This will be especially important if the diversification of feedstock patterns away from crude oil-based sources continues its momentum. Most forecasts (Malpas 1982) assume an important medium-term role for C_1-based technology relying on coal and methanol and on this basis downstream processes are unlikely to experience a fundamental change for some time. This does not exclude the continuing improvement of existing process technology by the development of better solvents and the enhancement of recovery and pollution-control techniques.

It is in the longer term scenario of shrinking fossil feed-based feedstocks where a shift to biochemical processing of renewable feedstock materials may become significant.

A wide range of renewable materials, such as starches, carbohydrates, glucose, cellulose, and lignins are potential renewable feedstocks for biochemical conversions. The major constraint (Porter 1982) of such routes is the dilution of the final broth mixture and the consequent high-cost water-removal problem which would conventionally involve distillation.

A few major organic chemicals are currently produced by fermentation. Citric acid (Roberts 1970) and ethanol (as gasohol) are two primary examples.

The microbiological techniques for the production of many other organic molecules already exist but exploitation is inhibited by economic constraints while fossil fuel-based feedstocks abound. The exception is the use of ethanol produced by fermentation as a fuel extender (as gasohol) in certain localities such as Mexico and Brazil. Ethanol is thought to have an important future role as an intermediate chemical which can be readily converted to ethylene by straightforward vapour phase catalytic dehydration.

5.2 Nuclear processes

[See also Cox Chap. 1, Batey Chap. 2]

5.2.1 Introduction

The development of new technology in solvent extraction processes in the nuclear industry is governed to some extent by rather more acute constraints compared to other areas of application. Two overiding considerations are likely to continue to dominate future developments, namely safety and environmental protection and process economics.

Significant changes in the public perception of the nuclear industry generally in the recent past have given rise to a much increased emphasis on improvement in waste management strategies, particularly with regard to the level of routine radioactive discharges and to the storage of long-lived radioactive waste. These new waste management objectives have spurred applications for novel solvent extraction processes and have given an additional impetus to the improvement of existing fuel separation processes. Such objectives include the separation and concentration of long-lived actinide or transuranic elements from active waste streams prior to discharge or storage. The desirability of salt-free processing in order to achieve greater concentration of liquid wastes is also a highlight of recent developments and the importance of this philosophy has been readily grasped by those involved in fast reactor fuel cycle development.

Changes in the economic climate, particularly with reference to the uranium market, will influence the direction of new developments in uranium extraction flowsheets. Naden *et al.* (1985) indicate how the price of uranium has fallen from \$44/lb in the late 1970s to less than \$20/lb in 1985. This fall not only leads to a restriction in the installation of new capacity but also stimulates innovatory techniques to improve processing. Recent methods of recovering uranium as a by-product of existing processes are looking more desirable compared to new mining and concentration operations where the mining and beneficiation costs can be 70 per cent of the total.

Developments in the application of solvent extraction processes in the fast reactor fuel cycle continue to be of importance in the nuclear programmes of many countries, and in light of the long lead times involved in these projects, it appears likely that this emphasis will continue. The basic economics of the fast reactor fuel cycles determines that reprocessing is an essential ingredient of the process. Improvements in existing solvent extraction flowsheets, in order to ensure reliable plant performance and product quality with an ever expanding requirement to fulfill tight waste management criteria, are likely to continue.

5.2.2 Uranium ore processing

5.2.2.1 *Solvents/Diluents*

Tertiary amines are the most widely used solvents for extraction of uranium from sulphuric acid leach liquors. Co-extraction of molybdenum results in major contamination but separation by carbonate stripping of loaded solvent is now a recommended route. Vanadium is also separated with molybdenum (Sialino 1977).

Solvent extraction is also employed as an adjunct to ion exchange where chloride eluates are obtained (Ritcey *et al.* 1981) from an ion exchange cycle and may be treated by alkyl phosphate amines or trioctyl-phosphine oxide (TOPO). The latter solvent is promising on account of favourable kinetics and there is potential application for uranium extraction from sulphuric acid solutions.

A major finding in recent years is the role of the diluent in determination of extraction performance (Blumberg and Gai 1977). Aromatic/aliphatic ratios, napthene content, and solubility parameters of the diluent all show possible effects on solvent performance (Ritcey 1985).

5.2.2.2 *Product stripping*

Product stripping appears to have been of lower research and development priority compared to primary extraction of uranium into solvent. A number of stripping agents are used in uranium processing including sodium carbonate, ammonium sulphate, sodium sulphate, ammonium chloride, and ammonium carbonate.

Ammonium compounds are favoured because they are often a by-product of ADU precipitation but poor pH control is sometimes unavoidable in the solvent extraction circuits leading to precipitate formation and resultant interfacial cruds and solvent losses.

Ritcey (1985) lists compatibility criteria against which the stripping technique should be judged: extractant degradation, final desired product, product specification, precipitation or hydrolysis of co-extracted impurities, ease of pH control, matching phase separation characteristics with design of equipment, and economics of stripping solution recycle after precipitation of the product.

5.2.2.3 *New processes*

The main incentive for development of new processes is the reduction of environmental impact, particularly in relation to the management of Ra^{226}. A second but less demanding incentive is to reduce the amount and hence the cost of solid–liquid separation. This may account for up to 50 per cent of the capital and operating costs of a typical uranium mill. A third aspect of new process development has been the desire to exploit the recovery of valuable by-products such as arsenic, nickel, and vanadium, a point especially important when seen against a background of falling uranium prices.

The key to tailings clean up lies in the ability to solubilize the Ra^{226} at an early stage in the process. The widely used sulphuric acid leach process does not do this and alternatives include nitric acid leaching (Skeaff 1975), hydrochloric acid leaching (Ritcey *et al.* 1973), and a chlorination roast-leach process (Ritcey 1985) (Lapage and Marriage 1973) (Skeaff and Laliberte 1980).

An appraisal of these process possibilities is given by Ritcey (1985) and may be summarized as follows. The use of nitric acid leach significantly improves solubilization of radionuclides and potentially reduces the Ra^{226} content of tailings by 75 per cent. The level is still considered to be high compared to that potentially achievable by other processes.

The chlorination roast leach process consists of a two-stage chlorination at 400°C to volatize uranium chloride followed by a second stage at 600°C to achieve conversion of radium to radium chloride (Suzuki *et al.* 1968).

The major portion of the uranium as uranium chloride would be condensed and purified conventionally. The roast solid arisings are leached with dilute HCl to solubilize radium chloride and any remaining uranium and rare earth metals. This liquor is treated to isolate the radium and the uranium sent to the main uranium recovery circuit. Tailings residues from this process are a factor of 20 leaner in Ra^{226} compared to those arising from conventional sulphate leach.

The HCl leach process involves leach in HCl with the addition of sodium chlorate as an oxidant. The tailings contain a 1/50th of the level of Ra^{226} compared to sulphate leach. Uranium recovery is achieved by solvent extraction into TOPO followed by stripping with H_2SO_4/H_2O_2. The uranium is precipitated as uranyl peroxide using H_2O_2. Radium removal and isolation are achieved by ion exchange.

A further possibility is the modification of the existing H_2SO_4 process (Fig. 5.1). Leaching is conducted under oxidizing conditions at pressure allowing conversion of sulphides to H_2SO_4 which facilitates uranium leaching. Radium is also dissolved under these conditions. Uranium is recovered at an early stage by solvent-in-pulp extraction. *In situ* extraction from the leach slurry by this process, it is believed, will allow scope for significant reductions in the cost of solid–liquid separation. The extracted uranium is

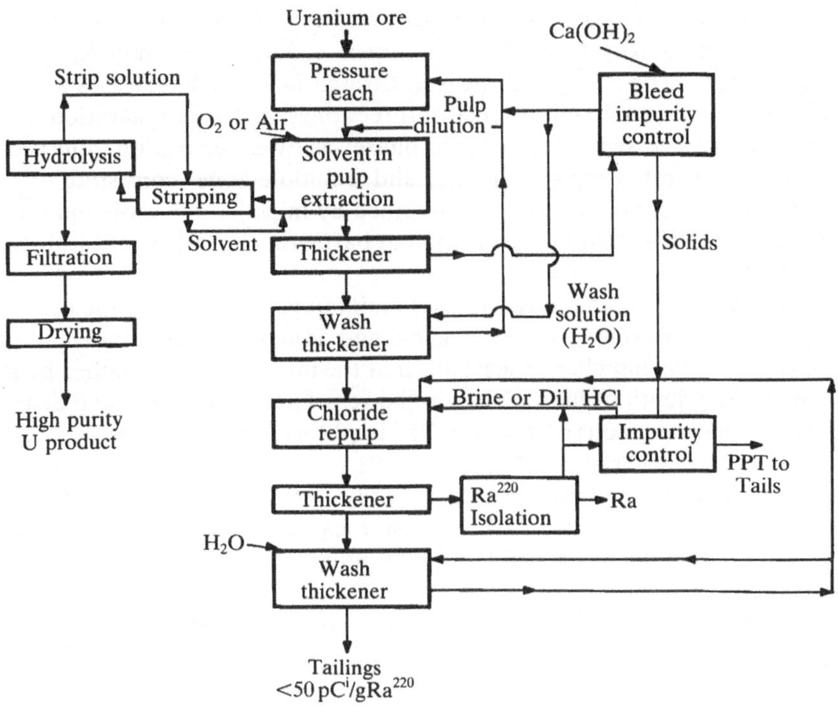

Fig. 5.1. Future uranium flowsheet incorporating solvent in pulp leach, strip hydrolysis and solution recycle. (From Ritcey 1985.)

stripped and converted by hydrolysis. It is claimed that this process offers may advantages not only in terms of cleaned-up tailings but offers higher uranium recovery values compared with other processes together with improved prospects for by-product recovery.

5.2.3 Equipment

Mixer settlers of both the gravity type design and the pulse mix type dominate the scene in uranium extraction plants. While simplicity of construction and operation are advantageous, conventional mixer settlers suffer from a number of disadvantages. These include: large size per unit throughput, tendency to encourage emulsion formation due to high shear rates required for good contact efficiency, and requirement of a high degree of feed clarification, high solvent inventory.

Two developments in this area are worth mentioning. The Davy McKee combined mixer settler (Scuffham 1981) (CMS) is designed to combine the advantages of reduced overall size per unit throughput with elimination of minority phase recycle and a consequent reduction of impeller power requirements. The CMS concept is illustrated in Fig. 5.2 and uses a single vessel which allows the coexistence of three zones: an upper separation zone, consisting of the separated organic phase; a lower separation zone, consisting of separated aqueous phase; and a middle zone, consisting of the mixed phase. The dispersion is maintained by impeller agitation and phase separation is facilitated by the location of baffles at either edge of the dispersion band.

The aqueous : organic phase ratio is maintained according to the position of the settled interface between organic and aqueous phases relative to the centre line of the impeller. The position of the interface is determined by the densities and depth of both phases in the vessel and by the level and density of liquor in the aqueous offtake leg. The implication of this behaviour is that the effective phase ratio within the middle zone is independent of the feed phase ratio and it can be controlled simply by adjustment of either outlet weir. Thus costly recycle control provision can be avoided and elimination of phase recycle means a smaller overall unit compared with conventional mixer settlers.

Successful incorporation of CMS into a commercial uranium extraction process has been achieved. A further benefit of significant commercial importance is the tolerance of the unit to levels of suspended solids in feed solutions. A particularly beneficial aspect of performance under these conditions is that the level of solids in the dispersion does not continually increase but reaches a state of dynamic equilibrium. Thus under a constant feed of

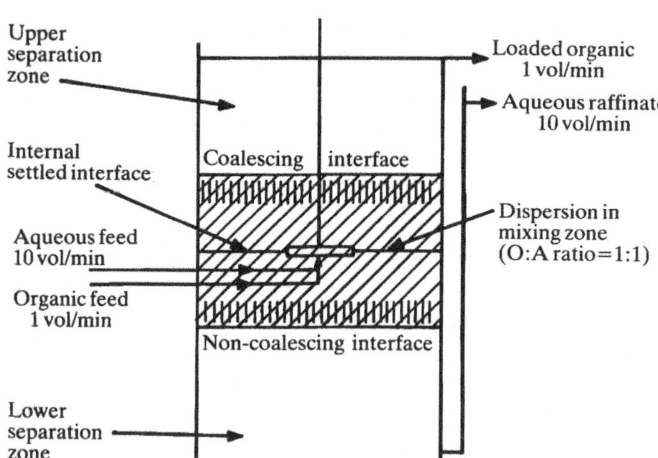

Fig. 5.2. The Davy McKee combined mixer settler. (From Schuffham 1981.)

suspended solids the solids content of feed and exit streams attain parity. It is reported that solid contents in the range 100–300 ppm can be accommodated without accumulation.

Reduction in the organic phase inventory, the reduction in organic/air interfacial area, and the ease of containment also lead to lower fire risks.

The use of continuous (differential) contact equipment is long established in the nuclear industry in the context of relatively small throughput, high-value uranium fuel reprocessing plants. Their use in uranium ore processing has been less widespread due to the greater fabrication and control costs compared to mixer settlers. Desson *et al.* (1984) examined the possibility of using pulsed plate columns in high throughput environments and found two major drawbacks. Firstly, column diameters in excess of 2 m led to a sharp increase in HETS values. Secondly, the cost of perforated plate fabrication at large diameters becomes excessive. Replacement of perforated plates with baffle plates (see Fig. 5.3) showed that consistent transfer efficiencies could be maintained over a wide range of column diameters. The extraction of uranium into 30 per cent TBP-dodecane from nitric acid solutions in pulsed

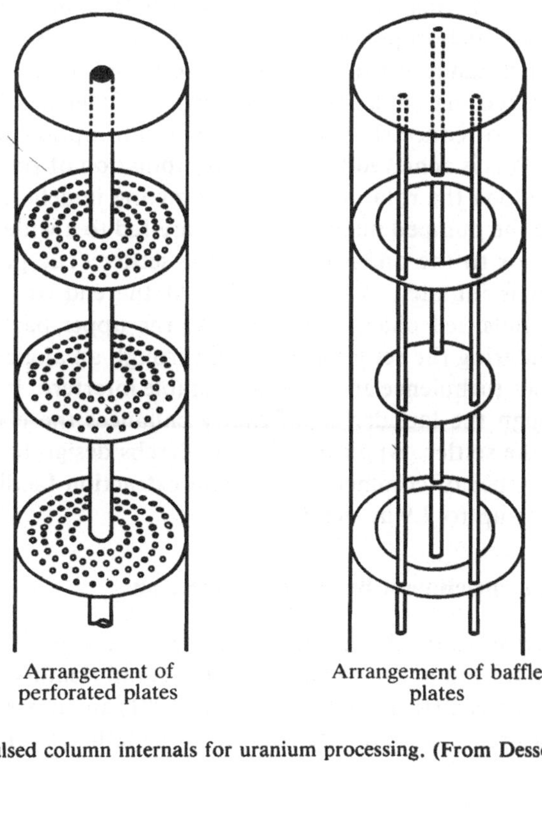

Fig. 5.3. Pulsed column internals for uranium processing. (From Desson *et al.* 1984.)

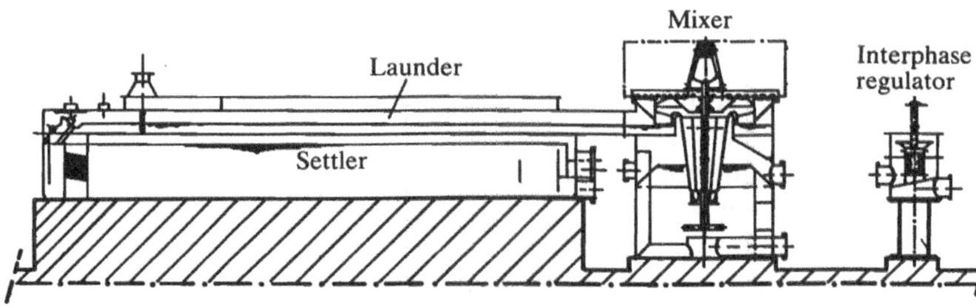

Fig. 5.4. The Krebs mixer-settler. (From Sonntag *et al.* 1984.)

columns fitted with baffle plates compared very favourably with extraction conducted in mixer settlers. Use of these columns for plants processing uranium-bearing solutions at feed rates up to 170 m^3/h is recommended and it is claimed that the design shows significant promise for scale-up to feed rates of 200 m^3/h.

Improvements in the design of mixer settler units for uranium processing have also continued, particularly to increase coalescence efficiency and to improve mixer performance. The Krebs system (Sonntag *et al.* 1984) is a good example of such improvements. The Krebs mixer/settler is shown in Fig. 5.4 and it is made up of three main sections: a mixing chamber; a launder; and a settling chamber. The mixing chamber is a high efficiency device designed to reduce short circuit flow and to optimize mass transfer performance. This is achieved by a novel combination of rotating and fixed cones located along the rotating axis of the mixing impeller. The dispersed phases leaving the pumped mixing chamber are fed to a wide launder which directs the liquor to the opposite end of the settler. The process of phase disengagement is initiated in the launder. At the end of the launder the organic and emulsified phases are routed to the upper part of the settler. Both phases entering the settler are routed through an angled plate distributor to cut down turbulence and produce a stable interfere. It is claimed that pre-flow through the launder significantly enhances the efficiency of the separation in the settler. Application of the Krebs design has been demonstrated at a number of commercial uranium extraction facilities at specific liquid flowrates up to 7.9 $m^3/h\,m^2$.

5.2.4 Isotopic enrichment by solvent extraction

The enrichment of natural uranium to increase the percentage of U^{235} is a major part of uranium fuel manufacturing. Traditionally, this has been achieved by gaseous diffusion (Benedict *et al.* 1981) and more recently by gas centrifugation processes. The use of chemical methods for isotopic sepa-

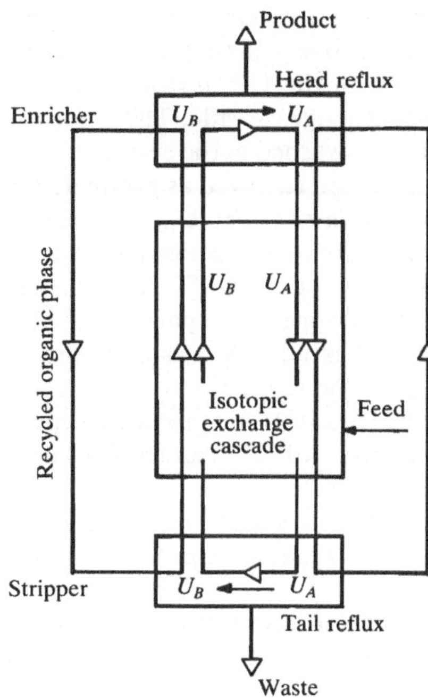

Fig. 5.5. Isotopic separation of uranium 235 by the Chemex process. (From Lerat *et al.* 1984.)

rations has been considered unfavourable because of slow kinetics and poor isotopic selectivities. In 1977 the CEA released details of a new process based on solvent extraction (Frejacques *et al.* 1977) in which the possible use of crown ether compounds as selective extractants for U^{235} was suggested. The process which was developed, known as CHEMEX (Coates 1982) (Lerat *et al.* 1984), is shown in principle in Fig. 5.5.

The process is based on a selective isotopic effect during the exchange reaction between two ionic species of uranium represented thus:-

$$U_A^{235} + U_B^{238} \rightleftharpoons U_A^{238} + U_B^{235}$$

The magnitude of the separation factor for the U^{235}/U^{238} separation is reflected in the magnitude of the reaction equilibrium constant. The valence states of the species U_A, U_B shown to exhibit a satisfactory separation factor are U_{III} and U_{IV} respectively. The enriched reaction products are separated from the depleted reactants by solvent extraction which is necessarily rapid to avoid the species present undergoing further reaction. The overall rates of the exchange are governed by reaction kinetics and by the rate of species transfer across the liquid–liquid interface. The enrichment is achieved by extracting the U_{IV} species selectively into an organic solvent and on the basis of the above reaction equilibrium, the U_{IV} species so extracted will be enriched in the U^{235} isotope. This primary extraction is achieved in the isotopic exchange cascade. The loaded solvent existing from the top of the cascade is backwashed in the enricher section and a proportion of the resulting uranium stream removed as product. A reflux stream is subjected to electrolytic reduction of U_{IV} to U_{III} and acidification whereupon it is returned to the isotopic exchange section. At the lower end the exiting aqueous stream is subject to oxidation of U_{III} to U_{IV} which is extracted into the recycled solvent in the Stripper. The partially loaded recycled solvent is returned to the exchange cascade. A waste stream, depleted in U^{235}, is rejected as tail reflux from the stripper.

Scale-up of the process to industrial scale is anticipated to use pulsed column contactors up to diameters of 1.2 m and heights up to 24 m. A flowsheet for a basic CHEMEX 500 000 s.w.u/annum industrial plant module is shown in Fig. 5.6. Such models could provide a basic building block for enrichment complexes of capacities up to 3×10^6 s.w.u/y. (Note: 1 s.w.u. = separative work unit.)

The economics of the solvent extraction processes based on CHEMEX bear favourable comparison with enrichment by gaseous diffusion.

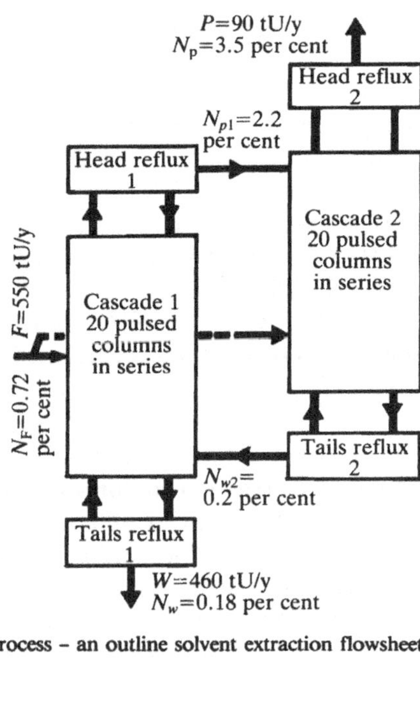

Fig. 5.6. The chemex process – an outline solvent extraction flowsheet. (From Lerat *et al.* 1982.)

5.2.5 Nuclear waste treatment processes
[See also Batey Chap. 2]

The concentration and conversion of wide categories of active wastes arising from fuel reprocessing operations involve wet chemical processes. With conventional solvent extraction flowsheets the efficiency of waste treatment is partially judged by the volume reduction factor based on the ratio of feed solution volume to that of the strip solution. Carley-Macauly (1985) indicates that this ratio seldom exceeds a factor of 10 and thus concludes that solvent extraction is of limited use for the treatment of low activity aqueous wastes.

The significant potential area of application of solvent extraction for rad waste clean up appears to be for the removal of actinides from highly active liquid wastes prior to solidification and storage. Actinides of importance here are the trivalent species of Americium and Curium, in addition to smaller amounts of Neptunium. Environmental pressure to achieve separation of these long-lived actinides from shorter-lived fission products has given rise to significant developments in separation techniques. New hazard data on the radioactivity of neptunium 237 (Guillaume *et al.* 1984) are of particular interest here. The transfer factors from gastrointestinal tract to blood for neptunium as ($Np.O_2^+$) are higher than previously thought and make it one of the more dangerous long-lived nuclides. Due to its ionic form, it exhibits possible earlier migration in underground environments which makes the radioactivity hazard more pertinent.

One recently developed technique is the TRUEX process which uses a modified solvent in a PUREX type solvent extraction flow scheme (Horwitz and Schulz 1985). Reprocessing of irradiated nuclear fuel by the PUREX process results in highly active raffinate solutions containing the majority of actinide waste. The TRUEX process (Fig. 5.7) is a flexible actinide removal process and is applicable to a wide range of waste solutions containing nitric acid salts and fission products. The feed solution is introduced into a conventional extract/scrub system in which the scrub solution contains oxalic acid to suppress extraction of zirconium and molybdenum. In the extraction section, the majority of the americium, plutonium, and uranium, together with a major fraction of the neptunium, are extracted. The solvent is TBP in an appropriate diluent such as kerosene or carbon tetrachloride modified with a small amount of oxtyl (phenyl)-N, N-diisobutyl carbamoylmethylphosphine oxide (CMPO). The CMPO additive enables extraction of trivalent actinides in addition to the readily extracted tetra and hexavalent species of the actinides. Almost 100 per cent of the fission products, oxalic acid, and inert species are rejected in the raffinate stream. The loaded solvent goes forward to strip unit 1 in which the americium and any technetium are backwashed into aqueous phase with 0.05 m nitric acid. The neptunium and

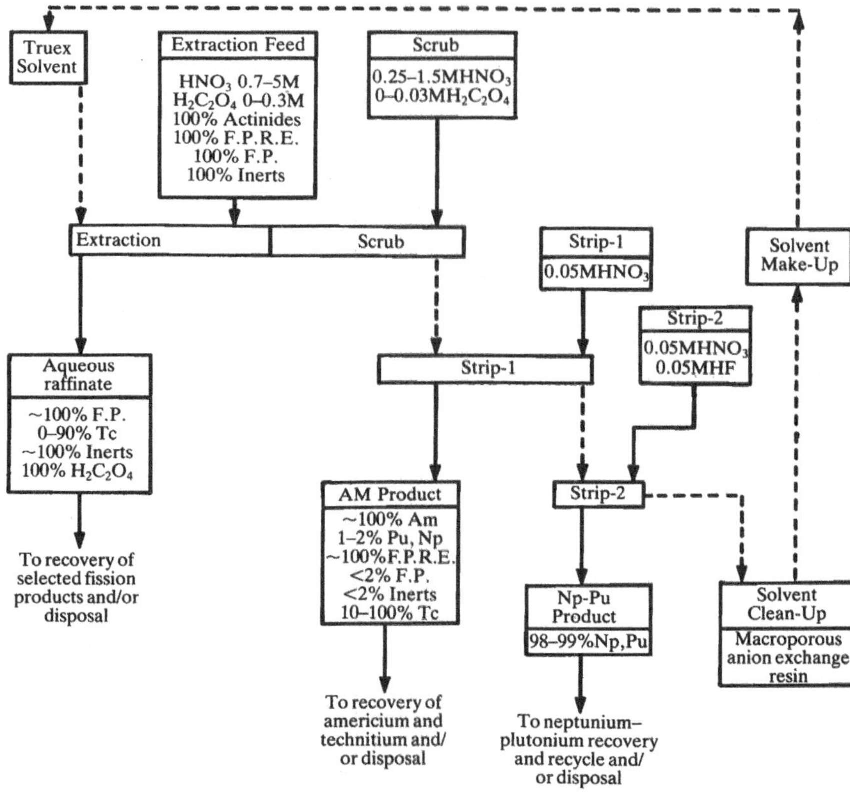

Fig. 5.7. Generic truex process flowsheet. (From Horwitz 1985.)

plutonium are backwashed separately in strip 2 with a mixture of 0.04 m nitric acid/0.05 m HF. Solvent clean up is achieved using ion exchange uptake of solvent degradation products followed by conventional sodium carbonate washing.

Decontamination factors of 5×10^4 for americium, $\gg$ 10 for uranium and plutonium, and 6 for neptunium have been demonstrated with a total reduction of actinide levels by a factor of 150.

The major advantage of this process is the similarity to the PUREX process with similar flow configurations and solvent characteristics. Minimum requirement of feed acidity adjustment makes process control relatively easy. This latter feature contrasts favourably with more complicated schemes put forward elsewhere. For example, Cecille *et al.* (1979) suggest an actinide removal (Fig. 5.8) process which requires denitration of the feed in order to reduce acidity levels to 0.1 m HNO_3. This is followed by extraction

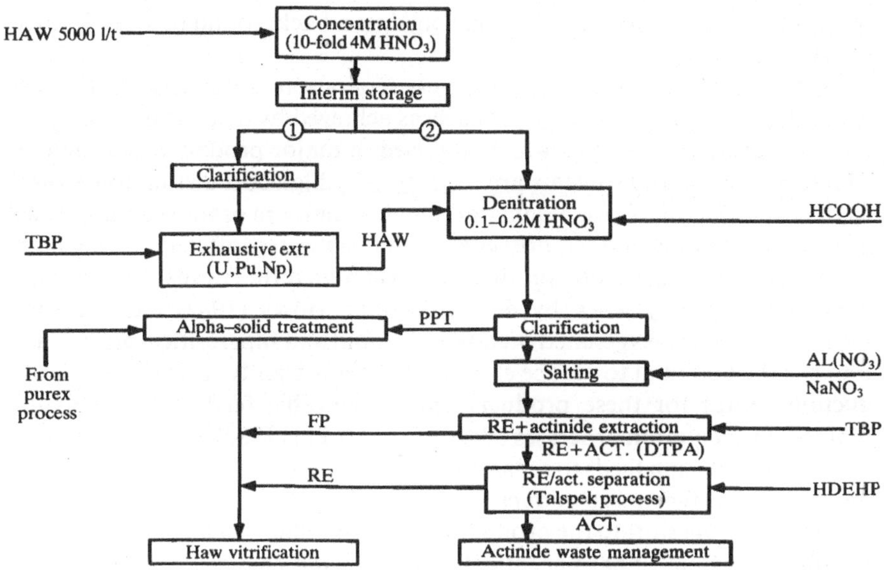

Fig. 5.8. Actinide removal from nuclear wastes by solvent extraction. (From Cecille *et al.* 1979.)

of rare earth species and actinides into bis-2-ethylhexyl phosphoric acid (HDEPH). Subsequently, americium and curium are backwashed selectively with a solution of di-ethylene triamine pentacetic acid (DPTA) and any remaining plutonium and neptunium stripped with oxalic acid solution.

5.3 Extraction using liquid and supercritical gases
[See also Verrall Chap. 3, Hamm Chap. 4]

5.3.1 Introduction

The use of liquified and supercritical gases as extracting solvents has attracted interest in recent years again as a commercial tool for the recovery of a wide range of products. There are two main reasons for this resurgence of interest. Firstly, many of the systems which demonstrate potential for application are essentially non-toxic and do not adversely contaminate the product being recovered. The second reason is that economic comparison of gas extraction processes with conventional distillation yields favourable results when based on energy consumption costs. Both of these benefits are of some importance when seen against a background of increasing interest in the manufacture of speciality products often requiring separation from sensitive environments and when seen against a background of significantly rising energy costs.

In the chemical processing industries, separation and purification from reaction products in a majority of cases is achieved by distillation. Many of the basic organic chemicals which are used in major product areas such as fibres, polymers, and solvents are recovered by high cost distillation. Some of these, such as alcohols, aldehydes, and ketones, are produced initially as an aqueous solution and require separation from water, often involving high latent heat costs and azeotropic distillation at high reflux ratios. Preliminary comparative survey work by de Filippi and Moses (1982) indicate that recovery of water-oxygenated organics by liquid and supercritical gas extraction has the potential to reduce energy costs significantly. Longer term production routes for these products from renewable feedstocks involving biomass fermentation processes (Porter and Jenkins 1982) will result in even more dilute solution feeds to separation processes, increasing the demand for more energy-efficient techniques.

The increasing shift to the production of high-value products such as food additives, flavours, pharmaceutical products, and enzymes has demanded greater consideration be given to product sensitivity and denaturing. Good product recovery processes must in many cases yield a clean product which is potable and does not produce undesirable interactions or side effects in its end-use environment. Use of organic extracting solvents may not be always acceptable if trace amounts of solvent remain with the product. Liquid–supercritical gas extraction systems involving non-toxic species such as carbon dioxide also have potential attractions in this respect.

5.3.2 Principle of liquid/supercritical gas extraction

The principle upon which supercritical gas extraction is based is well known. The concept of a supercritical state may be qualitatively described in terms of the critical temperature, T_c, which may be defined as the temperature above which it is not possible to liquify a gas regardless of the magnitude of the applied pressure. The critical pressure, P_c, may be defined as the pressure required to liquify the vapour or gas at the critical temperature. Examination of critical data for a range of gases (Nelson 1979) reveals that a wide range of operating conditions are available using different solvents (see Table 5.1).

Consideration of a pressure-temperature phase diagram for CO_2 (see Fig. 5.9) shows the typical range of conditions which allow the existence of both supercritical liquid and subcritical liquid CO_2. The range of temperatures in this case is large and the choice of operating conditions requires

TABLE 5.1 *Operating condition for different solvents*

	P_c (bar)	T_c (K)
Pentane	33.75	569.8
Butane	37.97	425.2
Propane	42.57	370.0
Ethane	48.84	305.4
Ethylene	51.17	283.1
Carbon Dioxide	73.82	304.2
Sulphur Dioxide	78.84	430.7
Ammonia	112.77	405.5
Water	221.19	647.4

examination of the manner in which solvent power of the fluid varies with temperature and pressure.

Liquid CO_2 (subcritical) exists between the temperature of −55°C and +31°C and within the pressure range of 5–74 bar (Brogl 1982). As such, liquid CO_2 can be handled as any other liquid and is miscible with most hydrocarbon solvents such as alcohols and alkanes. The solubility of water

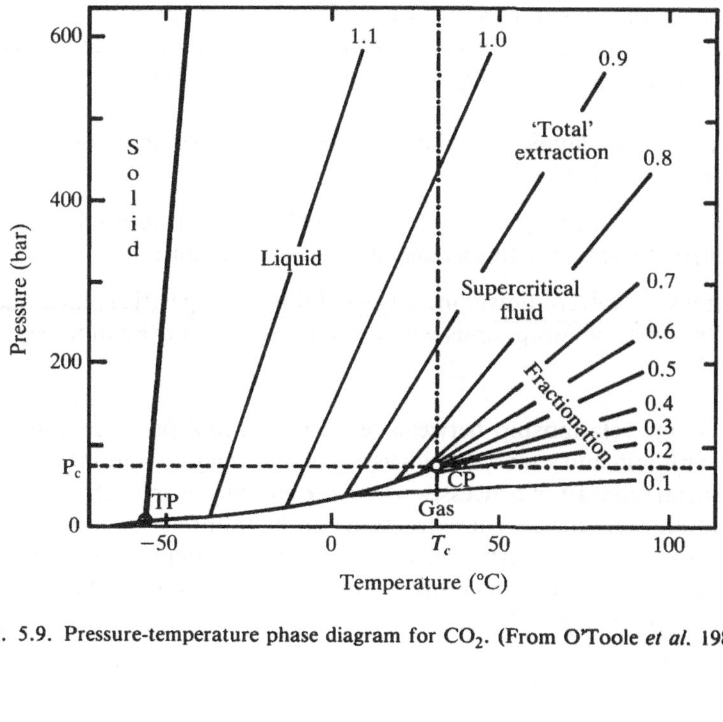

Fig. 5.9. Pressure-temperature phase diagram for CO_2. (From O'Toole *et al.* 1986.)

in liquid CO_2 is approximately 0.1 per cent, which indicates the potential of the solvent for the recovery of organic compounds from aqueous solutions.

Examination of the liquid region of the phase diagram shows that the liquid can be vaporized by heating at constant pressure or by reducing pressure at constant temperature; in other words, the physical behaviour is analogous with that of any other liquid and vaporization or distillation could be used as a technique for solute recovery following extraction.

In the supercritical region use of vapour–liquid separation as a recovery technique is no longer possible since, by definition, liquid and vapour phases no longer coexist above the critical point. Apart from this, supercritical fluid behaviour as an extracting solvent is similar to that of subcritical liquid. Brogl summarizes the general rules which govern the extraction behaviour of supercritical solvents:

(1) solvent power of a supercritical solvent increases with density at a given temperature;

(2) solvent power of a supercritical solvent increases with temperature at a given density.

These rules have been validated for pressures up to 100 bar and temperatures up to 100°C and it is also shown from experience of sub/supercritical fluids that, as a general rule, there is no step change in physical properties during the transition from sub to supercritical state or vice-versa.

Liquid and supercritical CO_2 therefore behaves in principle as any non-polar liquid solvent according to the following general rules:

(1) physical and chemical similarity of solvent and solute is consistent with higher solubility;

(2) for a range of compounds of comparable polarity, solubility in a given solvent decreases with increasing molecular weight;

(3) increase in solvent power increases solubility of specific solutes and also the number of compounds which exhibit selectivity (Brunner and Peter 1983).

The strong relationship between temperature and pressure and solvent power for liquid and supercritical gases has significant potential application to industrial separation processes where the conditions in both recovery and purification cycle of solvent extraction can be tailored to optimize overall performance. There is particular interest in application to the separation of fractions of compounds in complex mixtures; for example, those encountered in flavours and perfumes.

The transport properties of fluids in the subcritical and supercritical regions are also favourable for the kinetics of extraction. The diffusional properties of these fluids are about an order of magnitude greater than for normal liquids (De Filippi and Moses 1982). The diffusivity in high-pressure liquids or supercritical fluids is less than in the gas phase but is much higher than for normal liquids. The implications for contactor size and stage efficiency values are very favourable.

Other thermophysical properties of fluids in the region of the critical point are also favourable particularly when product recovery from loaded solvent is considered. For supercritical fluids, latent heats are zero and are low for subcritical liquids. This allows product recovery by distillation at low energy cost compared to conventional solvent distillation. Heat capacities in the supercritical regions are several-fold greater than for liquids and have a positive influence on heat transfer coefficients in reboilers and condensers.

Lower solvent densities, compared to conventional organic solvents (Hellyar and De Filippi 1982), can mean more efficient phase separation in contact equipment. Generally, hydraulic and mass transfer performance is enhanced by favourable density and viscosity values. In the case of CO_2-based systems, photographic studies of droplet behaviour indicate a similarity to gas bubble behaviour.

5.3.3 Applications

Two basic categories of separation lend themselves to the use of liquids or supercritical fluids as extraction agents. Processes based on conventional

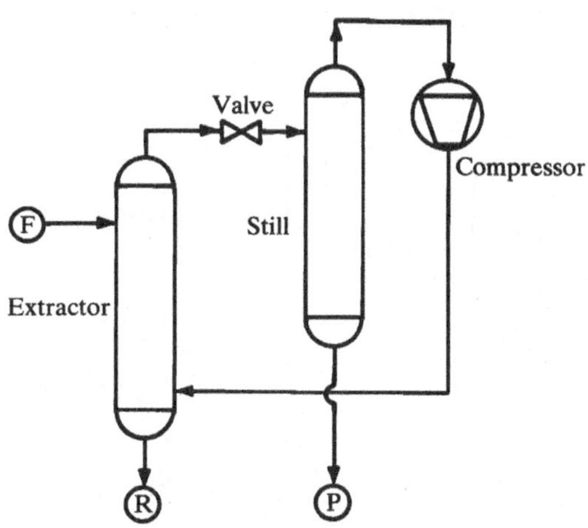

Fig. 5.10. Liquid–Liquid extraction with liquified gas and supercritical fluid solvents. (From Hellyar and De Filippi 1982.)

liquid–liquid extraction flowsheets and equipment have a wide range of potential applications. A typical outline flowsheet based on liquid extraction followed by product recovery is shown in Fig. 5.10. The liquid feed is pumped to the extractor unit which may be either a stagewise or a differential device and is contacted in countercurrent mode with the liquid extractant. For most liquid gas extractants this extraction will be conducted at relatively high pressure (see Table 5.1). The recovery of the product is achieved on the basis of high volatility differences between solvent and product and either flash distillation or conventional fractionation is used. The schematic process shown in Fig. 5.10 shows that the loaded extract is throttled to a lower pressure prior to distillation, reducing the volatility of the product to a much lower value compared to the solvent. The solvent is recompressed and recycled to the extractor.

The second type of separation is akin to a leaching process in which the liquid gas–supercritical fluid is contacted typically with a fixed bed of solid feed which contains the desired product incorporated in the matrix. Figure 5.11 shows an outline flowscheme for such a process and illustrates the two-stage nature of the extraction and product recovery cycle. The loaded solvent exits from the bed of solid feed, passes through a pressure reduction valve and into a gas–solid separator. The solid product is removed and the solute lean gas is recompressed and recycled to the extraction vessel. This type of process relies heavily upon the pressure dependence of solubility referred to previously. It is generally true (Hellyar 1982) that the solubility of some products is a strong function of pressure and temperature for certain supercritical fluids and that in the case of extraction from solids this effect is even stronger. Thus, recovery of a solid product is more favourable for liquid–solid extractions but is also possible for liquid–liquid systems. Thus, the

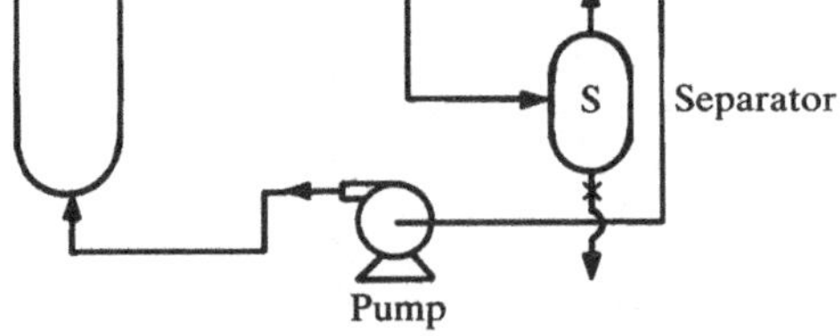

Fig. 5.11. Liquid–solid extraction with liquified gas and supercritical fluid solvents. (From Hellyar and De Filippi 1982.)

system conditions in the extraction vessel and the separator are heavily dependent upon the solubility–pressure–temperature relationship. Generally, the extraction is conducted at high pressure, corresponding to high solubility, and precipitation of the solid product achieved by pressure reduction. In many cases pressure reduction entirely within the supercritical region is adequate to effect the desired fall in solubility thus avoiding phase changes.

The specific applications of liquid–supercritical fluid extraction of significant potential cover a wide range of systems. Those of current commercial interest cover a more specific range of products. There are three types of system in which liquid gas extractions could be of importance:

(1) recovery of organic chemical products from dilute aqueous solutions such as those resulting from fermentations. These are liquid–liquid systems in general;

(2) flavour extraction and concentration which involves solid–liquid extraction;

(3) waste clean-up processes and regeneration.

5.3.3.1 *Organics extraction from aqueous solutions*

Much of the work in this area has been reviewed extensively by De Filippi and Moses (1982), Victor (1983), De Filippi and Moses (1982), and Randall (1982).

Extraction of organic solutes from aqueous solutions by direct extraction is attractive in terms of energy utilization compared to separation by conventional distillation. The recovery of a range of organic solutes is feasible using liquid–supercritical fluids and CO_2 is a major candidate. The recovery of the organic species, which may be numerous, is achievable by extraction. Separation and purification are achieved by CO_2 distillation involving heats of vaporization for CO_2 typically at the level of 65 Btu lb^{-1}. This compares to heats of vaporization for aqueous phases of the order of 1000 Btu/lb. Thus, in principle, energy requirements can be much reduced, this reduction being particularly favourable for feeds containing relatively low concentrations of organics in aqueous solutions.

Many of the fundamental data for these systems were published by Francis (1952, 1954). Francis (1954) published a comprehensive list of solubilities of both paraffinic and aromatic compounds in liquid CO_2 at 25°C and 65 atm. In addition, many data for ternary equilibria for organic-liquid CO_2-water systems are available.

For example, ternary data for the systems ethanol–water–carbon dioxide, isopropanol–water–carbon dioxide and secondary butanol–water–carbon dioxide are shown in Fig. 5.12. These show the limiting solubilities of each

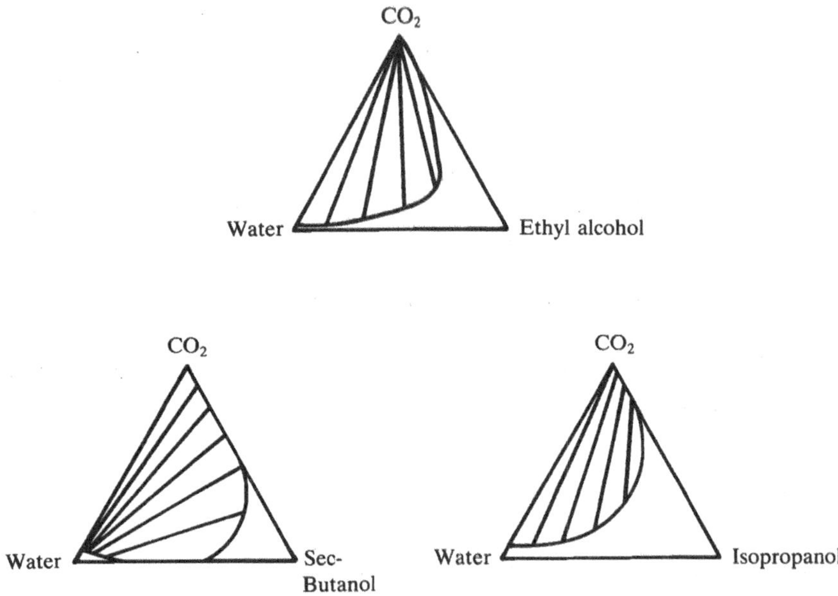

Fig. 5.12. Ternary equilibrium data with liquid CO_2 extractant. (From Francis 1954.)

component in each other and indicate the distinction between single-phase and two-phase regions. Beneath the phase envelope in the two-phase region, the slope of the equilibrium tie lines indicate the magnitude of the distribution coefficient. In the ethanol–water–CO_2 system, the CO_2-rich phase contains lower ethanol concentrations than the aqueous phase and distribution coefficients are less than unity. The diagram for sec.butanol–water–CO_2 shows different behaviour, with the tie line data indicating higher distribution coefficients and the phase envelope restricted by two further points of limiting solubility on the sec.butanol–water axis.

Comparison of the three diagrams shows that the affinity of the CO_2 phase for the organic component increases as molecular weight. De Filippi and Moses (1982) confirmed experimentally that generally the distribution coefficient increases for a homologous series as molecular weight increases. This is entirely consistent with an increasing affinity between CO_2 and organic components as the hydrophobicity of the organic molecule increases.

In terms of process economics the magnitude of distribution coefficient is inversely proportional to the minimum solvent rate required. Thus, on the basis that capital and running costs are approximately proportional to the solvent : feed ratio, the application of liquid gas–supercritical gas extraction will be economically more favourable where hydrophobic species are to be recovered.

Although no commercial processes for liquid phase organics recovery are reported, successful pilot plant scale demonstration of liquid gas extraction of ethyl alcohol involving extraction and product recovery has been described by De Filippi and Moses (1982). The flowscheme used is shown in Fig. 5.13 and the essential elements comprise a countercurrent extraction column, a carbon dioxide distillation column, and a gas compressor. The aqueous feed solution is pumped into the top of the extractor which is based on a perforated plate design. The extracting CO_2 stream is pumped countercurrent up the column and is discharged loaded with ethanol at the top. The aqueous raffinate is discharged following pressure let down. The loaded CO_2 stream, which leves the column at between 1000–2000 p.s.i.g., also undergoes pressure reduction to approximately 750 p.s.i.g. At this pressure a three-phase mixture of CO_2 vapour, CO_2 liquid solution and an aqueous solution is formed. The CO_2-rich phases go forward for distillation and recovery of the ethanol-rich product at the heavy end of the distillation column and recovery of the more volatile CO_2 for recompression and recycle.

Pilot scale runs on this flowsheet showed successful operation with feeds containing 14 wt per cent ethanol–water achieving a final product of up to 89 wt per cent ethanol. Successful operation of the process was also demonstrated using a corn broth feed containing up to 3 per cent solids with few performance problems in the extractor. On the basis of energy consumption

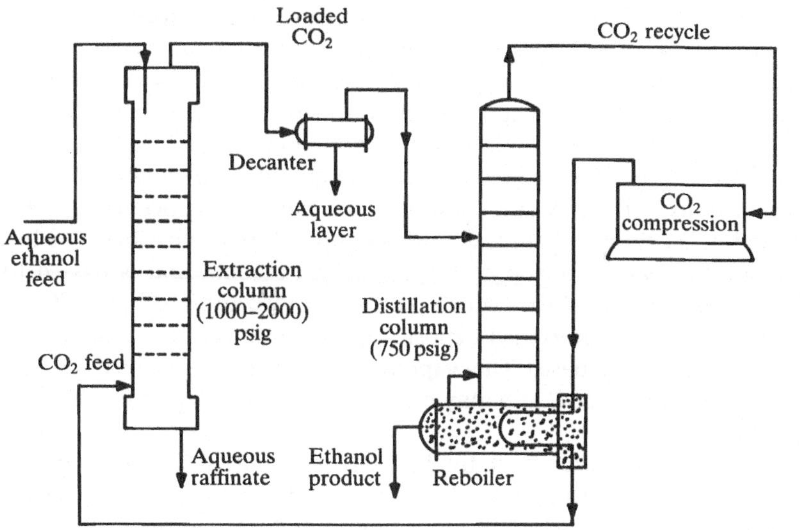

Fig. 5.13. Recovery of ethyl alcohol by liquid CO_2 extraction. (From De Filippi and Moses 1982.)

figures cited for ethanol produced by distillation and produced by extraction from a 5 wt per cent solution, savings in energy utilization may be of the order of ten. For higher alcohols which show a higher affinity for liquid CO_2 this saving will be greater.

Other fluids apart from CO_2 have potential as extractants for organic molecules from dilute aqueous solutions. Victor (1983), for example, describes the use of liquid propylene as an extractant for ethanol. A flow-sheet similar to that using CO_2 is proposed with an additional membrane system to enhance separation of propylene from the ethanol product.

5.3.3.2 *Hops extraction*
[See also Hamm Chap. 4]

Extraction and concentration of flavours from natural products for use in large-scale processing and marketing operations has been of considerable importance since the 1930s. Hops flavour extraction is a good example of a recently developed extraction process which has reached full scale demonstration. The use of hops in beer making has the dual objectives of imparting a bitter taste to the beer and also of adding the unique hop flavour. Addition of the flavour as a concentrate in a controlled manner facilitates the manufacture of a uniform product which the market demands. Grimmett (1981) summarizes the use of liquid CO_2 for flavour extraction and describes the hops extraction process. Approximately 15 per cent of the dried hop core consists of involatile resinous material containing α-acids which impart the bitter taste. Between 0.5 and 2.5 per cent of the core consists of steam volatile essential oils which are responsible for the unique flavour. A pilot scale process is described (see Fig. 5.14) which uses liquid CO_2 as extractant. This process is a typical model upon which commercial scale processes have been based Gardner (1982). Liquid CO_2 is supplied at 850 p.s.i. and is chilled to less than 7°C prior to being fed to a batch extractor loaded with finely divided hops. The maximum solubility of the α-acids occurs at 7°C and extraction is conducted as close to this temperature as possible. The loaded CO_2 solvent leaves the top of the extraction column and is passed to an evaporator held at 40°C which allows vaporization of the CO_2. The relatively non-volatile extract deposits upon the walls of the evaporator and is collected at the base. The vapour CO_2 stream is cleaned by carbon adsorption, is reheated, and may be recompressed for recycle to the extraction column.

The rate of extraction of individual flavour components in the extraction column varies and thus individual flavour fractions may be removed at various heights in the column according to end use requirement.

Use of CO_2 compares very favourably with other extraction processes based on conventional organic solvents. Liquid carbon dioxide does not

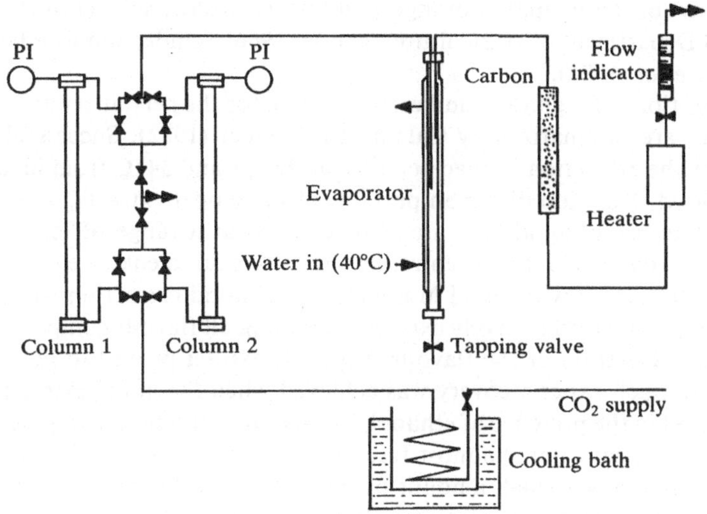

Fig. 5.14. Hop flavour extraction using liquid CO_2. (From Grimmett 1981.)

extract unpleasant smelling plant pigments and, because of the relatively mild processing conditions, most of the volatile essential oils are not lost during recovery processes.

Use of supercritical CO_2 also offers scope for further commercial exploitation. Vollbrecht (1982) lists the main advantages of CO_2 for hops extraction.

1. The process is free of potentially harmful organic solvents.
2. The hop extract products are virtually free of pesticides used during hop cultivation.
3. The extract is produced under inert conditions, thus allowing preservation of almost all aroma components.
4. Extract can be stored for a considerable length of time due to inert processing regime.
5. Only the resin fractions and a portion of the hard resins and the aroma oils are extracted. The alpha acids remain in their natural state in the extract with no evidence of isomerization.

5.3.3.3 *Other flavours*

[See also Hamm Chap. 4]

Use of liquid and supercritical CO_2 for the extraction of a wide range of flavours has been the subject of many experimental studies. Extraction of fruit flavours from apples, oranges, and pears was the subject of studies by the US Department of Agriculture (Schultz 1966; Schultz and Randall 1970; Schultz *et al.* 1974).

Extractions of high value materials suitable for the perfumery and flavour industry are summarized by Calame and Steiner (1982). Successful use of CO_2 for the extraction of lilac perfume at 90 bar and 34°C from lilac leaves is reported. Isolation of lemon peel oil, black pepper extracts, and almond flavouring using liquid CO_2 are also reported at a range of temperatures and pressures. Use of added entrainers are also said to enhance the recovery of the product. It was found in the case of almond extraction with CO_2 at 600 bar and 40°C that an oily extract containing all flavour components was obtained. Isolation of the flavour from this extract proved to be very difficult. A much easier recovery was achieved when the initial extraction was conducted in the presence of ethanol. Following extraction, two phases were obtained, a heavy phase almost free of flavours but rich in oil, and a light phase containing a small amount of oil and all the flavour components.

Generally, the use of entrainers (Brunner *et al.* 1979) increases the solubility of non-volatile compounds by influencing thermodynamic equilibria and also by shifting the critical point of the mixture. Decaffination of coffee provides a good example where the use of an entrainer is essential to the feasibility of the extraction.

This is a good example of a large-scale application of supercritical gas extraction. In Germany, Hag AG have a large supercritical extraction system based on four extractors of 40 m^3 capacity each. The decaffeination process requires use of a small amount of water as a cosolvent (Grimmett 1981).

Other applications in the context of consumer products include removal of off flavours from fruit concentrates and use of liquid CO_2 to expand tobacco leaves as an extension technique.

5.3.3.4 *Transformer oil decontamination*

Use of gas extraction in waste clean-up processes is well illustrated by this very interesting application described by de Filippi (1982). Until recently, polychlorinated biphenyls (PCBs) were used as a non-conducting dielectric fluid in electrical transformers. In view of the recently discovered carcinogenicity of PCBs, replacement of these liquids became necessary. Due to the complex geometry of the system, complete removal of the liquid by drainage was impossible. In addition, substitute oils are required to undergo a process of PCB decontamination prior to use. There is clear evidence that supercritical CO_2 offers favourable selectivities for PCBs which will allow their complete removal from transformer systems.

5.4 Liquid membranes

[See also Verrall Chap. 3]

5.4.1 Introduction

A liquid membrane system is a specific type of liquid–liquid system in which the droplets of a dispersed phase are contacted with a continuous phase. The droplets themselves are agglomerations of even smaller droplets dispersed in a carrier (see Fig. 5.15). Such a system is formed by emulsification of two immiscible liquid phases followed by redispersion of this in a third immiscible liquid phase. The small droplets held in the larger droplets are known as the encapsulated phase, and the continuous phase forming the larger droplets is known as the membrane phase. A stable system is possible only if the membrane phase is immiscible with both the encapsulated phase and the continuous phase. Usually the encapsulated phase and the continuous phase are miscible. Experience has shown (Li 1981) that the inner encapsulated droplets display diameters of the order of 1 μm and the outer membrane phase diameters are in the range 0.1–0.2 mm. The actual droplet sizes obtained in practice will depend upon the usual criteria of emulsion viscosity, pressure of surfactants, and the intensity of mixing. Achievement of a high effective membrane area is straightforward, thus enabling high rates of mass transfer between the encapsulated phase and continuous phase to be achieved.

The principle of the liquid membrane technique is that selective transport of one or more species from the continuous phase into the droplets of encapsulated phase is achieved by the maintenance of a concentration driving force across the membrane phase. Recovery of the product from the encapsulated phase is achieved by separation of the emulsion and continuous phase. This can be done in an analogous way to conventional liquid–liquid

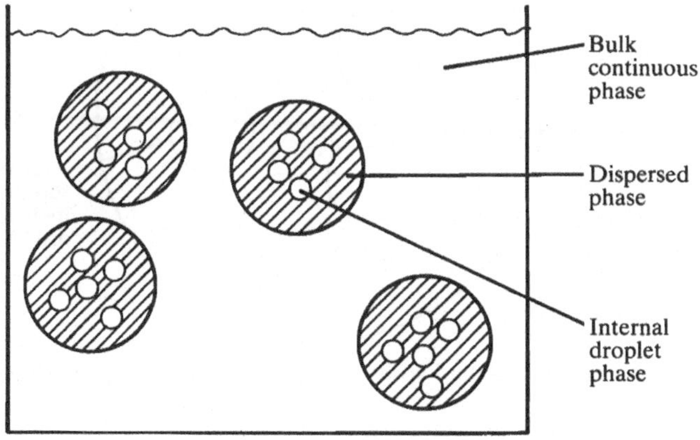

Fig. 5.15. Liquid membrane system.

Phase disengagement. The emulsion phase is then broken separately to result in separate encapsulated phase and membrane phases.

Li identifies two types of mechanism which are possible for the transport of solute across the membrane phase and which are designed to maximise flux and capacity. Firstly, the concentration gradient across the membrane phase can be maximized by adjusting the chemistry of the encapsulated phase to ensure reaction of the mobile species upon its arrival at the membrane–encapsulated phase interface. The reaction product should, if possible, be essentially non-diffusing in the membrane phase. An example of this mechanism is the liquid membrane extraction of phenol from an aqueous continuous phase, through an oil membrane into an aqueous solution of sodium hydroxide as encapsulated phase. With reference to Fig. 5.16, the phenol diffuses through the membrane to react with sodium hydroxide to form non-diffusing sodium phenolate in the encapsulated phase.

The second type of transport involves the incorporation of carrier species into the membrane phase which complex with the product species at the interface between continuous and membrane phases. Diffusion of the complex occurs through the membrane to the interface with the encapsulated phase whereupon the complex breaks down and the product species is released. The extraction of copper (Cahn and Li 1978) provides a good example of this mechanism.

At the continuous phase/membrane interface extraction occurs according to the complex reaction:

$$2RH + Cu^{2+} \rightleftharpoons R_2Cu + 2H^+$$

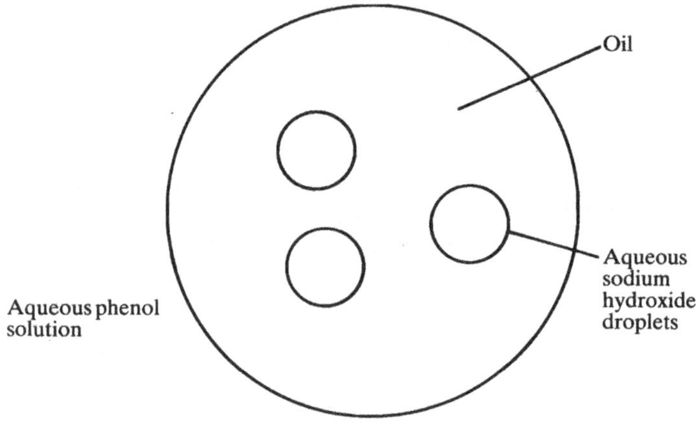

Phenol+Sodium hydroxide ⟶ Sodium phenolate

Fig. 5.16. Extraction of phenol in a liquid membrane system.

This is followed by the reverse reaction at the membrane-encapsulated phase interface:

$$2H^+ + R_2Cu \rightleftharpoons Cu^{2+} + 2RH$$

This latter step corresponds to a conventional stripping operation.

Use of a selective solvent as a permeable liquid membrane in this way eliminates an upper equilibrium loading of solvent which limits solvent utilization in conventional liquid–liquid extraction. Thus there is significant scope for reducing solvent inventory and its associated hazard and cost limitations.

In applying the liquid membrane concept to industrial scale separation flowsheets, formulation of the emulsion is one of the critical factors. The emulsion must maintain stability during the extraction stages and exhibit resistance to shear forces, temperature effects, and concentration changes. At the same time, the emulsion must be of a form which allows easy separation and recovery of the encapsulated phases when required.

The stability of the system will be affected to some degree by the formulation of the mixture and the type and amount of emulsifier which are used. The presence and distribution of the solute species, the pH and ionic strength of the system are all significant factors. Retardation of coalescence of internal aqueous droplets can be achieved by the use of mixed emulsifiers and the establishment of a complex liquid–crystalline layer at the oil–water interface (Brodin *et al.* 1978).

Droplet size and stability are also affected by osmotic gradient and in some cases the membrane phase can act as a semi-permeable membrane allowing significant water fluxes leading to system breakdown. Guarantee of stability of the system is important during uptake or delivery of solute and particularly so in the case of *in vitro* applications.

5.4.2 Applications

[See also Cox Chap. 1]

5.4.2.1 *Metals extraction*

In terms of commercial processes for using liquid membranes as an alternative to conventional solvent extraction, the dilute feed liquors emanating from metal ore leach processes render the application economically attractive. Two examples have been selected for consideration, uranium from wet process phosphoric acid liquors, and copper extraction from acid leach solutions. Published research data indicate that there is significant potential for commercialization of liquid membrane extraction applied to wet chemical recovery of zinc, tungsten, vanadium, cobalt, and nickel.

5.4.2.2 *Uranium extraction*

The potential use of liquid membrane systems for uranium extraction has been examined in some detail in the context of wet process phosphoric acid production (Dickens *et al.* 1984; Dickens and Davies 1983; Bock and Valint 1983). Extraction of uranium as a by-product of phosphoric acid plants has been well established and recovery is achieved by conventional solvent extraction using organo-phosphorous based solvents. Dickens proposes that liquid membrane systems, formulated on the same solvent chemistry, can be used and involve very much lower solvent inventories and can operate effectively on dilute feed solutions. The system proposed is a water/oil/water system using the extraction solvent DEHPA–TOPO dissolved in ESCAID 100 as the organic membrane phase.

Referring to Fig. 5.17, the uranium is complexed into the membrane phase in the hexavalent state as UO_2^{2+}. The mobile complex diffuses across the membrane and at the interface between membrane and encapsulated phase undergoes stripping to U^{4+}. This sequence of steps is identical to conventional solvent extraction but occurs simultaneously.

The process flowsheet proposed is shown in Fig. 5.18. The aqueous feed solution containing phosphoric acid (up to 54 per cent P_2O_5 maximum) and uranium as UO_2^{2+} (up to 100 ppm maximum) is contacted with the liquid membrane emulsion in mixer-settler contactors. The emulsion is formed by mixing and homogenizing of the DEHPA–TOPO solvent with 5 m H_3PO_4 in a solution of 5 g/l Fe^{2+}. The emulsion phase leaves the settler loaded

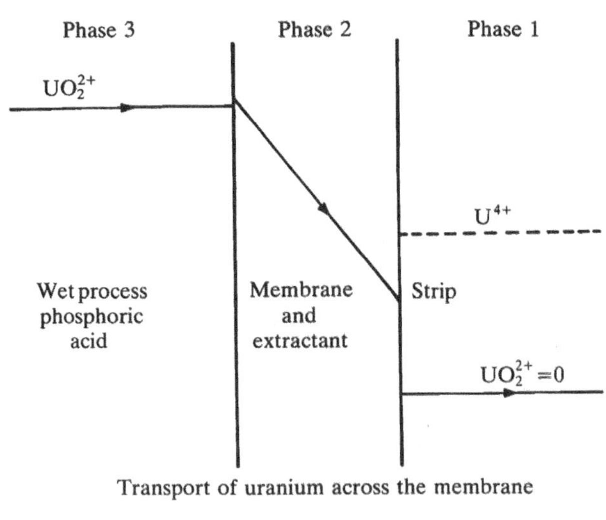

Fig. 5.17. Uranium extraction from wet process phosphoric acid liquors using a liquid membrane system. (From Dickens *et al.* 1984.)

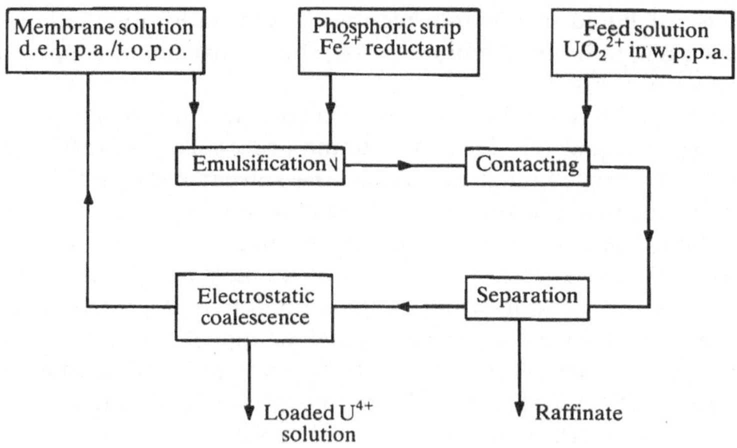

Fig. 5.18. Flowsheet for liquid membrane extraction of uranium from wet process phosphoric acid liquors. (From Dickens *et al.* 1984.)

with uranium which is recovered by breakage of the emulsion in an electrostatic coalescer.

Pilot scale demonstration of the process revealed two shortcomings. Firstly, it was found that the uranium concentration in the encapsulated phase following emulsion breakage was almost identical to that in the feed; selective extraction was thus achieved with minimal concentration. Secondly, the separation of loaded emulsion from the aqueous raffinate proved to be inefficient with very significant entrainments observed. This latter problem was solved by the use of fibre bed coalescers which reduced entrainment from a maximum of 200 ppm, v/v, to a maximum of 15 ppm, v/v.

It was concluded that, with the present state of knowledge and at current uranium prices, economic use of liquid membrane extraction in this context would require a doubling of uranium price.

5.4.2.3 *Copper extraction*

The application of liquid membrane extraction to copper recovery processes as an alternative to conventional solvent extraction has been proposed by Li *et al.* (1983), Frankfield *et al.* (1981), and Cahn *et al.* (1978). Two schemes have been examined in detail, one involving extraction by the liquid membrane emulsion followed by emulsion breaking for product recovery, the other by the use of the emulsion phase as a conventional solvent with no breaking and reformation of the emulsion phase.

Copper recovery from dilute leach liquors by solvent extraction using liquid ion exchange (LIX) type solvents employs the extraction and stripping steps shown previously.

The use of liquid membranes requires incorporation of the liquid ion exchange agent into the membrane phase; the mechanism of copper extraction is shown in Fig. 5.19.

A major difference between conventional extraction of copper into a selective solvent and the liquid membrane process is the function of the extractant. In conventional extraction, the solvent phase acts both as a separation, and as a storage, medium. The maximum storage capacity of the extract phase depends on the total amount of complexing agent available and on the equilibrium capacity of the solvent at ambient conditions. The liquid membrane system has only a small copper hold-up in the membrane phase and the total loading capacity depends mainly upon the acid concentration of the internal droplet phase. Li *et al.* (1983) present batch uptake data for minewater at 0.5 g/l copper at pH 2.5 in contact with (1) a homogeneous solvent, LIX64N, in kerosene; (2) an emulsion with membrane phase comprising 7.5 per cent LIX64N, 2 per cent surfactant in Isopar M, internal phase 20 g/l and 10 g/l Cu^{2+}; (3) an emulsion with membrane phase 7.5 per cent LIX64N, 2 per cent surfactant in Isopar M, internal phase 151 g/l H_2SO_4. The data reproduced in Fig. 5.20 show the change in uptake of copper by the fixed extracting phase contacted with successive batches of aqueous feed solution. In the case of LIX64N in kerosene, there is a very rapid decrease in copper uptake after the second uptake. With first emulsion containing a low acid internal droplet phase, the total loading of copper was doubled, and in the second emulsion the copper loading was raised by a factor of five compared to the homogeneous solvent.

These data are very important and clearly demonstrate that in terms of solvent utilization the loading capacities of the liquid membrane emulsions are much higher than that of the conventional solvent. The loading capacity of the emulsions for copper can be increased by raising the volume fraction of

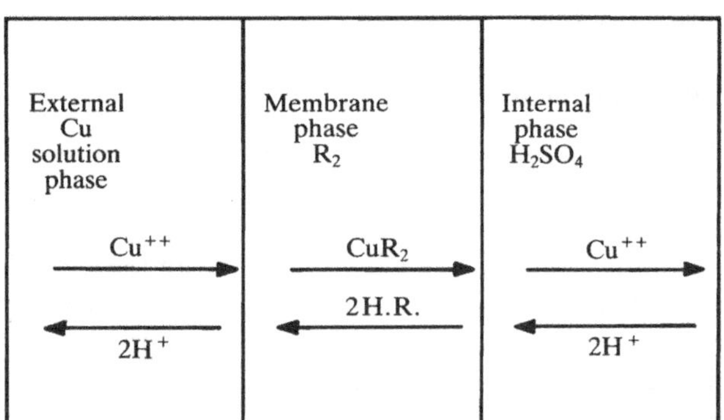

Fig. 5.19. Liquid membrane extraction of copper from aqueous solution.

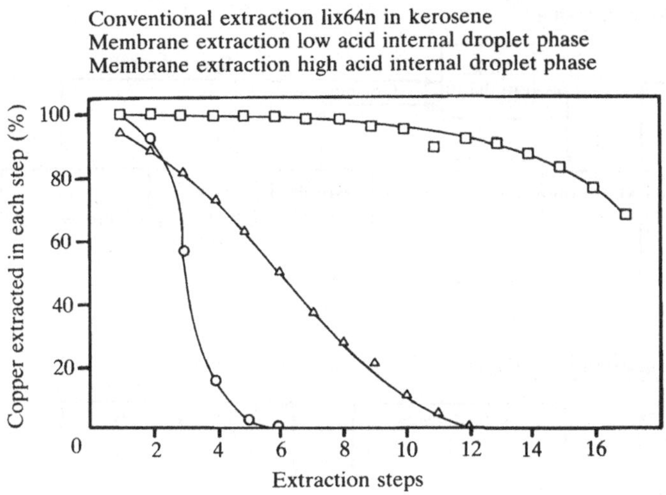

Fig. 5.20. Uptake data for copper extraction into LIX64N. (From Li *et al.* 1983.)

the internal aqueous phase and by increasing its acidity.

In both the proposed processes for copper recovery, the first major step is the extraction of copper ion from the external aqueous phase, which typically would be an ore leach solution, minewater, or a recycle stream. According to the mechanism shown in Fig. 5.19, the copper ion is loaded into a high acid internal droplet phase. Recovery of the copper can be achieved by two possible routes. The extracted copper can be removed directly from the internal droplet phase by breaking the emulsion into an organic phase and a copper-rich aqueous phase and subsequent phase separation by conventional means. Alternatively, the integrity of the emulsion can be maintained and the copper recovered by a sequence of steps similar to a conventional solvent extraction process. Here the emulsion phase would be separated from the external aqueous phase conventionally, and contacted with a stripping solution which would recover the copper from the internal droplet phase by diffusion through the membrane. A comparison of these two schemes is shown in Fig. 5.21.

In scheme 1 emulsion breakage is required to achieve recovery of the copper product and several methods are feasible. These include electrostatic coalescence which is already commercially demonstrated, chemical breakdown using de-emulsifiers and centrifugation followed by high shear. Of these three methods Li *et al.* (1983) observe that electrostatic methods represent the simplest and cheapest.

In scheme 2, uptake of copper ion into the emulsion phase is achieved by

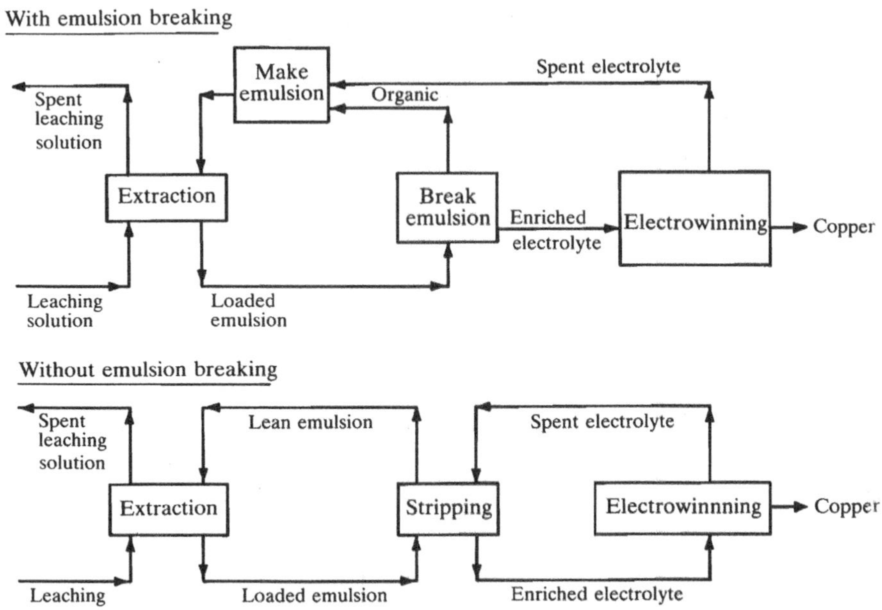

Fig. 5.21. Copper extraction and recovery using liquid membrane emulsions with and without emulsion breaking. (From Li *et al.* 1983.)

a similar mechanism to scheme 1. The re-extraction of copper from the internal droplet phase into an external stripping solution is achieved by ensuring a positive concentration driving force with respect to copper between internal and external phases, and by ensuring a suitably high acidity gradient. This latter condition, which would typically involve an external acidity seven to eight times that of the internal droplet phase, is of sufficient importance that successful re-extraction can be achieved even in the absence of a significant copper concentration driving force.

Economic comparison of these two liquid membrane-based extraction processes with conventional solvent extraction process for copper recovery indicated that in terms of both capital and operating cost the liquid membrane-based process with emulsion breaking was the best. The liquid membrane re-extraction process proved cheaper in terms of capital cost compared to conventional solvent extraction, but significantly more expensive compared to the cementation process. Operating costs proved lower for both liquid membrane processes.

5.4.2.4 *Liquid membrane encapsulation*

Liquid membrane systems have another important application in addition to separation. The liquid membrane can also be used to encapsulate reactive components in the internal droplet phase. Thus the membrane can be used to isolate the reactive component from its end use environment until desired or the membrane can be formulated to allow a controlled release of the reactive component into an external environment.

General applications of membrane encapsulation include formation of a reaction mixture in which, initially, one of the reactants is encapsulated within the membrane phase, which itself is in contact with another phase (typically a liquid or a solid) containing the other reactants. Emulsion breakdown allows ultimate contact of the reactants and hence the reaction proceeds. A schematic of this process is shown in Fig. 5.22. Application of pesticides and fungicides to crops could be a suitable area here, with emulsion breakdown being achieved simply by evaporation, allowing contact of the appropriate active component at the desired site and avoiding unnecessary wastage and pollution during application.

Liquid membrane encapsulation has also been used medically where controlled release of drugs into the gut is desirable. Here the membrane phase is chosen such that the compound to be released has a slight solubility in it, thus allowing slow permeation from the encapsulated phase into the external environment.

Permeation in the reverse direction can also be employed to remove toxins from the human gut by encapsulation of material and drugs which reacts with, or catalyses the breakdown, of the component (Frankfield *et al.* 1978; Asher *et al.* 1975).

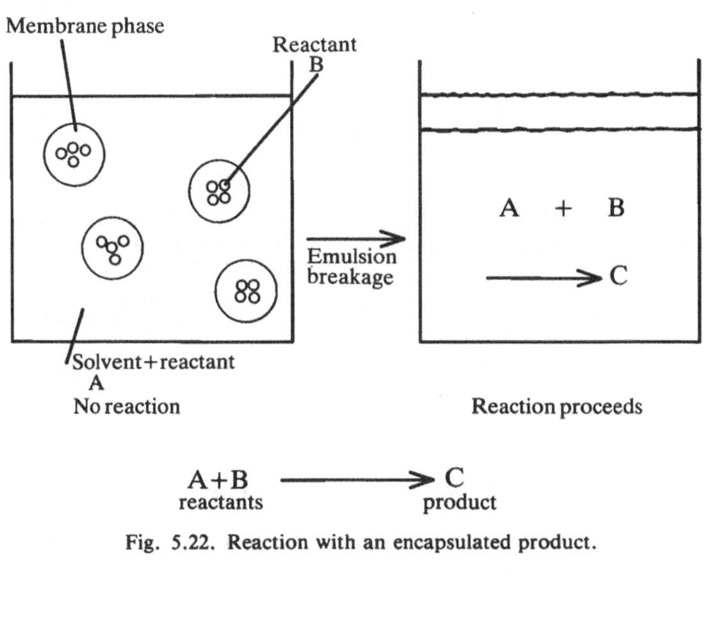

Fig. 5.22. Reaction with an encapsulated product.

Encapsulation of enzymes and whole living cells into liquid membrane emulsions has also been successfully demonstrated (Mohan and Li 1974, 1975) and has the potential for providing an effective enzyme immobilization technique. The principle behind it is illustrated in Fig. 5.23. The liquid membrane emulsion comprises an organic hydrocarbon phase encapsulating the inner aqueous droplets containing the enzyme, usually used in a stable buffer solution. The emulsion is contacted with an aqueous solution of the substrate (that is the reactant) which diffuses through the membrane to the interface with the encapsulated phase. Reaction occurs in the internal droplet and reaction products diffuse out of the emulsion into the external substrate solution.

Mohan and Li (1974) report successful implementation of this concept for the enzymic reduction of nitrate and nitrite ions in waste water. The enzyme-rich phase is produced by a cell extraction process, in this case of *micrococcus denitrificans* whereby intracellular enzymes from specific bacteria are recovered into an aqueous buffer solution and suitably separated from cell debris material. The resulting cell-free extract was emulsified with a mixture of Span 80, Enjay 3029, and S100N, these being surfactant, additive, and hydrocarbon diluent respectively. The substrate solution comprised 0.1 m phosphate buffer at pH 6.5 and 10 μg ml^{-1} NO_3^-. The reduction of NO_3^- to NO_2^- was clearly demonstrated together with further enzymic reduction of NO_2^- to nitrogen and water. It was shown that encapsulated enzymes do not

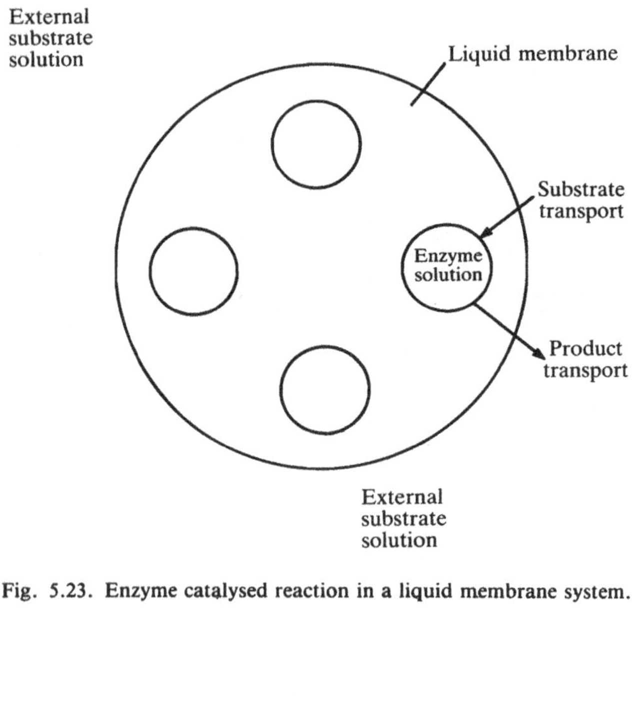

Fig. 5.23. Enzyme catalysed reaction in a liquid membrane system.

leach but can act on externally applied substrates which are capable of diffusional transport across the membrane phase. The rate data generated showed the importance of the diffusional resistance offered by the membrane both to substrate transport into the emulsion and to product transport out.

Another example with potential medical application is the enzymic oxidation of urea (May and Li 1972; Asher *et al.* 1975). In this system the enzyme urease is encapsulated in the internal droplet phase. Urea from the external feed phase diffuses through the membrane and undergoes catalytic oxidation to CO_2 and NH_3 upon contact with the enzyme.

A further development is the encapsulation of whole cells into liquid emulsions which allows microbiological reactions to be conducted in environments which would normally be hostile to live cells. Mohan and Li (1975) report the encapsulation of whole cells of *micrococcus denitrificans* into a liquid membrane system which was successfully used to effect nitrite reduction. Again, the rate of reaction was reduced due to the diffusional resistance offered by the membrane but maintenance of cell activity in the presence of toxic species in the external substrate–non-environment was confirmed. This represents a significant breakthrough in the potential use of microbiological reactions for waste clean-up in chemically hostile environments.

Pharmaceutical applications of liquid membrane systems in the context of drug administration are summarized by Davies (1981). Liquid membrane emulsions were found to be superior to simple water–oil emulsion systems for direct drug injection. They were easier to inject and produced a more favourable antibody response. The drug is contained in an encapsulated aqueous phase and is released slowly through the membrane layer to give a sustained release system. Attention has recently been focussed on the use of liquid membrane emulsions for the delivery of anti-cancer drugs and some clinical successes have been reported. Uptake of the drug at specific sites in the body can be greatly enhanced.

Davies proposes three possible mechanisms governing rate of drug release (see Fig. 5.24):

(1) diffusion of the unionized drug through oil;

(2) diffusion of ionized and/or unionized drug through thin oil–water lamellae;

(3) coalescence of the internal aqueous phase and subsequent rupture of the oil droplet.

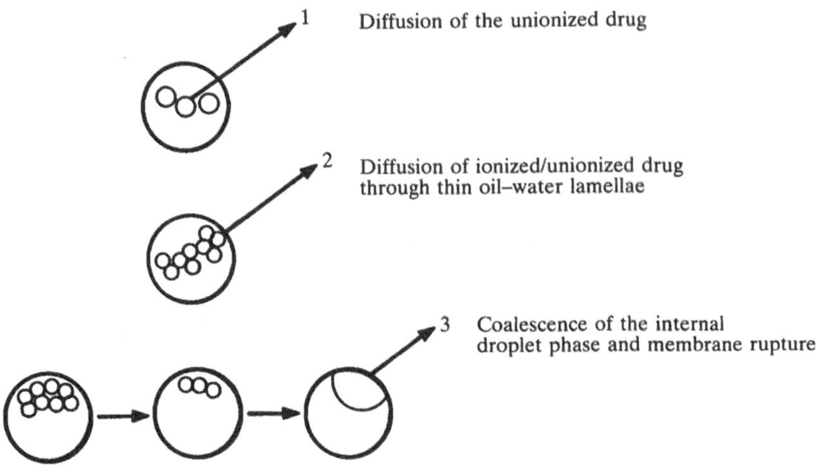

Fig. 5.24. Drug release mechanisms from liquid membrane emulsions. (From Davis 1981.)

5.5 Biochemical processes
[See also Verrall Chap. 3]

5.5.1 Introduction

The production of new chemicals and the development of routes for the manufacture of existing products from renewable feedstocks will initially be related to the successful exploitation of biochemical processes. Atkinson and Sainter (1982) identify three factors upon which this proposition is based:

(1) raw materials will come increasingly from renewable sources;

(2) biotechnological processes appear likely to be economical compared to the chemical processing of vegetable materials;

(3) a wide range of possibly valuable products is being defined via both traditional biological routes and genetic manipulation.

The potential input of biological processes to the chemicals market covers three areas which are related to the type of product: large-, medium-, and small-scale divisions

(1) *large-scale manufactured products*: petroleum and coal which are used mainly as fuels or as primary feedstocks for bulk industrial chemicals;

(2) *medium-scale*: commodity chemicals and their substitutes;

(3) *small-scale*: small-volume, high-value speciality chemicals where biological production routes are the only ones available.

Currently, with the exception of one or two specific geographical locations, large-scale bulk chemicals manufacture by biological means is unlikely to be economic. The same applies to medium-scale products. The major current interest lies in the cost-effective small-scale production of high-value products. In many of these cases the cost of production is not a critical factor, rather whether or not the product can be manufactured at all at the desired specification.

Synthesis of chemicals by biological fermentation processes often convert only a small fraction of the feedstock into the desired product requiring separation from a dilute and often complex mixture. For example, fermentation of sugar to alcohol results in 20–25 per cent of the feed carbon component being converted to CO_2, 40–50 per cent converted to ethanol, and the remainder either unfermentable or converted to higher alcohols and other organic species. Maximum conversions with other products may be much lower, usually in the order of 10 per cent for penicillin and as low as 1 per cent for some enzymes. Product inhibition of microbiological activity, together with the above factors, combine to produce a feedstock for separation which is dilute and can contain significant amounts of solid material.

Typical product concentrations (Atkinson and Sainter 1982) prior to separation are shown in Table 5.2. In addition to low concentration the liquors themselves are often physically unattractive, having high viscosities, and containing high levels of surfactant material (Stageman 1984). Such properties add to the difficulties of using liquid extraction processes where emulsion formation and low rates of mass transfer are frequently observed. Another problem encountered in the use of liquid extraction for biochemical separations is the denaturing effect which many common organic solvents and diluents can have on sensitive molecules.

TABLE 5.2 *Typical product concentrations leaving fermenters (after Atkinson 1982)*

	g/l
Single cell protein	30–50
Ethanol	70–120
Acetone–butyl alcohol–ethanol mixture	18–20
Organic acids (e.g. citric, lactic)	40–100
Antibiotics (e.g. penicillin G)	10–30
Lipids	10–30
Enzyme protein	2–5
Vitamin B_{12}	0.02
Riboflavin	10–15

In spite of these difficulties, liquid extraction offers the combined advantages of a process in which selectivity can be controlled chemically and which can be conducted at ambient temperatures, thus avoiding thermal degradation of relatively unstable molecules. The relative importance of solvent extraction is confirmed by a brief examination of some of the products and processes to which it has been applied.

5.5.2 Ethanol production and recovery

Ethanol and acetic acid are two organic chemical intermediates for which the technology of production by fermentation is well established. This is mostly on account of historical background to the time prior to cheap crude oil-based processes introduced in the 1950s. However, in the early 1980s, particularly in North America, there was a resurgence of interest in production of ethanol from biomass both for use as a chemical feedstock substitute and as a fuel extender. There has been strong interest in developing technology for separating ethanol from dilute solutions arising from fermentation processes (Munson and King 1984). This has occurred because of the energy-intensive nature of azeotropic distillation processes, and the costs involved can represent a substantial fraction of the value of the product ethanol as a fuel constituent or chemical intermediate. Thus there has been a significant appraisal of liquid extraction as a low-energy intensive process for ethanol production.

Apart from energy cost considerations, production of ethanol by fermentation is also subject to the restriction of total product yield imposed by inhibition of yeast activity as the ethanol concentration rises during the course of the reaction. At ethanol concentrations above approximately 12 wt per cent, yeast activity is suppressed to an unacceptably low level, thus virtually placing an upper limit on final ethanol concentration. A second area of investigation has concentrated on the removal of ethanol by liquid extraction as it is produced thereby extending the total yield of product available from a given batch of feedstock.

In both contexts the choice of extracting solvent is a primary consideration and there are three desirable characteristics of a solvent. These are high capacity for ethanol, a high selectivity for ethanol over water, and non-toxic interactions with fermentation liquor. The capacity to extract ethanol is characterized by the distribution coefficient, K_D. The magnitude of K_D is critical in determining the solvent to aqueous flow ratio required for a viable extraction process. High distribution coefficients mean low solvent circulation rates and thus lower energy costs for solvent pumping.

Munson and King (1984) list experimental distribution coefficient data for ethanol–water–organic mixtures and these are summarized in Table 5.3. These data confirm the earlier conclusion of Roddy (1981) that the order of

TABLE 5.3 *Distribution coefficient data for ethanol–water–organic mixtures (after Munson and King 1984)*

Solvent	K_D	Temperature (°C)	Aqueous feed concentration (wt% E + OH)
MIBK	0.5	30	0.78
DIBK	0.19	30	0.78
DIBK*	0.11	24	4.6
TBP	0.8	30	0.78
TBP*	0.56	25	4.6
1-Hexanol	1.0	25	0.78
2-Ethylbutanol	0.97	30	0.78
2-Ethylbutanol*	0.83	25	4.6
1-Octanol	0.64	25	0.78
1-Octanol*	0.61	25	4.6
1-Decanol	0.57	30	—

*After Roddy (1981).

extractability of ethanol was hydrocarbon = halocarbon < ether < ketone < amine < ester < alcohol = phosphate.

In addition to affinity for ethanol a second vital criterion for solvent choice is separation factor with respect to water. Roddy (1981) concludes that a limited number of solvents are available with distribution coefficients ~0.5 for ethanol and separation factors ~10 from aqueous solutions. The highest coefficient measured was for 2 ethyl-1-butane at 0.69. Tributyl phosphate exhibited a lower value of distribution coefficient (0.54), but may be preferable as a candidate solvent since it is available at high purity, has a very low solubility in water, and exhibits a low vapour pressure.

The application of liquid extraction processes for ethanol recovery during fermentation has attracted a considerable amount of interest. In particular, in-line extraction as a means of improving the overall conversion of sugars to alcohol by removal of the product during fermentation has featured in recent literature. The major problems include undesirable effects of organic solvents on microbial growth and product formation (Wang *et al.* 1981), and serious reduction in mass transfer fluxes during extractions in the presence of cell debris and other biological material (Crabbe 1983).

In order to avoid the depression of cell activity caused by direct contact of organic solvent and unfiltered fermentation liquor, Wang *et al.* (1981) proposed an in-line extraction loop (see Fig. 5.25), in which the ethanol-rich fermentation broth is filtered in a series of graded filters which retain the majority of yeast cells. The clarified broth goes forward for extraction; in this case the long-chain aliphatic alcohols, butanol, hexanol, and decanol,

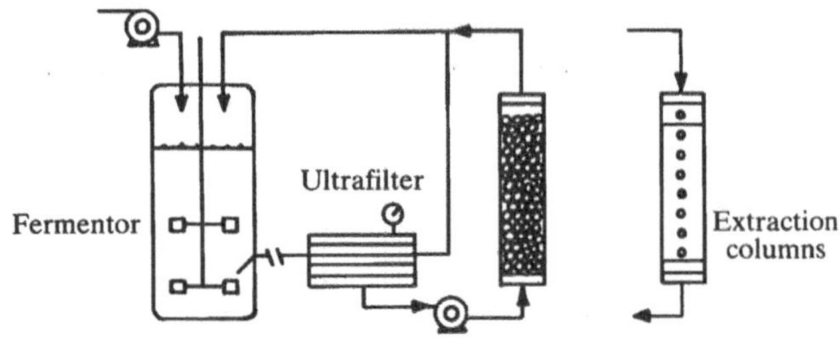

Fig. 5.25. In-line ethanol extraction during fermentation. (From Wang *et al.* 1981.)

are proposed as extractants. It is proposed to remove residual organic solvent from the raffinate by carbon adsorption treatment prior to recycle to the fermenter. The yeast biomass would be recycled to the fermenter as a thickened stream from the ultra filter.

Minier and Goma (1980) demonstrate an alternative approach to *in-situ* extraction of fermentation solutions using an immobilized cell fermenter. Their system comprised a column equipped with perforated plates between which porous brick was packed (see Fig. 5.26). The porous brick acted as an

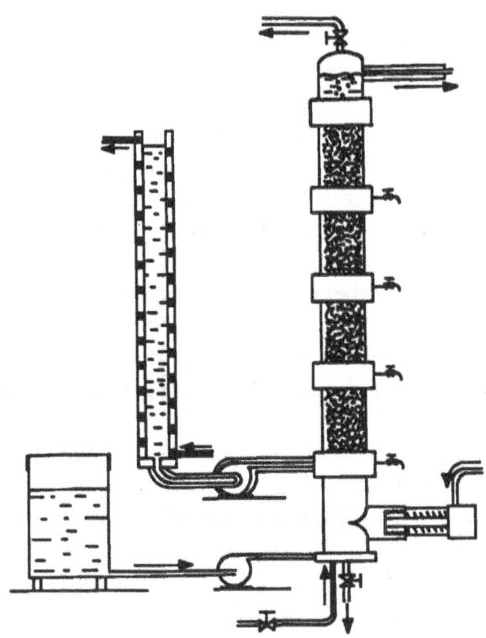

Fig. 5.26. Ethanol production by combined fermentation and extraction. (From Minier and Goma 1980.)

immobilization medium to which yeast cells could adhere. The feed to the base of the column comprised an aqueous nutrient stream containing sugar and an extractant organic stream of dodecanol. Results of performance trials showed that ethanol productivity was enhanced by a factor of up to five in the presence of the extracting solvent. It is claimed that the extractive fermentation contactor developed satisfies design criteria for plug flow, high liquid–liquid interfacial area, low carbon dioxide hold-up, and compatability with a continuous process environment.

Use of ethanol produced from biomass as a motor fuel extender is well established in certain specific geographical locations where the economics of the process are favourable. Generally the major constraint to economic viability of this application is the cost of separation by distillation. A novel process for the manufacture of 'Gasohol' (that is an ethanol–gasoline mixture) which combines distillation and solvent extraction is described by Leeper and Wankat (1982). The proposed flowsheet is shown in Fig. 5.27 and comprises a predistillation section in which the dilute fermentation solution is concentrated to approximately 90 wt per cent ethanol under relatively inexpensive conditions. The ethanol-rich product from predistillation is then contacted in a liquid–liquid contactor with pure gasoline. The gasoline extracts some of the ethanol to yield a 'gasohol' product containing some 10 wt per cent ethanol. Aqueous raffinate enrichment and gasoline recovery are conducted separately in a gasoline recovery distillation column.

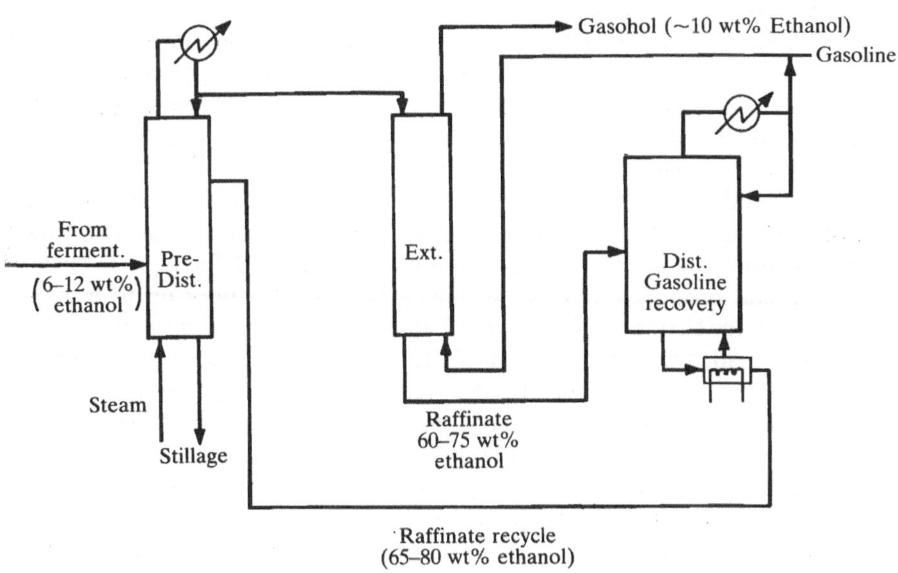

Fig. 5.27. Gasohol production using solvent extraction. (From Leeper and Wankat 1982.)

5.5.3 Citric acid

Citric acid is one of the few bulk products which has been produced by fermentation for some considerable time. Two fermentation processes have been used (Sodeck *et al.* 1981): firstly by a surface fermentation process in which the mould, *Aspergillus niger*, forms a mycelium layer on the liquid surface of an aqueous solution of sucrose and effects conversion of the sucrose into citric acid by enzymic reaction. Upon completion of fermentation the mycelium layer is removed and the citric acid recovered. Currently, however, approximately 80 per cent of the citric acid is produced by a second process known as submerged fermentation in which the mould *Aspergillus niger* is cultured aerobically beneath the surface of a solution of molasses.

Both processes require an important downstream separation to yield the citric acid in a pure marketable form. Currently citric acid separation is achieved by precipitation of calcium citrate from the fermentation broth. Following separation of the mycelium, the broth is filtered and a lime suspension is added to precipitate calcium citrate. After filtration and washing the calcium citrate is dissolved in sulphuric acid to precipitate calcium sulphate in a solution of citric acid. The solution is separated, cleaned, and evaporated to yield crystalline citric acid.

Simplification of the citric acid recovery and purification process using solvent extraction is proposed by Wennersten (1980) and (1983). Citric acid is readily extracted into a number of organic solvents, with high molecular weight aliphatic amines and phosphorous-bonded oxygen-donor solvents exhibiting particularly good selectivities. Three important criteria have been established for solvent selection: (1) good distribution coefficient (but < 10) for citric acid uptake; (2) easy product backwashing with water; (3) low tendency to emulsion formation.

A reasonable value of distribution coefficient reduces the number of extraction stages required and reduces the required solvent/aqueous flow ratio. However, excessively high coefficients lead to difficulties in backwashing and product recovery. Wennersten suggests values up to a maximum of 10 for citric acid. Easy water backwashing of the citric acid product from the loaded solvent is essential. If a basic backwashing solution is employed the product is removed as citrate and thus precipitation and solid–liquid separation operations of the type used in the conventional process become unavoidable, negating the main advantage of using extraction.

Solvents which exhibit a strongly temperature-dependent distribution coefficient with respect to citric acid offer the most attractive solution to the problem of solvent selection. For example, Wennersten presents data for the system water–citric acid-25 per cent TBP in Shellsol A which illustrates the potential of a temperature controlled extraction (see Fig. 5.28).

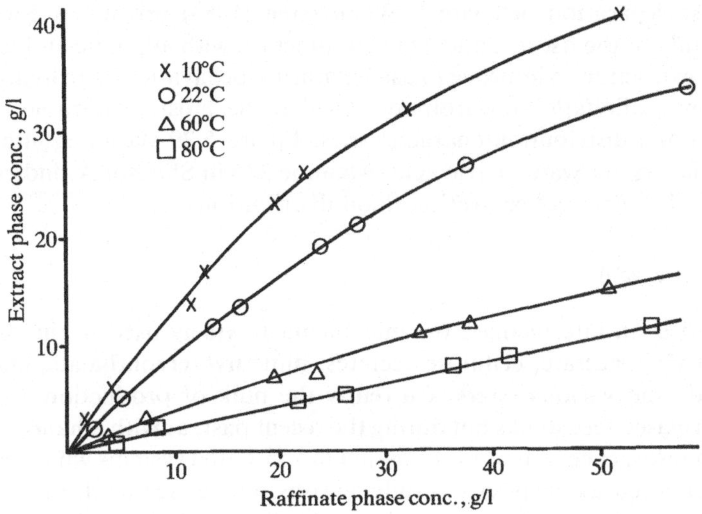

Fig. 5.28. Extraction of citric acid into TBP/shellsol A as a function of temperature. (From Wennesten 1983.)

Uptake of citric acid from the broth would be efficiently achieved at temperatures less than 20°C and product backwashing achieved at around 60°C.

Long chain aliphatic amines also exhibit a suitable temperature/distribution coefficient relationship but their use in the pure state as extractants is restricted by a strong tendency to form stable emulsions and consequent

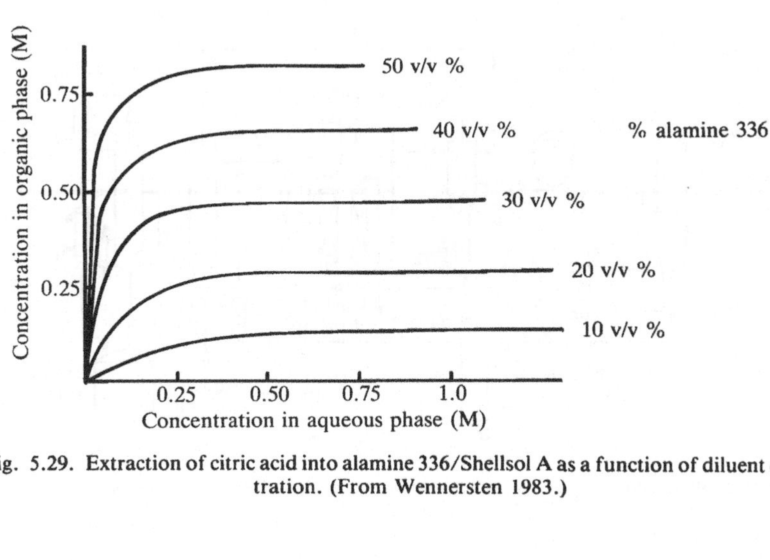

Fig. 5.29. Extraction of citric acid into alamine 336/Shellsol A as a function of diluent concentration. (From Wennersten 1983.)

poor phase separation behaviour. Wennersten (1983) presents a comprehensive study of the use of diluents in conjunction with aliphatic amines for citric acid extraction to improve phase separation behaviour. Pure hydrocarbon diluents, aliphatic and aromatic, yielded the most favourable phase separation and distribution characteristics. Figure 5.29 shows equilibrium data for the system water–citric acid–Alamine 336 in Shellsol A, indicating the effect of diluent concentration upon distribution.

5.5.4 Acetic acid

Acetic acid is an intermediate organic chemical widely used in the manufacture of vinyl acetate, cellulose acetates, dimethyl terephthalate, chloroacetic acid, and various esters. Currently the bulk of production is from petroleum-based feedstocks but during the recent past, as with ethanol, there has been a resurgence of interest in fermentation-based systems which would become economic as the price of crude oil rises. These include, for example, bioconversion of corn starch hydrolysates and hydrolysates of lignocellulosic biomass (Busche *et al.* 1982) into acetic acid solution. Economic exploitation of this production route would rely upon the introduction of solvent extraction into a substantial proportion of the separation processes. A combination of liquid extraction and azeotropic distillation is proposed by Busche (1983) (see Fig. 5.30). Extraction of the dilute feed liquid by diethyl ester is proposed. The energy-intensive section of the process centres around the azeo-still and heat recovery by vapour recompression at the top of the still is proposed. Acetic acid, being a relatively high boiling point material,

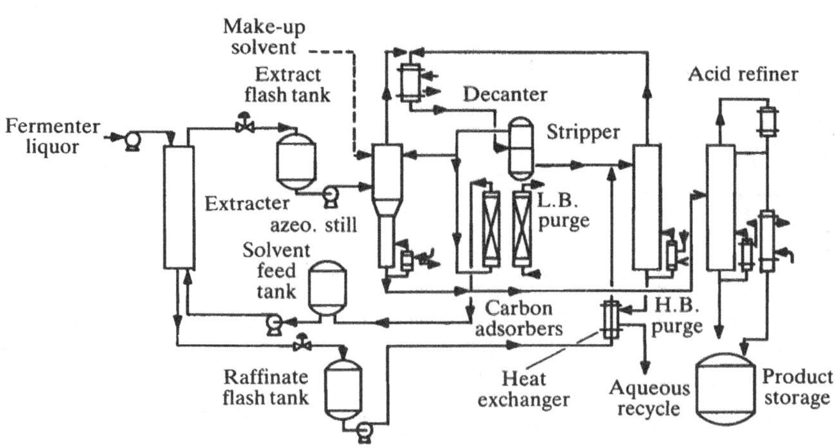

Fig. 5.30. Acetic acid recovery by solvent extraction. (From Busche 1983.)

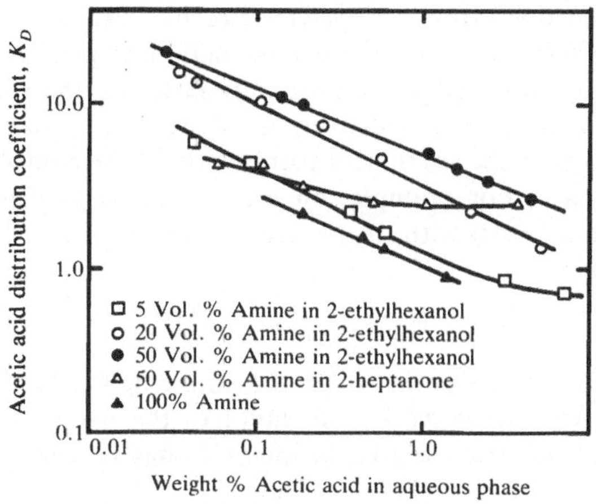

Fig. 5.31. Extraction of acetic acid by alamine 336 in various diluents. (From Ricker *et al.* 1979.)

is recovered at considerable expense by distillation, especially when the feedstock contains a large proportion of water.

Ricker *et al.* (1979) show that organic bases, including a variety of amines and trioctyl phosphine oxide (TOPO), can exhibit much higher distribution coefficients for extraction of acetic acid compared with conventional solvents. The effect of diluent upon distribution coefficient is critical. Figure 5.31 shows the effect of diluent (type and concentration) upon the extraction properties of Alamine 336, in the acetic acid–water system and illustrates the increase in distribution coefficient exhibited in the presence of the diluent.

5.5.5 Penicillin extraction

The use of liquid extraction for the recovery of penicillin, as penicillin G, from fermentation liquors is long established and currently large-scale operations are conducted according to a process now more than 40 years old. Typically, pen G is extracted in its undissociated free state at a pH of 2–2.5 into *n*-benzyl acetate. On account of the relative instability of undissociated pen G at ambient temperatures, and at low pH, the cell-free fermentation liquor is chilled to 5°C and contacted with the solvent in short residence time centrifugal contactors. Both the distribution coefficient and stability of pen G are strongly dependent upon pH. Thus, pen G used in aqueous solutions at low pH is highly extractable but also fairly unstable. The use of refrigeration and energy-intensive liquid–liquid contact are expensive features of conventional pen G plants, and, because of stability problems, plant losses can be as high as 15 per cent.

The use of a reactive extraction system (Reschke and Schugerl 1984) for pen G offers the possibility of efficient extraction of pen G at higher pH values in the aqueous phase, thus drastically reducing the rate of decomposition. Reactive extraction effectively involves a neutralization reaction between the extracting species in acid form, with a base component carried in the organic phase. For example, an organic amine A dissolved in the organic phase would react with the penicillic acid anion P^{-1} and proton in the aqueous phase:

$$A_{org} + P^-_{aq} + H^-_{aq} \rightleftharpoons AHP_{org}$$

This is an ion pair extraction in which the amine acts as a carrier for the desired species. An alternative system involves the use of a quarternary ammonium salt as a carrier to give an ion-exchange reaction system:

$$NR_4^+Cl^-_{org} + P^-_{aq} \rightleftharpoons NR_4^+P^-_{org} + Cl^-_{aq}$$

Comparative extraction data for pen G extraction into *n*-butyl acetate in the presence of amine carriers are shown in Fig. 5.32 after Reschke and Schugerl (1984). The ability to achieve a great improvement in extraction uptake at high pH is clearly demonstrated in the presence of the secondary amine, dioctylamine. The degree of extraction is defined as the fraction of the total penicillin present which is extracted.

Comparison of the performance of primary, secondary, and tertiary amines as carriers reveals that primary amines are not particularly suitable because of the moderate partition coefficients of their penicillin complexes and because of their relatively high solubility in water. The secondary amines exhibited the best extraction performance. Those studied were dioctyl amine and a mixture of secondary amines Amberlite LA-2. It is concluded that

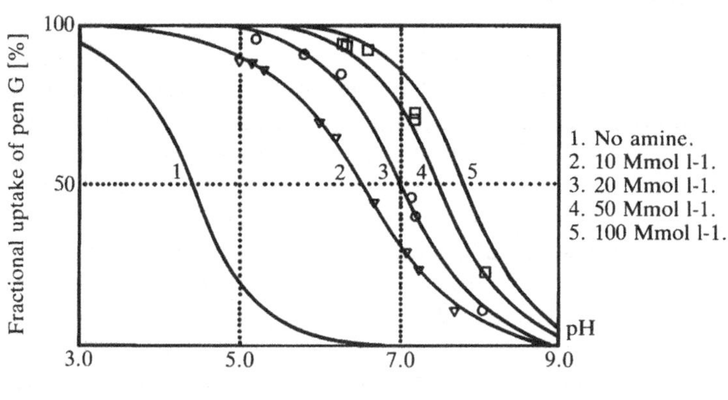

Fig. 5.32. Extraction of penicillin G into n-butyl acetate in the presence dioctylamine as a function of pH.

amine carriers facilitate penicillin extraction in the pH range 5–7 where pen G is most stable. The stability of pen G in the organic phase is slightly reduced in the presence of carriers but in practice this is insignificant.

Ion-exchange extraction also facilitates pen G extraction at neutral pH values but there are problems during re-extraction for product recovery.

5.5.6 Aqueous two phase systems and enzyme extraction

[See also Verrall Chap. 3, Hamm Chap. 4]

Sales of industrial enzymes worldwide were valued in 1986 at £560 million (Godfrey 1986), with the many sectors of the market including enzymes for reducing starch to sugars, additives for detergents, rennets and pectinases used in fruit wine and vegetable processing. Enzymes are essentially natural proteins which promote many different chemical reactions, that occur in living cells. In recent years there has been a continuing interest in harnessing their properties for the processing of a wide variety of materials.

The number and range of enzymes of commercial importance has increased very significantly in recent years but the vast majority (Kroner *et al.* 1982) of enzymes are intracellular, which has made recovery on a commercial scale difficult. New processes based on liquid–liquid extraction are showing significant potential in overcoming some of the recovery problems associated with traditional protein precipitation methods. To avoid enzyme denaturating associated with conventional organic solvents, the use of hydrophilic two-phase systems in which both phases contain substantial proportions of water has been developed for enzyme recovery. Kula *et al.* (1982)

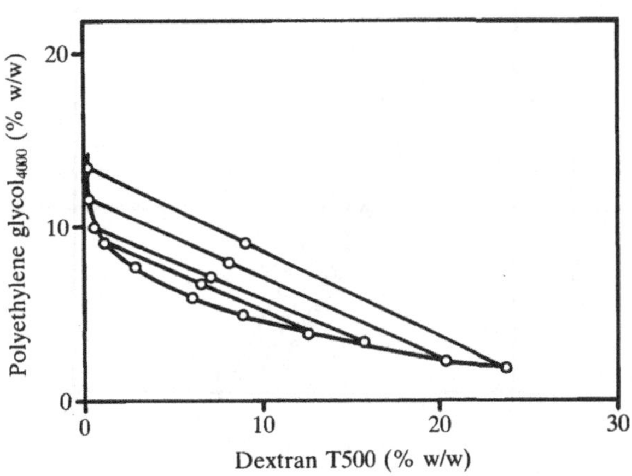

Fig. 5.33. Phase diagram for the system polyethylene glycol 400/water/dextran T-500. (From Kula *et al.* 1982.)

review in depth the application of this extraction technique to enzyme recovery.

Typically an aqueous two-phase system is made up by mixing a hydrophilic polymer such as polyethylene glycol (PEG) with certain salts, for example potassium phosphate in the presence of water. Mixtures of polyethylene glycol and dextran in aqueous solutions produce similar two-phase behaviour. Phase diagrams for two of the commercially used systems, PEG-water-dextran and PEG-water-potassium phosphate, are shown in Figs. 5.33 and 5.34. The high water content of both phases makes the systems suitable for extraction of sensitive molecules such as enzymes from broken cells. Table 5.4 lists some of the intracellular enzymes which can be potentially recovered in aqueous two-phase systems together with their applications.

Enzyme partition in aqueous two-phase systems can be controlled by a number of parameters which include concentration of polymer, mean molecular weight of the polymer, type and concentration of added salts, pH and temperature. Figure 5.35 (Kula *et al.* 1982) illustrates the effect of polymer concentration and type upon the partition of the enzyme fumarase and total protein in a PEG-phosphate system. An important property exhibited in these data is the difference in the change of partition coefficient of the single enzyme fumarase relative to total protein. Lowering the mean molecular weight of the PEG is readily achieved to increase the total partition coefficient for a particular enzyme. The mean molecular weight of the PEG is easily manipulated by mixing appropriate amounts of polymer of known molecular weight.

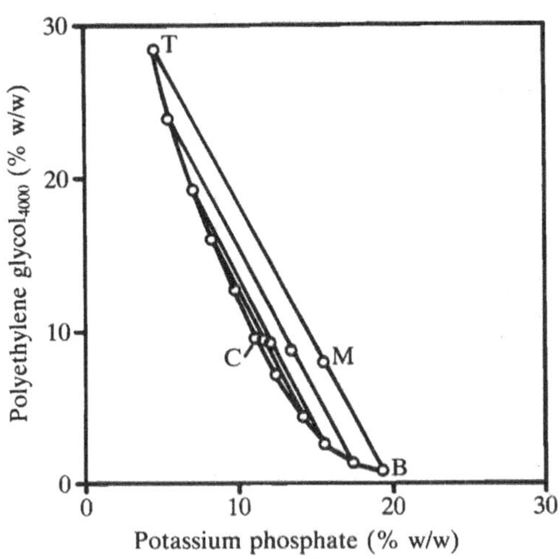

Fig. 5.34. Phase diagram for the system polyethylene glycol 400/water/potassium phosphate. (From Kula *et al.* 1982.)

TABLE 5.4 *Enzymes with favourable partition properties in systems based on PEG–dextran–salt solutions*

Enzyme	Applications
Alpha-amylase	Glues/food ingredients
Glucoamylase	Corn starch/glucose conversion: starch/glucose conversion
Alpha-glucosidase	Maltose/glucose conversion
Glucose-6-phosphate dehydrogenase	Medicinal indicator
Formate dehydrogenase	Oxalate/formate determination
Formaldehyde dehydrogenase	Aldehyde/alcohol conversion
Catalase	Cold milk sterilization
Pullunanase	Starch/maltose conversion
Glucose isomerase	Glucose/fructose conversion
Beta-glucosidase	Food processing
Interferon	Pharmaceutical applications

The effect of salt type and concentration is shown in partition data for the affinity extraction system glucose-6-phosphate dehydrogenase in the presence of PEG 6000-blue shown in Fig. 5.36.

Other systems exhibiting similar behaviour are listed by Albertson (1971) and include dextran–ficoll–water, ficoll–polyethylene glycol–water, polyvinyl alcohol–dextran–water, dextran–methyl cellulose water.

A further tool for modification of partition behaviour in aqueous two-phase systems is affinity partition. The use of biospecific interactions of enzymes with substrates, products, inhibitors, or antibodies appears to offer a rational method for specific separation of a single protein from a complex mixture. Affinity agents are usually a polymer derivative of one of the solvent components; for example, PEG-NADH and triazine dyes have been successfully used to adjust enzyme partition.

Another important characteristic of aqueous two-phase systems is the selectivity of one or other of the phases, usually the heavy phase, for cell debris. Use of this property can, in some cases, preclude the use of expensive solid–liquid separation processes.

Demonstration of scale-up to commercial operation of enzyme recovery processes based on hydrophilic liquid–liquid extraction to date is limited. The potential of these systems has been demonstrated at intermediate scale for the recovery of the enzyme formate dehydrogenase (Kroner *et al.* 1982). Formate dehydrogenase is an important enzyme used in the regeneration of the enabling agent NADH (the hydrogen form of nicotiamide adenine

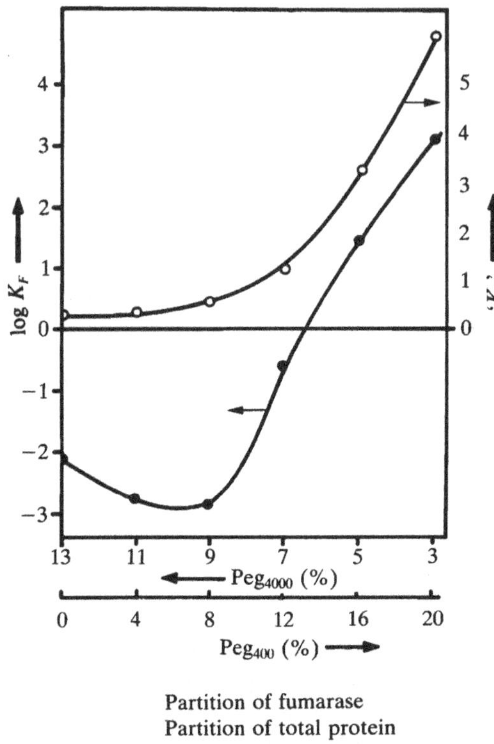

Fig. 5.35. Effect of molecular weight of polymer upon partition of fumarase into polyethylene glycol. (From Kula *et al.* 1982.)

dinucleotide) which is a very high value material associated with the metabolism of lipids, carbohydrates, and proteins.

Formate dehydrogenase (FDH) is isolated from the yeast *Candida boidinii* grown on methanol. A typical sequence of production steps includes cell breakage, solid–liquid separation and a sequence of protein precipitations each with an associated centrifugation step in order to achieve a final product (see Fig. 5.37). The final product yield based on the feed from the cell disruption stage is typically of the order of 50 per cent at a FDH product concentration of 2.2 units/mg. Use of liquid extraction based on hydrophilic solvents raised the product yield to over 70 per cent at a product concentration of 2.1 units/mg. A summary flowsheet is shown in Fig. 5.37. Cell disruption, which is essential for efficient release of the intracellular enzyme, is conducted in a glass bead mill in the presence of water. A heat denaturation process is employed to raise enzyme activity and is followed by the first partition step in a single stage contact with a solution of PEG 400,

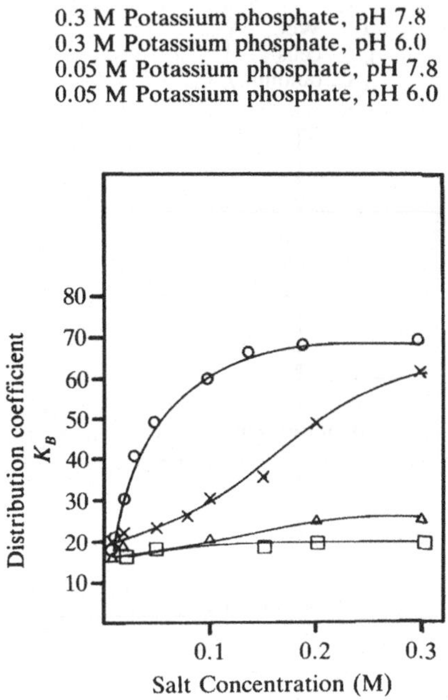

Fig. 5.36. Effect of different salts upon the partition of glucose-6-phosphate dehydrogenase in the system polyethylene-glycol-blue. (From Kroner *et al.* 1982.)

PEG 1550 in potassium phosphate solution at pH 7.8. At this stage the cell concentration is approximately 20 per cent. Following equilibration, the phases are separated in a centrifugal nozzle separator resulting in an upper extract phase containing the bulk of the FDH and a lower raffinate phase containing most of the cell debris and lysis. The extract goes forward for the next contact with a 10 wt per cent solution of potassium phosphate and potassium chloride and a further two-phase system is formed. The enzyme remains in the light phase and a majority of nucleic acids and polysaccharides are rejected with the lower raffinate phase. At the next stage, partition C, a fresh solution of heavy phase of 4 wt per cent PEG 400, 2 wt per cent PEG 1550, and 18 wt per cent potassium phosphate is added to the enzyme-rich light phase and subjected to a single stage contact. Under these conditions the FDH product enzyme is preferentially extracted into the heavy phase, leaving the majority of other proteins in the light phase which is rejected. Final product concentration is achieved in partition D by addition of 10 per cent potassium phosphate solution. This effects transfer of the enzyme back to the PEG-rich phase, at the same time providing a polishing effect of removing traces of residual polysaccharides, nucleic acids, and small amounts of protein.

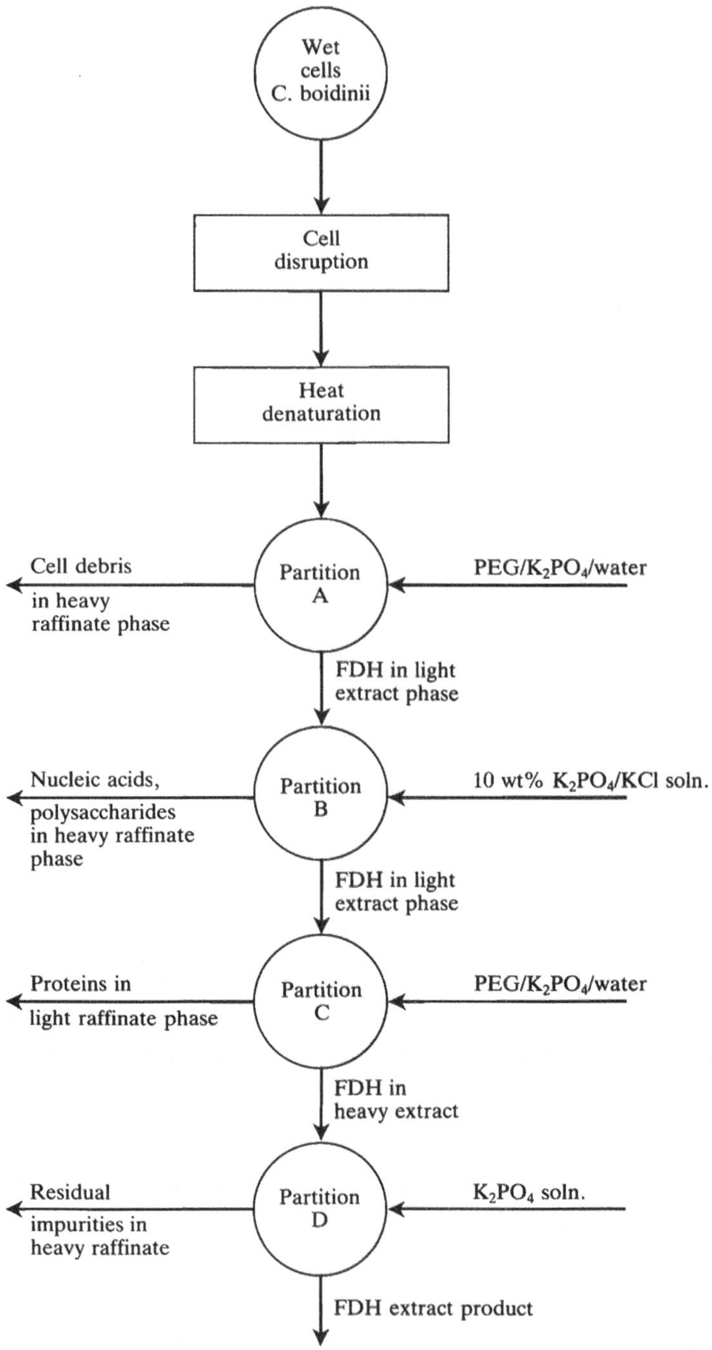

Fig. 5.37. Flowsheet for purification of formate dehydrogenase using aqueous two-phase extraction. (From Kula *et al.* 1982.)

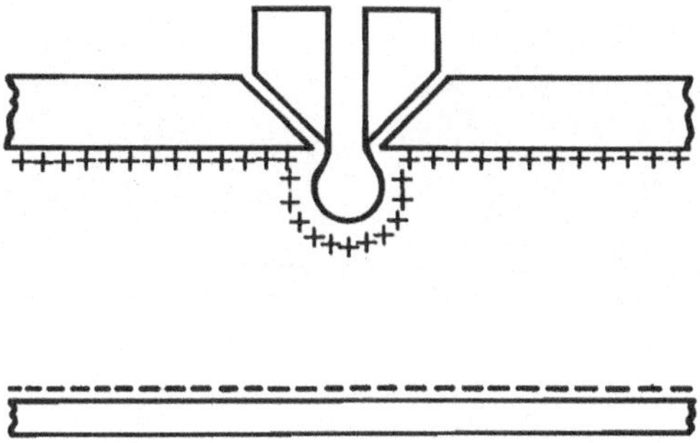

Fig. 5.38. Charge induced on a conducting droplet in a uniform field.

There are several attractive features of the liquid extraction-based enzyme recovery process. The final concentration of enzyme over the process is enhanced by about ten fold with only a very small reduction in specific enzyme activity. Compared to the conventional precipitation-based process, liquid extraction achieved a higher yield of enzyme and at a production rate approximately ten-fold better. Batchwise centrifugation processes and chromatographic separation steps are avoided, leading to substantial reductions in residence time. This latter feature leads to the conclusion that substantial economic savings are anticipated (Kula 1982) both in terms of labour cost and energy cost. Higher product activity and improvement in production rate more than compensate the additional cost of solvent inventory and added reagents.

5.6 Electrically enhanced extraction
[See also Batey Chap. 2]

5.6.1 Introduction

A number of important industrial separation processes which use liquid extraction demand that the desired product be recovered as quickly as possible to maximize yield. For example, the recovery of certain antibiotic products in the pharmaceutical industry which are highly unstable is conducted in short residence time contact equipment. Enhancement of rates of extraction is most commonly derived by introduction of mechanical energy into the system in the form of stirrers, rotating discs, or by the use of centrifugal contactors. The major drawback of this approach is that only a small percentage of the applied energy is actually utilized in the process of enhancement and the majority is dissipated either as heat or as mechanical vibration.

Solvent extraction processes may also be enhanced by direct utilization of electrical energy to achieve higher specific rates of extraction within a par-

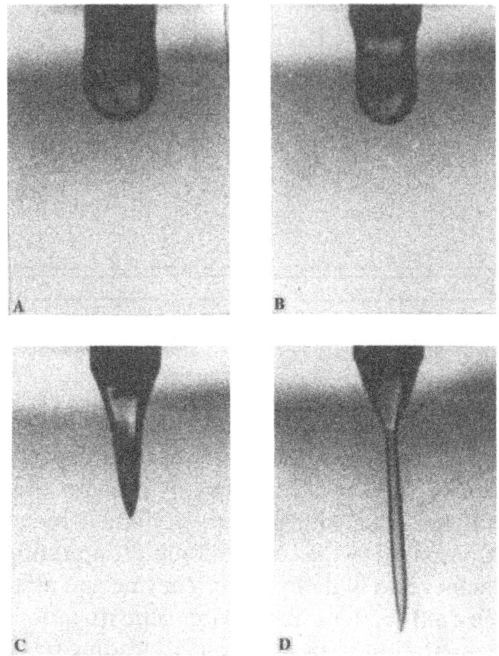

Fig. 5.39. Droplet formation at a charged nozzle at different field strengths for furfural dispersed in n-heptane. (A) Zero field, (B) 0.5 KV/cm, (C) 1.0 KV/cm, (D) 1.5 KV/cm. (From Bailes and Thornton 1971.)

ticular contactor. The technique of electrically enhanced solvent extraction using electrically charged drops was demonstrated initially by Thornton and Stewart (1967) and Thornton and Brown (1967). The basis of the method relies on the ability to place an electrostatic charge on droplets of dispersed phase as they are formed and to accelerate the droplets through the continuous phase under the influence of a potential gradient (see Fig. 5.38). The ability to produce charged droplets at the point of formation depends to a large extent on the electrical and physical properties of both dispersed and continuous phases. Providing that droplets be produced which retain their charge and if the continuous phase is essentially non-conducting, enhancement of interphase mass transfer may be achieved in three ways:

(1) small oscillating droplets are produced with a high specific interfacial area;

(2) droplet oscillation is maintained in spite of the small droplet size, giving greatly improved dispersed phase mass transfer coefficients;

(3) electrostatic acceleration of dispersed phase droplets through the continuous phase gives rise to enhanced interfacial shear and thus improved continuous phase film mass transfer coefficients.

5.6.2 Droplet behaviour

The mechanism of droplet formation and the charge acquired by the droplets exert a vital influence upon their subsequent behaviour in the contactor. Referring to Fig. 5.38 where we see droplet formation pictured under somewhat idealized conditions, the forces tending to detach the droplet at the nozzle comprise gravitational forces due to the weight of the droplet and electrostatic forces which depend upon the electrical properties of the system and the magnitude of the applied field.

Upon detachment from the nozzle, if the droplet retains its acquired charge, it can be accelerated through the continuous phase, enhanced by the potential gradient applied between the nozzle and the lower electrode. The resulting terminal velocity of the droplet will be determined from the force balance between gravitational and electrostatic forces acting and the drag forces. Under electrically enhanced conditions the terminal velocity is thus increased.

A further effect of the charge on the droplet relates to the force interactions at the liquid–liquid interface. The surface charge on the droplet acting normally outward will tend to oppose the inward acting interfacial tension forces and thus the effective interfacial tension is lower. This latter effect raises the value of Weber number and since droplet oscillation generally occurs at values of Weber number in excess of 3.5 and at Reynolds numbers greater than 300 (Thornton 1976), the likelihood of droplet oscillation is much enhanced.

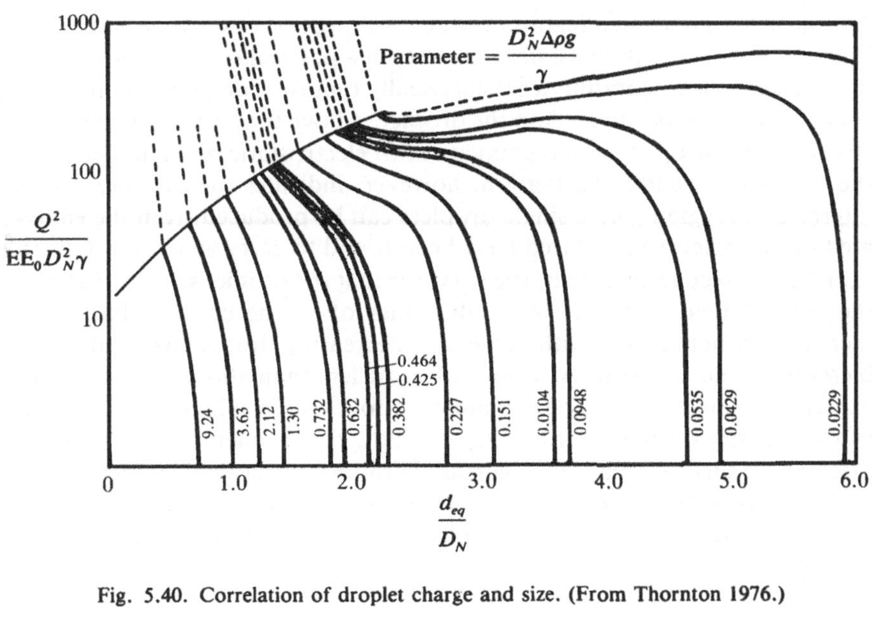

Fig. 5.40. Correlation of droplet charge and size. (From Thornton 1976.)

The droplet size and thus the specific interfacial area is also strongly influenced by the mechanism of formation at the nozzle. Bailes and Thornton (1971) illustrate clearly two types of droplet formation behaviour depending upon the applied feed strength (see Fig. 5.39). At low field strength a pendant droplet will increase in size until the combined magnitude of gravitational and electrostatic forces exceed the restraining interfacial tension forces at the nozzle–droplet interface. The droplet then detaches, retaining a charge, the magnitude of which depends upon the charge density at the nozzle. This process is shown in A and B of Fig. 5.39.

At high field strengths, the pendant droplet at the nozzle elongates considerably and electrostatic forces dominate the droplet formation process (see Fig. 5.39, C and D). As the charge acquired by the forming droplet increases, the interface becomes inherently unstable and a continuous stream of very small, highly charged droplets emanates from the tip of the elongate. Data presented by Thornton (1976) for the system water dispersed in *n*-heptane are reproduced in Fig. 5.40 and show the experimental correlations between droplet diameter, nozzle diameter, and droplet charge.

Three dimensionless groups were used to correlate droplet size data, namely:

$$\text{Charge group } \pi_1 = \frac{Q^2}{\epsilon\epsilon_0 D_N^3 \gamma} \tag{5.1}$$

$$\text{Bond group } \pi_2 = \frac{D_N^2 \Delta\rho g}{\gamma} \tag{5.2}$$

$$\text{Ratio group } \pi_3 = \frac{d_{eq}}{D_N} \tag{5.3}$$

Thornton highlights two features of the resulting curves shown in Fig. 5.40. Firstly, each curve exhibits a discontinuity at a particular value of (d_{eq}/D_N). The locus of the discontinuities theoretically represent an upper limit of the charge which can be induced on the droplet. Above the limit, at higher field strengths, the droplet should elongate until electrostatic breakup occurs at the tip. Experimental observation, however, indicates the existence of an intermediate region where single droplets can be produced from the elongated tip. Droplets so produced have been found to carry an extremely high charge. The second feature of the curves is that the droplet size decreases as the applied field increases. At high values of π_2 the droplet charge, Q, increases monotonically with respect to decreasing droplet size and hence increasing field strength. At values of π_2 of less than about 0.1 the curves exhibit maxima. This can be explained as follows. In spite of the fact that charge density increases with field strength, the droplet charge, which is affected by surface area, may decrease with increasing field strength if the surface area is decreasing faster than the charge density is increasing.

The significant reduction in droplet size for a given nozzle diameter is highly beneficial in promoting a high rate of mass transfer. The acquisition of a significant charge by the droplets enhances the terminal velocity and thus enhances continuous phase mass transfer. The parallel electrode arrangement shown in Fig. 5.38 ensures that, following detachment of the droplet from the nozzle, the subsequent motion of the droplet is influenced by the electric field. For a droplet of total charge Q, moving in an electric field of strength E_0, parallel to the direction of motion, the accelerative force acting on the droplet, F, is given by

$$F = \underbrace{V\Delta\rho g}_{\text{Force due to gravity}} + \underbrace{QE_0}_{\text{Force due to electrical field}} \tag{5.4}$$

In theory the droplet will lose some of its charge during its passage between the electrodes and field strength is also position dependent. Under experimental conditions it has been shown (Bailes 1981) that these effects can be neglected and the droplet accelerates to a terminal velocity with force balance achieved by the combined opposing forces of viscous and form drag. A significant improvement in droplet velocity is possible (see Fig. 5.41). The

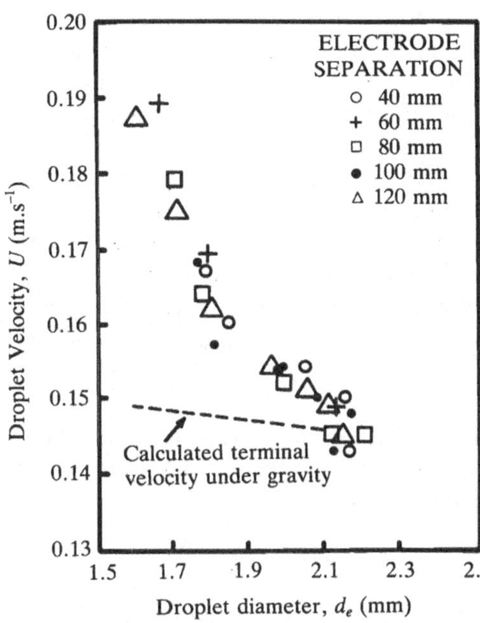

Fig. 5.41. Charged droplet velocity as a function of droplet size for the furfuraldehyde/N-heptane system. (From Bailes 1981.) (Nozzle diameter – 1.10 mm.)

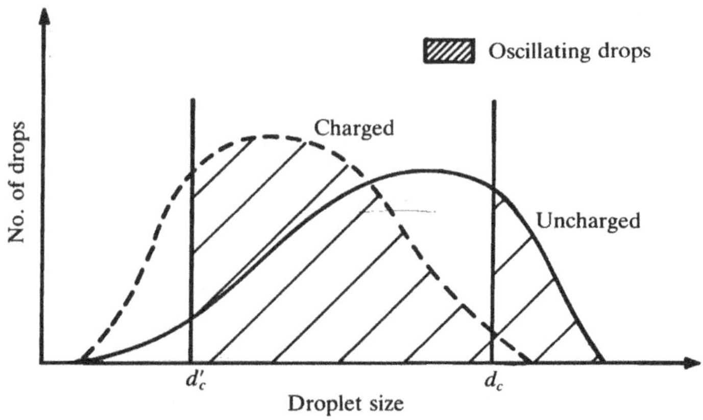

Fig. 5.42. Effect of charge upon critical droplet size.

comparison between velocities of droplets of different sizes under enhanced conditions and under gravity-only conditions clearly demonstrates the great improvement in velocity which is possible, especially at the smaller droplet sizes.

A third important feature of charge acquisition upon droplet behaviour is the enhancement of droplet oscillation. The important role played by droplet oscillation in the promotion of dispersed phase mass transfer is well known (Rose and Kintner 1966; Angelo *et al.* 1966) and thus the interrelation between electrical field conditions, droplet size and velocity, and oscillation behaviour is of some considerable importance.

Prediction of the oscillation frequency for charged droplets was successfully developed by modifying the interfacial tension term in the established relationship for uncharged droplets validated by Schroeder and Kintner (1965). Thus, after Stewart and Thornton (1967), the oscillation frequency, for charged droplets, is given by the expression:

$$\omega^2 = \frac{192b\gamma'}{(3\rho_d + 2\rho_c)d_{eq}^{\ 3}} \tag{5.5}$$

where b is a modified amplitude correction factor equal to $0.725\,(100\,d_e)^{0.23}$.

The influence of electrical charges is taken into account in a separate expression for interfacial tension, thus:

$$\gamma' = \gamma - 0.037\left(\frac{Q^2}{64\pi^2\epsilon\epsilon_0 r_0^3}\right) \tag{5.6}$$

The reduction in interfacial tension caused by the electrical charge on the droplet and the electrically enhanced acceleration of the droplets lead to increases in both Weber number and Reynolds number respectively, where:

$$We = \frac{\rho U^2 d_{eq}}{\gamma'} \tag{5.7}$$

and

$$Re = \frac{\rho U d_{eq}}{\mu} \tag{5.8}$$

With reference to Fig. 5.42, the effect of charge acquisition upon droplet size distribution can be related to critical droplet size. Qualitatively, the effect on

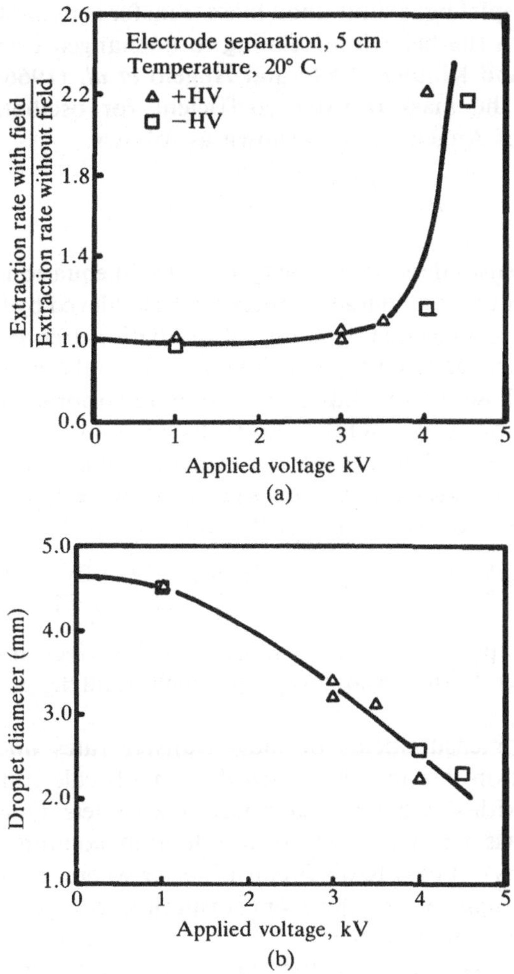

Fig. 5.43. Electrically enhanced extraction of copper into L I X64N in escaid 100. (From Bailes 1977.)
(a) Effect of applied voltage upon relative extraction rate.
(b) Effect of applied voltage upon average droplet size.

a hypothetical drop size distribution is to shift the value of critical droplet size, d_c, above which oscillation occurs, to a smaller value.

By appropriate manipulation of field conditions it should be possible to reduce the critical droplet size to a value where effectively all the actual droplets produced at a nozzle are oscillating and experiencing internal eddy diffusional mass transfer.

5.6.3 Mass transfer enhancement

The effect of droplet behaviour upon mass transfer is theoretically analysed with reference to the behaviour relating to unchanged oscillating droplet systems. Rose and Kintner (1966) and Angelo *et al.* (1966) developed an expression for the mass transfer coefficient for oscillating uncharged droplets. That of Angelo *et al.* is shown as follows:

$$k_d = \left(\frac{4D_d\omega(1 + \lambda_0 + \frac{3}{8}\lambda_0^2)}{\pi}\right)^{\frac{1}{2}} \tag{5.9}$$

where λ is a dimensionless amplitude factor. Use of equation (5.9) to predict the mass transfer rate for charge droplets can be achieved by using equations (5.5) and (5.6) to determine the value of oscillation frequency ω.

Both models suffer from the disadvantage that a knowledge of droplet deformation and oscillation frequency are required prior to implementation. Bailes (1977) found that a slightly modified version of an empirical correlation by Skelland and Wellek (1964) for dispersed phase film coefficients in oscillating droplets predicted experimental extraction efficiencies of charged droplets with an error of ± 7 per cent, namely:

$$\frac{k_d d_{eq}}{D} = 0.42\left(\frac{d_{eq}U\rho_c}{\mu_c}\right)^{0.68}\left(\frac{\gamma^3\rho_c^2}{g\mu_c^4\Delta\rho}\right)^{0.1}\left(\frac{4\ \mathrm{Dt}}{d_{eq}^{\ 2}}\right)^{-0.14} \tag{5.10}$$

This expression appears to take no account of the effect of oscillation frequency but in fact the dimensionless groups include all the parameters which describe it.

Experimental measurements of mass transfer rates under electrically enhanced conditions are mainly confined to small-scale continuous equipment equipped with single-nozzle dispersed phase inlets. Quantitative interpretation of mass transfer data from enhanced column experiments is difficult since it is not easy to distinguish exactly where, during the passage of a droplet through the column from formation to final coalescence, the major enhancement takes place.

Bailes (1981) suggests that the dominant 'end effect' contribution to mass transfer is probably at formation rather than at final coalescence. This contribution still appears to be rather less important than the enhancement of extraction rate which occurs during droplet motion due to the field and its effect upon levels of turbulence and eddy diffusion. These conclusions are based upon single nozzle extractions conducted in a small column in which *n*-heptane was transferred into droplets of furfuraldehyde from a bulk *n*-heptane phase saturated with furfuraldehyde. By taking mass transfer rate measurements in columns of different lengths using droplets of the same size in each case, the effective rate of mass transfer in the falling drop region was separated from that occurring due to the end effects by a process of subtraction.

The overall magnitude of the enhancements which are possible in the presence of electric fields are amply demonstrated for extraction in the copper-LIX system (Bailes 1977) in Fig. 5.43 and for the extraction of ethanol into n-decanol (Weatherley 1987) shown in Fig. 5.44. Figure 5.43(a) shows the effect of negative and positive applied potentials on the relative extraction rate of copper into LIX 64N in Escaid 100 in a 6 cm diameter column at an electrode separation of 5 cm. Very considerable enhancements are possible at applied voltages in excess of 4 kV. Figure 5.43(b) shows the effect of the applied voltage on the average droplet size and at voltages in excess of 5 kV an electrostatic spray was formed with droplet diameters of the order of 0.1 mm.

The data shown in Fig. 5.44 for ethanol extraction from fermentation liquor illustrates another important potential area of application for electrically enhanced extraction. These data, which compare rates of mass

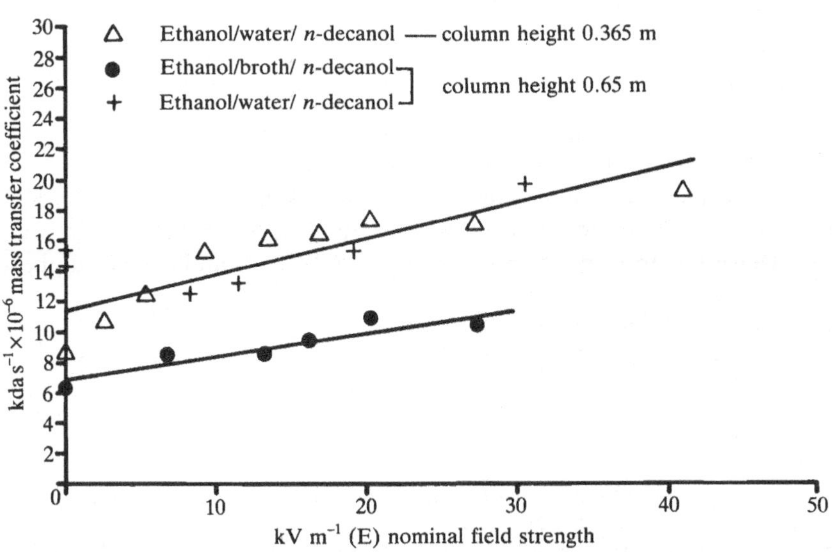

Fig. 5.44. Electrically enhanced extraction of ethanol into *N*-Decanol from fermentation liquor.

transfer in a solid free system to those in the presence of fermentation debris, show that electrical enhancement could be used to counteract the reduction in extraction rates exhibited in whole broth systems. The reduction in droplet size effected by the electric field means that efficient processing of liquors containing substantial amounts of solid material through large diameter nozzles becomes a possibility.

Designs for large capacity electrostatic extractors have not yet been investigated in any detail but the use of radial nozzle trays has been proposed for larger diameter columns (Thornton 1989).

Nomenclature

D Molecular diffusivity
D_N Nozzle diameter
d_c Critical drop diameter for onset of circulation
d_c' Critical drop diameter for charged drop
d_{eq} Equivalent spherical drop diameter
E_0 Electric field in vicinity of drop
g Acceleration due to gravity
k_d Dispersed phase film coefficient
Q Charge on drop
r_0 Equivalent spherical drop radius
t Drop contact time
U Drop velocity
V Voltage; Drop volume in eqn. 5.4

Greek symbols

γ Interfacial tension
γ' Effective interfacial tension
ϵ Dielectric constant (Defined by E in Fig. 5.40)
ϵ_0 Permittivity of free space (Defined by E_0 in Fig. 5.40)
ρ Density (subscripts c and d refer to continuous and dispersed phases respectively)
$\Delta\rho$ Density difference between phases
ω Drop oscillation frequency

References

Albertson, P. A. (1971). *Partition of Cell Particles and Macromolecules* (2nd edn). Wiley Interscience, New York.

Angelo, J. B., Lightfoot, E. N. and Howard, D. W. (1966). *AIChE J.* **12**, 751.

Asher, W. J. *et al.* (1975). *Kidney International*, Vol. 7, S-409-412.

ATKINSON, B. and SAINTER, P. (1982). *Chem. Eng.* Nov., pp. 410–19.

BAILES, P. J. and THORNTON, J. D. (1971). *Proc. Int. Solvent Extr. Conf.* **2**, 1431–9.

BAILES, P. J. (1977). *ISEC77 Proc. Int. Conf. Solvent Extr. Conf.*, Vol. 21, pp. 233–41. Canadian Institute of Mining.

BAILES, P. J. (1981). *Ind. Eng. Chem. Proc. Des. Dev.* **20**, 564–70.

BARNES, J. E. and EDWARDS, J. D. (1982). *Chem. Ind.* p. 151.

BENEDICT, M., PIGFORD, T. and LEVI, H. (1981). *Nuclear Chemical Engineering* (2nd edn). McGraw-Hill, New York.

BLUMBERG, R. and GAI, J. E. (1977). *Proc. ISEC 77, Toronto*, Paper 1b.

BOCK, J. and VALINT, P. L. (1982). *Ind. Eng. Chem. Fundam.* **21**, 417–22.

B. P. (1984). *B. P. Briefing Paper.* Ref. 5/84.

BRODIN, A. F., KAVALUNIAS, D. R. and FRAUH, S. G. (1978). *Acta Pharmac. Suecica* **15**, 1–11.

BROGL, H. (1982). *Chem. Ind.* 19 June 1982, pp. 385–90.

BRUNNER, G. *et al.* (1979). Paper presented at the 6th AIR APT Conference (1977). *High Pressure Science Technology*, Vol. 1, pp. 565–72.

BRUNNER, G. and PETER, S. (1983). *Proc. Int. Solv. Extr. Conf. Denver*, pp. 558–9.

BUSCHE, R. M., SHIMSHICK, E. J. and YATES, R. A. (1982). *Biotechnol. Bioeng. Symp.* No. 12, pp. 249–62.

BUSCHE, R. M. (1983). *Biotechnol. Bioeng. Symp.* No. 13, pp. 597–15.

CAHN, R. P. and LI, N. N. (1978). *Metal Extraction by Combined Solvent and LM Extraction.* US Patent No. 4 086 163 (April 25).

CALAME, J. P. and STEINER, R. (1982). *Chem. Ind.* 19 June 1982, pp. 399–402.

CARLEY-MACAULY, K. W. (1985). *Solvent Extraction and Ion Exchange in the Nuclear Fuel Cycle*, pp. 128–35. SCI, London.

CECILLE, L., DIVORSCHEK, H., GIRARDI, F., HUNT, B. A., MANNONE, F. and MOURTY, F. (1979). *Am. Chem. Soc. Symp. Series.* No. 117, Section 30, pp. 427–40.

COATES, J. H. (1982). *Proc. AIChE Winter National Meeting, Orlando, Florida.*

CRABBE, P. G. (1983). *Proceedings of Eleventh Australian Conference on Chemical Engineering, Brisbane, 4–7 Sept.*, pp. 253–9.

DAVIES, S. S. (1981). *Chem. Ind.* pp. 683–7.

DE FILIPPI, R. P. and MOSES, J. M. (1982). *Biotechnol. Bioeng. Symp.* No. 12, pp. 205–19.

DESSON, M., DUCHAMP, C., HENRY, E., JOUIN, J. P. and MICHEL, P. (1984). *AIChE Symp. Ser.* No. 238, 80, pp. 162–15.

DICKENS, N. and DAVIES, G. A. (1983). *E.E.C. Research Report* No. 4, 058/79/MPP.

DICKENS, N., ROSE, M. and DAVIES, G. A. (1984). *Proceedings of Extraction '84* EFCE Publication Series No. 43, pp. 69–81.

FLETT, D. S. (1981). *Chem. Eng.*, No. 370, pp. 321–4.

FRANCIS, A. W. (1952). *Solubilities of Inorganic and Organic Compounds* (eds. A. SEIDELL and W. F. LINKS). Supplement to 3rd edn. Van Nostrand, New York.

FRANCIS, A. W. (1954). *J. Phys. Chem.* **58**, 1099–114.

FRANKFIELD, J. W., CAHN, R. P. and LI, N. N. (1981). *Separ. Sci. Technol.* **16**(4), 384–402.

FREJACQUES, C., GELEE, M., MASSIGNON, D. and PLURIEN, P. (1977). Paper

36/FR/257 *Proc. Int. Conf. Nuclear Energy and Its Fuel Cycle*. International Atomic Energy Agency, Salzburg.
GARDNER, D. S. (1982). *Chem. Ind.* 19 June 1982, 402–5.
GODFREY, A. (1986). *Link-Up* April–June 1986, 22–25.
GRIMMETT, C. (1981). *Chem. Ind.* 16 May 1981, 359–62.
GUILLAUME, B., MOULIN, J. P. and MAURICE, C. (1984). *Inst. Chem. Eng. Symp. Ser.* **88** (Extr. 1984), 31–45.
HELLYAR, K. G. and DE FILIPPI, R. P. (1982). *Chem. Eng.* April 1982, 136–8.
HORWITZ, E. P. and SCHULZ, W. W. (1985). *Solv. Extr. Ion Exch. in the Nucl. Fuel Cycle.* pp. 138–44. SCI, London.
KRONER, K. H., CORDES, A., SCHELPER, A., MORR, M., BOCKMANN, A. F. and KULA, M. R. (1982). *Affinity Chromatography and Related Techniques* (ed. T. C. J. GRIBNAU, J. VISSER and R. J. F. NIVARD), p. 491. Elsevir, Amsterdam.
KULA, M. R., KRONER, K. H. and HUSTEDT, H. (1982). *Adv. Biochem. Eng.* **24**, 73–118.
LAPAGE, K. and MARRIAGE, J. W. (1973). *Inst. Mining Met. Trans.* Sect. C, 82, pp. C101–C102, June 1973.
LERAT, J. M., LORRAIN, C. H., SANTARY, G. and VENOT, P. (1984). *Nucl. Eng. Int.* January, pp. 25–6.
LEEPER, S. A. and WANKAT, P. C. (1982). *Ind. Eng. Chem. Process Des. Dev.* **21**, 331–4.
LI, N. N. (1981). *Chem. Eng.* No. 370, 325–7.
LI, N. N., CAHN, R. P., NADEN, D. and LAI, R. M. M. (1983). *Hydrometallurgy* **9**, 277–305.
MALPAS, R. (1982). *The Chemical Industry* (ed. D. SHARP and T. F. WEST), pp. 99–117. SCI, London.
MAY, S. W. and LI, N. N. (1972). *Biochem. Biophys. Res. Commun.* **47**(5), 1179–85.
MINIER, M. and GOMA, G. (1980). *Paper to International Conference on Energy from Biomass, Brighton, Nov. 1980.*
MOHAN, R. R. and LI, N. N. (1974). *Biotechnol. Bioeng.* **16**, 513–23.
MOHAN, R. R. and LI, N. N. (1975). *Biotechnol. Bioeng.* **17**, 1137–56.
MUNSON, C. L. and KING, C. J. (1984). *Ind. Eng. Chem. (Proc. Des. Dev.)* **23**(1), January, 109–15.
NADEN, D., PILOT, J. P. and RENNER, C. W. (1985). *Solvent Extraction and Ion Exchange in the Nuclear Fuel Cycle*, pp. 18–33. SCI, London.
OTA (1982). *OTA Report*. Congressional Office of Technology Assessment Report, Washington, USA.
PORTER, K. D. and JENKINS, J. D. (1982). *The Chemical Industry* (ed. D. H. SHARP and T. F. WEST), Chap. 41, pp. 544–64. SCI, London.
RANDALL, L. G. (1982). *Separ. Sci. Technol.* **17**, 1.
RESCHKE, M. and SCHUGERL, K. (1984). *Chem. Eng. J. Biochem. Eng. J.* **28** (1), Feb. B1–9; **28**(1), Feb. B11–20.
RICKER, N. L., MICHAELS, J. N. and KING, C. J. (1979). *J. Separ. Proc. Technol.* **1**(1), 36–41.
RICKER, N. L., MICHAELS, J. N. and KING, C. J. (1980). *J. Separ. Proc. Technol.* **1**(2), 23–30.
RITCEY, G. M., SLATER, M. J. and LUCAS, B. H. (1973). *Proc. Int. Symp. Hydrometallurgy, AIME Annual Meeting, Chicago, February*, pp. 419–74.

RITCEY, G. M., PRICE, K. T. and WONG, E. W. (1981). *Development of an Eluex Process for the Treatment of High Chloride Uranium Eluates*. Paper presented at the 1981 Annual Canadian Uranium Producers Metallurgy Committee Meeting, Saskatoon, May.

RITCEY, G. M. (1985). *Solvent Extraction and Ion Exchange in the Nuclear Fuel Cycle*, pp. 35–51. SCI, London.

ROBERTS, F. (1970). British Patent 1182983, March 4th.

RODDY, J. W. (1981). *Ind. Eng. Chem. Proc. Des. Dev.* **20**, 104–8.

ROSE, P. M. and KINTNER, R. C. (1966). *AIChE J.* **12**, 530–4.

SCHROEDER, R. R. and KINTNER, R. C. (1965). *AIChE J.* **11**(1), 5–8.

SCUFFHAM, J. B. (1981). *Chem. Eng.* **370**, 328–30.

SCHULTZ, W. G. (1966). *Liquid CO_2 for Selective Aroma Extraction*. Paper presented at the 26th Annual Meeting of the Institute of Food Technologists, Portland, Oregon (May 1966).

SCHULTZ, W. G. and RANDALL, J. M. (1970). *Food Technol.* **24**, November, 94–8.

SCHULTZ, W. G., SCHULTZ, T. H., CARLSON, R. A. and HUDSON, J. S. (1974). *Food Technol.* **28**(6), 32–6.

SIALINO, E., MIGNOT, C., MICHEL, P. and VIAL, J. (1977). *Proc. ISEC 77, Toronto, Canada*, Paper 18B.

SKEAFF, J. M. (1975). *Canmet Report MRP/MSL 75*–165 (TR) Energy, Mines and Resources Canada.

SKEAFF, J. M. and LALIBERTE, J. J. (1980). *American Institute of Mining Engineers Report* ERP/MSL 80–48, August.

SKELLAND, A. H. P. and WELLEK, R. M. (1964). *AIChE J.* **10**, 491–6.

SODECK, G., MODL, J. and SALZBRUN, W. (1981). *Process Biochem.* **16**(6), 9–11.

SONNTAG, A. A., GOODMAN, D. A. and KOHLER, K. C. (1984). *AIChE J.* Symp. Ser. (ed. R. AKERS and C. J. KING), No. 238, **80**, 102–14.

STAGEMAN, J. (1984). Private Communication. (Covering the Difficulties in the Solvent Extraction of Fermentation Liquors.)

STEWART, G. and THORNTON, J. D. (1967). *I. Chem. E. Symp. Ser.* (ed. J. M. PIRIE) No. 26, pp. 29–42.

SUZUKI, T., IZO, S., TOJANA, S., TOMIZAWA, S. and KODAMA, K. (1968). *Tokyo Kozyo Shikensko Hokoku.* **63**, 75–86, February 1968.

THORNTON, J. D. (1976). *Birmingham University Chem. Eng. J.* **27**(1), 6–13.

THORNTON, J. D. European Patent Specification No. 0356030AZ (filed 1989).

THORNTON, J. D. and BROWN, B. A. UK Patent 1205562 (Filed 1967)

VICTOR, J. G. (1983). *Proc. Int. Conf. Solvent Extraction, Denver*, pp. 511–12.

VOLLBRECHT, R. (1982). *Chem. Ind.* **19** June 1982, 397–99.

WANG, H. Y., ROBINSON, E. M. and LEE, S. S. (1981). *Biotechnol. Bioeng. Symp.* No. 11, pp. 555–65.

WEATHERLEY, L. R., MILLAR, M. W. and LAUGHLAND, G. J. (1987). *Proc. Extr. '87.* Institute of Chemical Engineers, London.

WENNERSTEN, R. (1980). *Proc. ISEC 80*, Vol. 2, Paper No. 80–63.

WENNERSTEN, R. (1983). *J. Chem. Technol. Biotechnol.* **33B**, 85–94.

WILKINSON, W. L. (1986). *Atom*, No. 352, February, p. 37. UKAEA, London.

INDEX

The manufacturer's authorised representative in the EU for product safety is Oxford University Press España S.A. of El Parque Empresarial San Fernando de Henares, Avenida de Castilla, 2 - 28830 Madrid (www.oup.es/en or product.safety@oup.com). OUP España S.A. also acts as importer into Spain of products made by the manufacturer.
Printed and bound by CPI Group (UK) Ltd, Croydon, CR0 4YY
30/09/2026
02225174-0005